OXFORD HISTORICAL MONOGRAPHS
EDITORS
M. H. KEEN P. LANGFORD
H. C. G. MATTHEW H. M. MAYR-HARTING
A. J. NICHOLLS SIR KEITH THOMAS

LATIN SIEGE WARFARE IN THE TWELFTH CENTURY

R. ROGERS

CLARENDON PRESS · OXFORD
1992

Oxford University Press, Walton Street, Oxford OX2 6DP
Oxford New York Toronto
Delhi Bombay Calcutta Madras Karachi
Petaling Jaya Singapore Hong Kong Tokyo
Nairobi Dar es Salaam Cape Town
Melbourne Auckland
and associated companies in
Berlin Ibadan

Oxford is a trade mark of Oxford University Press

Published in the United States
by Oxford University Press, New York

© R. Rogers 1992

All rights reserved. No part of this publication may be reproduced,
stored in a retrieval system, or transmitted, in any form or by any means,
electronic, mechanical, photocopying, recording, or otherwise, without
the prior permission of Oxford University Press

British Library Cataloguing in Publication Data
Data available
Library of Congress Cataloging in Publication Data
Rogers, Randall.
Latin siege warfare in the twelfth century/Randall Rogers.
p. cm. — (Oxford historical monographs)
Includes bibliographical references and index.
1. Attack and defense (Military science)—History.
2. Mediterranean Region—History, Military. I. Title. II. Series.
UG444.R64 1992 355.4'4'091822—dc20 92-8978
ISBN 0-19-820277-6

Typeset by BP Integraphics Ltd, Bath, Avon
Printed and bound in
Great Britain by Bookcraft Ltd
Midsomer Norton, Bath

For
Rosie and Séamus

PREFACE

In writing this book I have benefited from the assistance and kindnesses of a considerable number of teachers and friends. In acknowledging my most prominent debts, I hope that others will once again bear with me.

I owe the most to Professor Karl Leyser, whose erudition and patience has been instrumental. Dr Maurice Keen has maintained an encouraging interest in my work for a considerable period, sharing insights and giving guidance as well as occasional consolation. I am also grateful for the assistance and enthusiasm of Professor P. R. Hyams and Dr Henry Mayr-Harting. I owe much to Mr J. O. Prestwich and the late Dr R. C. Smail, who taught me a very great deal. Dr William Hamblin, Dr Jeremy Johns, and Dr Christopher Corrie each helped with subjects about which I remain all too unfamiliar. Dr Charles Everret, Dr Stuart Airlie, Mr D. N. Fishel, and Dr D. L. D'Avray also helped me greatly. I would also like to thank Professors Robert Brentano, Thomas Bisson, and Randolph Starn for directing my interests at an early stage. Anne Gelling and Hilary Walford of Oxford University Press have been enormously helpful in dealing with a difficult typescript.

My wife Susie has been much beleaguered by twelfth-century siege warfare and its pursuit. I would like to thank my father, J. G. Rogers, for first interesting me in military history and the experiences of those caught up in it.

R.R.

Baton Rouge
November 1991

CONTENTS

List of Figures x

List of Maps x

Abbreviations xi

Introduction 1

1. Latin Siege Warfare in the First Crusade 10
2. The Capture of the Palestinian Coast and the Development of Crusader Siege Technique 64
3. Siege Operations in the Establishment of Norman Authority in Italy and Sicily 91
4. Cities and Siege Warfare: Lombardy in the Twelfth Century 124
5. Siege Warfare in the Iberian *Reconquista* 154
6. Seaborne Siege Warfare: The Italian Maritime States and Latin Expansion 193
7. Towards Conclusions 237

   Appendices
   I. Evidence for Crusader Siege Towers 1099–1169 249
   II. Siege Engines: General Descriptions and Terms 251
   III. The Problem of Artillery 254
   Bibliography 274
   Index 289

LIST OF FIGURES

1. Drawing of a thirteenth-century stone carving at the Church of Saint-Nazaire in Carcassonne (Viollet le Duc, *Dictionnaire*, viii. 389; redrawn from Finó, *Fortresses*, 157)   268
2. Traction lever artillery piece with fixed counterweight (from 'Maciejowski Bible', Pierpont Morgan Library, New York M.638, fo. 23^v)   269
3. Traction lever artillery piece (from Peter of Eboli, *Liber ad honorem Augusti*, Burgerbibliothek, Berne)   271

LIST OF MAPS

1. The siege of Antioch, 1097–1098   27
2. The siege of Lisbon, 1147   184
3. The siege of Acre, 1189–1191   216

ABBREVIATIONS

AA	Albert of Aachen, *Historia Hierosolymitana* (RHC Occ. 4)
AC	Anna Comnena, *Alexiade*, ed. and trans. B. Leib (3 vols.; Paris, 1941–5)
AG	*Annali Genovesi di Caffaro e de' suoi continuatori*, ed. L. T. Belgrano and C. Imperiale di Sant'Angelo (5 vols.; FSI 10–15; 1890–1929)
AI	Caffaro, Oberto, and Ottobuono, *Annales Ianuenses*, in *AG*
BM	Bernard Maragone, *Annales Pisani*, ed. M. L. Gentile (RIS 6; Bologna 1930)
CDRG	*Codice diplomatico di Republica di Genova*, ed. C. Imperiale di Sant'Angelo (3 vols.; FSI 77–9; 1936–8)
CS	P. W. Edbury (ed.), *Crusade and Settlement: Papers Read at the First Conference of the Society for the Study of the Crusades and the Latin East and Presented to R. C. Smail* (Cardiff, 1985)
DC	Ibn al-Qalanisi, *The Damascus Chronicle of the Crusades*, ed. and trans, H. A. R. Gibb (London, 1932)
EHR	*English Historical Review*
FC	Fulcher of Chartres, *Historia Hierosolymitana*, ed. H. Hagenmeyer (Heidelberg, 1913)
FSI	*Fonti per la Storia d'Italia* (94 vols.; Rome, 1887 ff.)
MGH	*Monumenta Germaniae Historica* (1826 ff.)
MGH SS	*Monumenta Germaniae Historica*, Scriptores (32 vols.; 1826–1934)
MGH SRG	*Monumenta Germaniae Historica*, Scriptores Rerum Germanicarum in Usum Scholarum Separatim Editi (61 vols.; 1840–1937).
MGH SRG NS	Monumenta Germaniae Historica, Scriptores Rerum Germanicarum, new series (12 vols.; 1922–59)

OV	Orderic Vitalis, *Historia ecclesiastica*, ed. M. Chibnall (6 vols.; Oxford, 1967–80)
RHC	Recueil des Historiens des Croisades (Paris, 1841–1906)
RHC Occ.	Recueil des Historiens des Croisades, Historiens Occidentaux (5 vols.; 1844–95)
RHC Orient.	Recueil des Historiens des Croisades, Historiens Orientaux (5 vols.; 1872–1906)
RIS	Rerum Italicarum Scriptores, new series (34 vols.; Bologna and Città di Castelo, 1900 ff.)
RS	*Chronicles and Memorials of Great Britain and Ireland in the Middle Ages* (Rolls Series; London, 1858–96)
TNV	*Translatio Sancti Nicolai in Venetiam* (RHC Occ. 5)
WT	William of Tyre, *Chronicon*, ed. R. B. C. Huygens, identification des sources historiques et détermination des dates par H. E. Mayer et G. Rösch (2 vols.; Turnhout, 1986)

INTRODUCTION

THE importance of siege warfare in the twelfth century manifests itself in the political and military history of the period. As commentators have noted, sometimes in exasperation, much of the military activity of the age revolved around fortifications. In such an environment an ability to conduct and endure siege operations was fundamental. Blockade and attrition were the staples of siege warfare in this period as in most others. Many fortifications were stormed in onslaughts in which the number and ardour of attackers were paramount. Yet some positions were taken by assaults based on complex machinery which overcame stone and personnel defences. On one hand, castles and cities were either captured or compelled to seek a negotiated surrender. On the other, the failure of a major attack and the destruction of large and expensive siege engines on occasion brought an end to the siege.

Although sieges in which the latter methods were predominant were not unknown in the early Middle Ages, their number and scale increased significantly after 1050. While in part due to the considerable bellicose activity in Latin Europe from about 1060 to 1120, it was also a response to military problems. The twelfth century marked a transition from earth and wood to stone in castle construction, as well as growth and development in urban fortifications. It also encompassed a notable increase in complex siege operations.

The Mediterranean was profoundly affected by these developments during the twelfth century. The extension of Latin authority which characterizes the history of the Mediterranean during this period involved much siege warfare in the Levant, southern Italy, and Iberia. The establishment of Latin rule depended substantially on the conquest of the cities of these regions. Every method of siegecraft, including bribery, was employed in these endeavours. However, it was Latin ability in overcoming the formidable defences of great cities which proved decisive.

The impulses behind much of this military activity were similar to those which shaped the twelfth century. The Crusades and the growth of Latin maritime power brought diverse groups of warriors, sailors, and pilgrims before large and well-fortified cities. Although powerfully motivated by spiritual and often material concerns, these composite forces were assembled from disparate groups of crusaders with a rudimentary common organization. Moreover, their logistical support was limited and sometimes precarious. While centred initially on the Holy Land, such expeditions played a notable role in military endeavours throughout the Mediterranean. That such forces maintained their concentration outside major fortifications and organized themselves to capture them is a measure of Latin achievement in the twelfth century.

Those who served their leaders out of obligation or for pay also fought in Latin sieges. Crusading expeditions after the First Crusade usually operated in conjunction with a Latin power which recruited locally. Some rulers channelled fiscal and military resources into besieging units which contributed significantly to the growth and centralization of their political authority. Although the majority of these troops were assembled and financed on an *ad hoc* basis, some nascent states became adept in this kind of military organization. To a degree twelfth-century siege warfare is a manifestation of the 'growth of government' often associated with the period.

Another customary theme of the characteristic features of the twelfth century prominent in its siege warfare is urbanism. Although castles and dependent fortifications were the scenes of protracted and sometimes bitter fighting, decisive encounters in southern Europe and the Levant revolved around urban centres. While this was also the case in some areas of northern Europe, particularly as the century progressed, towns did not dominate siege warfare to such a degree. In the Mediterranean the measure of Latin siege technique was the reduction of cities.

In some areas of Italy urban communities were not only the objects of siege warfare but also some of its most effective practitioners. Communal militias assisted in defence and in the suppression of seigniorial depredations throughout Latin Europe. Moreover, Italian cities and particularly those of Lombardy conducted campaigns of siege warfare on a large scale. The personnel, material, and technological resources of cities made them significant in waging siege warfare.

Experts in the organization and technology of this form of conflict are notable in their involvement in twelfth-century sieges. While some were political and military leaders who developed abilities in conducting siege operations, others were clearly professional military engineers. Evident more in narrative sources than in fiscal records, their often briefly noted presence nevertheless demonstrates their significance. Although the means by which they acquired their esoteric knowledge remains largely conjectural, their role in Latin siege warfare is a reflection of some of the impulses of the twelfth century.

Unlike the role of heavy cavalry or developments in fortification, siege warfare has received relatively little attention from modern military historians. This contrasts with the twelfth century, in which siege warfare was understood to be an integral part of military experience and a suitable subject for historical writing. This is not to assert that modern analysts have been unaware of the importance of sieges in medieval military affairs. Lot provides an example of an influential historian who clearly noted the importance of siege warfare during the central Middle Ages.[1] Lot is also representative in that, despite his acknowledgement, he chose not to write systematically about siege warfare. Smail's model study of crusading warfare discusses the place of battle in a wider context, and among other points clearly establishes the relationship between field armies and fortifications in defending the Latin states.[2] However, Smail provides little concerning relationships between offensive forces and fortifications.

When siege warfare has been discussed, it is usually as an adjunct to the history of military architecture. This has reinforced a tendency to concentrate upon devices and methods by which besiegers overcame the fortifications of the Middle Ages. However, despite this sometimes acute interest, depictions of siege engines and their employment are often far from unclouded, and owe much of the accuracy they possess to accounts of ancient rather than medieval writers. This is primarily a result of the nature of the sources and particularly their use of terminology. Yet an element of this confusion stems from a desire to extract clarity from difficult material and the imposition of anachronistic conceptions on to the medieval period and evidence. While a number of examples could be noted here, the most pertinent comes from the work of General Köhler.

After more than forty years' service in the Prussian artillery, Köhler retired to write a thorough history of warfare from the mid-eleventh to

---

[1] F. Lot, *L'Art militaire et les armées au Moyen Âge en Europe et dans le Proche Orient* (2 vols.; Paris, 1946), i. 17.
[2] R. C. Smail, *Crusading Warfare* (Cambridge, 1956).

the mid-fifteenth centuries.[3] While laudable in its scope and range of material, his work was deficient in several regards and trenchantly criticized by several writers, including Hans Delbrück. Although Delbrück's comments did not focus upon Köhler's discussions of siege warfare, they could be applied with equal vigour to these sections of his work.[4] While a fuller discussion of Köhler and the problem of artillery follows in Appendix III, further comment is merited here.

Köhler's greatest limitation was an uncomplicated if thoroughgoing application of the assumptions and standards of the later nineteenth century on to the problems and sources of the Middle Ages. His treatment of siege warfare and particularly its technology was considerably affected by this tendency, and his own professional experience may have played a role in this. Of course, Köhler was not the only medieval military historian of his period whose work suffered from this problem. Smail lucidly demonstrates how profoundly beliefs and concerns of the nineteenth and early twentieth centuries shaped medieval military historiography.[5]

While many topics within military history have been placed in a clearer perspective by more recent writers, siege warfare and technology have fared less well. The frequency with which the fanciful imaginings of Viollet-le-Duc continue to be reproduced as illustrations of siege engines demonstrates this. Köhler's work continues to exercise an influence, particularly on English-speaking students, because it provided much of the basis for Sir Charles Oman.[6] Although Oman's account of siege warfare is not solely dependent on Köhler, his descriptions of machinery and particularly artillery are. While scholars have developed a more accurate understanding of medieval artillery since the time of Köhler and Oman, much of it has remained obscure, resulting in continued reliance on Oman.[7]

Another and probably more important reason for the confusion which surrounds the technology of medieval siege warfare is that an accurate understanding of machines, their capabilities, and limitations cannot be completely developed in the abstract. The employment of

---

[3] G. Köhler, *Die Entwickelung des Kriegwesens und der Kriegführung in der Ritterzeit von Mitte des II. Jahrhunderts bis zu den Hussitenkriegen* (3 vols.; Breslau, 1886–90).

[4] H. Delbrück, *Geschichte der Kriegkunst im Rahmen der Politischen Geschichte* (7 vols.), iii. *Das Mittelalter* (Berlin, 1907), 678–83.

[5] Smail, *Warfare*, 3–17.

[6] C. Oman, *A History of the Art of War in the Middle Ages* (2 vols.; 2nd edn., London, 1924).

[7] See Appendix III; J. Sumption, *The Albigensian Crusade* (London, 1978), 116.

machinery depended on the particular factors of a given siege as well as the technology of attackers. The choice of devices and methods brought against a position was influenced and sometimes dictated by topography. The design and strength of masonry of a fortress as well as the number and condition of its defenders were also important if not critical factors. The availability of materials, threats posed by field forces operating in support of the besieged, available supplies of food and forage, as well as weather conditions affected the choice of method as well as other aspects of a siege. The length of time allocated, as well as the willingness of attackers to contend with the tedium of blockade or the efforts and casualties of close assault, also affected operations. Tensions within composite besieging forces on occasion influenced methods, and the attackers' willingness to accept a conditional surrender. While these factors varied from siege to siege, their influence must be taken into account.

If establishing a proper context is important for poliorcetics—the art of conducting and resisting sieges—then it is essential in understanding the place of sieges in medieval warfare. It is a commonplace that patterns of conflict in several portions of the medieval period were dominated by sieges. Oman's work is representative of this view and his is one of the strongest interpretations that 'the military art' of the twelfth century suffered because of a predominance of defence over offence, leading to a preponderance of sieges over field battles.[8] It may be noted that this assessment also involves a considerable grafting of late nineteenth-century military opinions and conventions on to the twelfth century. For Oman, like many of his time, decisive battles were the consummate military experience, and periods in which such encounters were few were judged accordingly. Keegan's observations about the place of battles in military historiography, and particularly the tendency of English writers to perceive military affairs primarily in terms of great battles, is pertinent here.[9] While this is not the place for an extended critique, it should be noted that the nature of warfare in much of the Latin west renders this orientation largely irrelevant. For economic and political as well as military reasons, sieges largely displaced field battles and, to a degree, were the 'decisive encounters' of a campaign. It is probably not possible to resolve, and doubtless not worth debating, whether the sieges or field battles of the First Crusade

[8] Oman, *Art of War*, ii. 52.
[9] J. Keegan, *The Face of Battle* (Harmondsworth, 1976), 55–7.

were more important in the expedition's outcome. This might seem superfluous, except that this section of Oman's study remains a quarry.[10] As with other aspects of medieval military history, siege warfare needs to be examined in its own terms as much as is possible. In addition historians have seldom looked at the character of siege warfare within periods of the medieval epoch. Given the diverse ways in which armies were raised across the medieval centuries, and their varied strategic objectives, the intrinsic aspects of siege warfare within specific periods is significant.

Moreover, siege warfare in the twelfth century has particularly notable and distinctive characteristics which should be illuminated. Contamine's survey addresses this question by discussing siege warfare according to his divisions of the medieval period.[11] Important as this work is, it serves more as a beginning than as a detailed discussion of the character and notable features of siege warfare in the twelfth century.

This study hopes to further understanding of these matters by discussing the characteristic features of twelfth-century siege warfare in the context of the Latin expansion of the age. In this endeavour important or representative sieges will be examined as part of the campaigns and wars of which they were a part. In so doing it will be necessary to identify particular factors which shaped the course of notable actions. While this initially makes comparison difficult, it is hoped that this approach will reveal, not only differences, but also some broader patterns in the siege warfare of the twelfth century.

By way of introduction it is appropriate to mention something about the focus and scope of this enquiry. If nothing else this may serve as a partial explanation to readers who feel that the sieges in which they are most interested have been given short shrift, and that the choice of examples has been arbitrary.

In regard to poliorcetics, the twelfth century is a fitting period upon which to concentrate. Historians have rightly considered the time between the First and Third Crusades as one in which western Europeans matched the standards of the Romans and may have surpassed them, particularly in the development of fortification-destroying artillery. One commentator has even added poliorcetics to the ever-growing list of arts and endeavours which enjoyed a renaissance

---

[10] Bloch draws extensively on Oman in connecting literary motifs with military realities (R. H. Bloch, *Medieval French Literature and Law* (Berkeley, Calif., 1976), 81–97).
[11] P. Contamine, *War in the Middle Ages*, trans. M. Jones (Oxford, 1984).

during the twelfth century.[12] Some historians have maintained that this development was the result of a movement of advanced techniques from east to west in the wake of the Crusades. White characteristically reversed this assessment, arguing that the successes of the crusaders stemmed from a superiority in military technology, including poliorcetics.[13] A concentration upon the Latin siege warfare of the twelfth century may not answer questions about the diffusion of military techniques definitively, largely because they need to be examined from much broader geographical, cultural, and linguistic perspectives. However, it is possible to focus these problems in their proper perspective, and concentrate on diffusion and development within the Latin west.

Having set forth reasons for concentrating upon the twelfth century, it is appropriate to explain the Mediterranean focus of this study. In the first place, the primary sources are compelling. Men who witnessed or heard or read something about the great sieges of the Crusades and other campaigns wrote some of the most illuminating accounts of the Middle Ages. In some ways the concentrations of this study are determined by these writers and the evidence they provide.

Historians of fortification in twelfth-century England and France have worked material from these areas more thoroughly than has been done for the Mediterranean.[14] Consequently a concentration on Latin activities in the Mediterranean will provide material which may be compared with that unearthed concerning the wars of the Conqueror, his sons, Plantagenets, and Capetians.

Latin siege warfare in the Mediterranean was often waged on a scale far beyond that customary in north-western Europe. As Contamine has noted, the period in which feudal Europe experienced conflicts which consisted largely of very small-scale ambushes, skirmishes, burnings, and sieges also saw the conquests of the Normans in Italy, the Crusades, and the *Reconquista*.[15] There was, of course, much siege warfare waged in the Mediterranean comparable to that conducted north of the Alps. However, there was also a great deal of action which went far beyond the usual experience of Latin combatants. These actions were of the essence in what distinguishes the twelfth century in the history of the Middle Ages. While a focus on this activity may

[12] J.-F. Finó, *Fortresses de la France médiévale* (3rd edn., Paris, 1977), 149.
[13] See below, pp. 9–10.
[14] Finó, *Fortresses*; P. Héliot, 'Le Château-Gaillard et les fortresses des XIIe–XIIIe siècles en Europe occidentale', *Château Gaillard*, i (1962); R. A. Brown, *English Medieval Castles* (3rd edn., London, 1976); P. S. Fry, *British Medieval Castles* (London, 1974).
[15] Contamine, *War*, 31, 54–64.

eclipse more 'typical' twelfth-century siege warfare, it will take a measure of the twelfth-century Latin west.

It may be noted by way of introduction that this study began as an examination of siege warfare in the Holy Land Crusades. However, it became clear that crusader operations in the Levant were but a part of a larger phenomenon evident in other areas of Latin military expansion. At the least, it is hoped that this enquiry will locate twelfth-century crusading siege warfare in its wider context.

Readers comfortable in the parallels and ravellins of Vauban and Uncle Toby may find aspects of twelfth-century siege warfare and its organization unfamiliar. While a number of these differences are manifest in the pages which follow, two should be underscored here.

The first concerns the number of combatants involved in twelfth-century siege warfare. Accurately ascertaining the number of those who fought vexes almost all enquiries into medieval military affairs. Chroniclers are notoriously unreliable in estimating combatants mixed up in a conflict, and on occasion give considerably exaggerated figures. While this is true of some medieval historians of siege warfare, many otherwise accurate writers simply fail to mention the number of besiegers or defenders at a given siege. While figures have been given when possible, this information is often simply not recorded.

Moreover, there is another difficulty in estimating participants in siege warfare not common to other forms of conflict. Personnel who were normally non-combatants played a role in siege warfare. Cities mobilized all available personnel in manning fortifications, repairing damage, and sometimes in resisting attacks. The spectre of the sacking which inevitably followed a successful general assault by Latin forces concentrated the efforts as well as the minds of those attacked wonderfully.

This problem also affects measuring attacking forces, particularly those involved in some of the great sieges of the Crusades. Those expeditions motivated a significant number of 'non-military' personnel, including the aged and juvenile, as well as those whose station and gender rendered them unsuitable for the military life. Whatever their limitations in full battle or on the march, they played a role in siege warfare, and should be listed among those whose efforts facilitated the capture of notable fortifications.

A second and related problem is what might be termed the structure of forces involved. As mentioned above, defenders often included both garrison and citizens, whose interests were not always coterminous. Moreover, as in many large communities, not all inhabitants had the same interests, and some groups could accommodate a change in lordship with greater equanimity than others. However, this is a facet of urban siege warfare in many times and places, although perhaps heightened during the central Middle Ages.

What should be emphasized for the twelfth century is that attacking forces were on occasion equally if not more disparate. Some besieging forces were composed of coalitions of different and sometimes conflicting interest groups. A 'fair' distribution of plunder to all those who laboured and fought was often difficult. This problem was usually met by an agreement and/or common oath to distribute booty according to a pre-arranged formula. More critical were divergences in long- and short-term political and economic gains, which occasionally led to sharp divisions, especially over accepting a negotiated surrender. A considerable number of important twelfth-century sieges cannot be analysed as simply 'two-player games'—that is as contests between one attacker and one defender. The composition of besiegers as well as the besieged was often more complex.

Moreover, the possible outcome of operations cannot be seen in terms of whether one side or another enjoyed control of a given position. Besieged communities strove above all to survive. In some cases they preserved their own religion and legal customs under the lordship of an infidel, and in others counted themselves fortunate to be exiles. Some attacking groups sought plunder, which was often employed in further military ventures, and revenge for their fallen comrades. Others served a ruler seeking lucrative sources of taxation. Twelfth-century siege warfare was profoundly affected by these factors and their interplay.

Twelfth-century sieges differed in several notable respects from those of the seventeenth and eighteenth centuries. In regard to the organization of forces, objectives, and logistics, the two periods may appear more divergent than similar. This is equally, if not more so, for poliorcetics. No medieval military engineer is known to have boasted that the date of the fall of a fortification could be accurately predicted by an expert from its outset. Such forecasting depends not only on an accurate knowledge of opposing fortifications and the condition of their defenders, but also certainty regarding the resources and capabilities of attackers. Twelfth-century engineers and commanders were on occasion unsure of the latter concerns, and improvisation was a keynote of a number of major siege operations of the period. It is to these subjects—notable sieges of the twelfth century and the manner in which they were organized and conducted—that we now turn.

I

LATIN SIEGE WARFARE IN THE FIRST CRUSADE

SOME notable ingredients of the success of the First Crusade are manifest in that expedition's siege warfare. Complex sieges were waged in circumstances made arduous by hostile field forces and limited logistical support. Not only did supply problems affect war materials, but on occasion the provisioning of foodstuffs especially to the indigent became an overriding concern. While these difficulties were inherent in the first mass movement of the Crusades, they were exacerbated by the crusaders' *ad hoc* organization and occasional inability to agree on military and territorial objectives. Although these factors influenced many aspects of the armed pilgrimage, they brought particular complications to siege operations.

Moreover, the First crusaders are thought to have encountered particular difficulties in the design, production, and employment of siege machinery because of a Near-Eastern superiority in poliorcetics and fortification. Those historians who have commented on the relative sophistication of crusader and Near-Eastern siege technology have argued that the crusaders learnt much from their allies and foes. Runciman has stressed the importance of direct imperial Byzantine assistance in the production of crusader siege machinery. More recently, Prawer has argued that native Christian workmen took service with crusading leaders during their march through Syria.[1] White has reversed this traditional view, arguing that in critical areas of military technology, the crusaders enjoyed a superiority. Regarding siege warfare, he suggests that the crusaders' 're-introduction' of the mobile siege tower may explain their successes.[2]

While each interpretation is plausible, neither may be confirmed or

[1] S. Runciman, *A History of the Crusades* (3 vols.; Cambridge, 1951-4), i. 227-8; J. Prawer, *Histoire du royaume latin de Jérusalem*, trans. G. Nahon (2 vols.; Paris, 1969), i. 228 n. 5.

[2] L. H. White, 'The Crusades and the Technological Thrust of the West', in V. J. Parry and M. E. Yapp (eds.), *War, Technology, and Society in the Middle East* (London, 1975), 107.

even denied by available evidence. Moreover, each assumes that poliorcetics and the engineering of machinery were the principal challenges crusaders faced in waging siege warfare. While questions of engineering capabilities are obviously important, they must not be viewed primarily in abstract terms. Thus it is appropriate in this chapter to locate them in their proper military and logistical context. In outlining the problems which confronted the First crusaders, and the solutions they devised, we may obtain a clearer understanding of relationships between technology and warfare. We will also see how the crusaders functioned as a group in meeting these not inconsiderable challenges, and perhaps learn something of the interplay between ideology and organization on the Crusade. In focusing on the principal sieges of the expedition, a range of operations and types of sieges will also be illustrated. In this way we may take the measure of Latin siege technique at the beginning of our period of enquiry.

Sources

Because of its importance to contemporaries, the First Crusade inspired considerable historical writing during the twelfth century. This corpus of material presents difficulties of interpretation, not only on account of occasional disagreements over specific information, but also because of its nature and the manner in which it has survived. The authorship, originality, and interrelations of the sources for the First Crusade are long-standing problems with their own historical traditions. While this is partially due to the anonymity of important writers and a lack of original manuscripts, it also reflects something of the sources themselves. Although the letters and narrative histories of participants provide the fundamental material, later twelfth-century writers also drew on the oral accounts of returned crusaders and perhaps on written versions no longer extant. While the modern historian may consult a range of the available material, a critical attitude must be maintained in reconstructing and analysing the expedition's siege warfare. As it is not possible to review all of this material here, interested readers may consult more extensive expositions of the sources of the First Crusade.[3] However, it is important to discuss pertinent aspects of

[3] Runciman, *His. Crus.* i. 327–35; C. Cahen, *La Syrie du Nord à l'époque des croisades* (Paris, 1940), 1–100.

source criticism here, partially because accounts of siege operations have seldom been examined in this context. Moreover, the problems involved in this discussion are exemplary for similar concerns in other areas of Latin siege warfare.

Almost all of the Latin accounts of the First Crusade contain some information about the expedition's siege warfare. This information was obtained through a variety of means: personal observation and questioning, eyewitness reports, other written accounts, oral tradition, literary convention, and imagination. While there is no simple test for assessing the utility of a given account, internal consistency, narrative detail, the sequence of events, and comparison with other sources are important criteria in evaluating material. Occasionally it is clear that an account written somewhat after the events described and based on earlier written narratives owes much to an author's style and ingenuity. Such is the case with Robert the Monk's description of siege machinery employed at Antioch in his history of the First Crusade written in about 1110 and based principally on the *Gesta Francorum*. Robert's list of machinery built and employed unsuccessfully at Antioch occurs at the beginning of his account of the siege, lacks specific narrative details or reflections later in the text, and includes very many of the terms for machinery known in the west.[4] Robert's list also serves a stylistic purpose by emphasizing Antioch's impregnability.

However, some writers who added details to earlier accounts provide valuable information, such as Guibert of Nogent, whose *Gesta Dei per Francos* interpreted the significance of the First Crusade in the course of Christian history.[5] Although based largely on the history of Fulcher of Chartres, Guibert's work, written before 1108, incorporated considerable original material, particularly concerning the siege of Arqua. Ralph of Caen wrote a history of the deeds of Tancred in the Crusade soon after the latter's death in 1113, which also provides important information.[6] Although probably based on the work of Fulcher of Chartres, Ralph yields much knowledge about Tancred's considerable role in the expedition's siege warfare. While the works of Peter Tudebode, Albert of Aachen, and William of Tyre could be mentioned in this context, they are discussed below.

The most important descriptions of siege operations come from men who participated in the First Crusade. However, the different forms

[4] Robert the Monk, *Historia Hierosolymitana* (RHC Occ. 3; IV. i. 775).
[5] Guibert of Nogent, *Gesta Dei per Francos* (RHC Occ. 4).
[6] Ralph of Caen, *Gesta Tancredi in expeditione Hierosolymitana* (RHC Occ. 3).

and structures authors chose for their works occasionally complicate an understanding of the Crusade's siege warfare. Some writers so abbreviated or telescoped events that it is difficult to connect their sequences of events with those of other historians. While all writers sought to give a comprehensive history of the *exercitus Dei*, each invariably includes more information about one of the major contingents, usually because the author or his principal sources were members of that group. This occasionally obscures a comprehension of the particular crusaders responsible for machinery construction or major policy decisions. While this problem is encountered in all of our sources, it is most prominent in the history of Raymond of Aguilers.[7] Despite these difficulties and the author's bias in favour of the count of St Gilles and the Holy Lance found at Antioch, Raymond of Aguilers gives an important account of the organization of siege operations in the southern French contingent.

Some participant historians did not remain with the expedition for the whole crusade. Fulcher of Chartres's *Historia Hierosolymitana* is a key source for the Latin establishment in the east as well as for the First Crusade. However, Fulcher left the main force to join Baldwin of Boulogne at Edessa in 1097 after the siege of Nicea and was not present at subsequent actions of the First Crusade. Nevertheless, his account, compiled soon after the Crusade, is an important source.

Two crusaders whose letters provide valuable information did not complete the pilgrimage. Stephen of Blois abandoned the First Crusade in 1098, and Anselm of Ribemonte was killed at the siege of Arqua.[8] However, their accounts of operations at Nicea and Antioch are particularly useful.

Before ending this general discussion of the sources, it is appropriate to review recent scholarship not found in the works cited above.

The Anonymous Gesta Francorum *and the* Historia de Hierosolymitano itinere *of Peter Tudebode*

The relationship between the anonymous *Gesta Francorum* and the *Historia de Hierosolymitano itinere* of Peter Tudebode has long been a

---

[7] Raymond of Aguilers, *Le Liber*, ed. J. H. Hill and L. L. Hill (Paris, 1969). See also below, pp. 32–3.
[8] Stephen's letters to his wife Adela and Anselm's to Manasses II of Rheims are printed in Hagenmeyer's letter collection (H. Hagenmeyer (ed.), *Die Kreuzzugsbriefe aus den Jahren 1088–1100* (Innsbruck, 1901)).

topic of historical debate.[9] Until recently modern scholars were agreed that the crusader Tudebode, who wrote in approximately 1110, based his account closely on the *Gesta Francorum*, leaving it substantially unchanged except for the addition of personal reminiscences and stories which he had heard upon his return to France. Tudebode's most recent editors and translators, however, have argued differently, suggesting that both Tudebode's history and the *Gesta Francorum* were based on a now lost original source. They also argue that Tudebode may be closer to a putative original.[10]

While a new examination of the relationship of the First Crusade sources is needed, it is not clear that this recent argument is any more justified than the interpretation it seeks to replace. It is difficult to understand how the first nine books of the *Gesta Francorum*, which narrate events from the Crusade's beginning to the defeat of Kherboga, could have been written by anyone other than a follower of Bohemond. Moreover, not all of Tudebode's details, and especially those concerning the siege of Antioch, argue for the accuracy of his memory or putative written source. In any case, I have relied upon the *Gesta Francorum* as a source of fundamental importance, and followed R. Hill in attributing the work to a southern Norman of Bohemond's contingent.[11] Nevertheless, both writers provide important information, and this study tries to note details of siege operations which bear on the relationship between these two texts.

Albert of Aachen, William of Tyre, and a 'Lotharingian Chronicle'

While an established consensus holds that William of Tyre, who wrote in the second half of the twelfth century, based his account on Albert of Aachen's work composed between 1120 and 1130, it has been argued that both writers drew on a now lost original source. Knoch revived the debate arguing that both Albert's and William's accounts of the First Crusade were derived primarily from a now lost chronicle written soon after the capture of Jerusalem, and that William's less embellished

---

[9] *Gesta Francorum et aliorum Hierosolimitanorum*, ed. R. Hill (London, 1962); Peter Tudebode, *Historia de Hierosolymitano itinere*, ed. J. H. Hill and L. L. Hill (Paris, 1977).

[10] Following this argument is difficult, because the reader must consult the Hills' Latin text as well as their earlier English translation, which contains most of their arguments (Peter Tudebode, *Historia de Hierosolymitano itinere*, ed. and trans. J. H. Hill and L. L. Hill (Philadelphia, 1974), 1–12).

[11] *Gesta Franc.*, pp. ix–xvi.

history may be closer to the original.[12] If correct, Knoch's thesis allows us to read both accounts critically together as a 'Lotharingian' account of the First Crusade, focusing primarily on crusaders from Lorraine and the Rhineland, and pre-eminently on the deeds of Godfrey of Bouillon. While there has been some support for this interpretation, Mayer and others have commented that, attractive and cogent as Knoch's thesis is, one cannot argue from what is essentially silence, and that, until substantial manuscript evidence becomes available, Knoch's formulations must remain a hypothesis.[13]

Nevertheless, it is important to consult both twelfth-century historians in an examination of the Crusade and its siege warfare. A detailed analysis of the siege warfare described in both works compared with that found in other sources shows that, in their similar sections, Albert's and William's accounts contain considerable valuable material. Moreover, such an analysis underscores the importance of Albert's history in understanding the siege operations of the Lotharingian contingent. His account maintains a consistent perspective on the actions of Godfrey's followers, particularly during the assault on Jerusalem. While more narrow in its focus than the account of William of Tyre, which relates events from other contingents more fully, Albert of Aachen provides details and a clarity which William does not.[14] In any event, it is clear that Albert of Aachen's history is an important source for the siege warfare of one major contingent of the First Crusade. It may be noted that this perspective on source material and accounts of siege warfare can, on occasion, be reversed. Detailed examination of descriptions of operations and comparison with other accounts may shed light on the problems of the interrelationships of the sources of the First Crusade. In this regard, a study of siege warfare may also contribute to a clearer understanding of our sources.

Because the difficulties of certain sieges involved profound challenges to the cohesion and even ideology of the Crusade, a study of siege warfare inevitably leads into wider questions of organization and motivation. Yet this cannot be a history of the Crusade through its siege

[12] P. Knoch, *Studien zu Albert von Aachen* (Stuttgart, 1966).
[13] E. O. Blake and C. Morris, 'A Hermit Goes to War: Peter and the Origins of the First Crusade', in W. J. Shiels (ed.), *Monks, Hermits and the Ascetic Tradition* (Oxford, 1985), 91–2, 98–104; H. E. Mayer, 'Literaturbericht über die Geschichte der Kreuzzüge', *Historische Zeitschrift*, 3 (1969), 657, and 'Review of P. Knoch, *Studien zu Albert von Aachen*', *Deutsches Archiv für Erforschung des Mittelalters*, 23 (1967), 218–19; P. W. Edbury and J. G. Rowe, *William of Tyre: Historian of the Latin East* (Cambridge, 1988), 45–6 nn. 4, 5.
[14] See below, pp. 33, 58–61.

warfare, and the reader should turn to wider histories of the Crusade for a narrative of events and the larger context in which crusading siege warfare occurred.[15]

Nicea, 6 May–19 June 1097

Capturing Nicea from its recent conqueror, Turkish Sultan Kilij Arslan, was the Crusade's first major military undertaking. The crusaders' inexperience added to the difficulties presented by Nicea's fortifications and a Turkish field army. However, crusader unity of purpose and considerable Byzantine assistance facilitated the siege's success. The degree to which crusader siegecraft was dependent upon this assistance has never been clarified, and remains an important topic for discussion. Moreover, a detailed examination of this siege illustrates how the crusaders organized their efforts in the first major military venture of the expedition.

Establishing and Securing a Blockade

The siege began with the arrival of Lotharingian, Flemish, and southern Norman contingents on 6 May. The crusaders rapidly ringed their camps in close blockade, with Godfrey of Bouillon's and Robert of Flanders's troops facing the northern wall and the southern Normans the eastern. Positions facing the southern wall were vacant, as neither Raymond of St Gilles, Adhemar of Le Puy, Robert of Normandy, nor Stephen of Blois had arrived. The city's western approaches were lapped by Lake Ascanius.

Nicea's fortifications and topography combined to make the city's capture a challenging problem. While its western wall rested on Lake Ascanius, a double ditch surrounded the rest of the circuit walls. These were formidable, as they were buttressed by many towers allowing enfilading fire to be delivered all along them. Moreover, the city's masonry seems to have been particularly strong and durable.[16]

[15] Runciman, *His. Crus.* i; R. Grousset, *Histoire des croisades et du royaume franc de Jérusalem* (3 vols.; Paris, 1934), i; Prawer, *His. roy.* i; H. E. Mayer, *The Crusades*, trans. J. B. Gillingham (Oxford, 1972); H. Hagenmeyer, 'Chronologie de la Première Croisade', *Revue de l'Orient latin*, 6, 7 (1898, 1900), 214–93, 430–503; Smail, *Warfare*; J. S. C. Riley-Smith, *The First Crusade and the Idea of Crusading* (London, 1986).

[16] Raymond of Aguilers, *Lib.* 43; AA ii. 21, p. 314; WT iii. 1, pp. 111–12; A. M. Schneider and W. Karnapp, *Die Stadtmauer von Iznik-Nicea* (Berlin, 1938).

The crusaders attacked Nicea as immediately and directly as the situation allowed. While isolating the city in close blockade, they strove to overwhelm the defenders with assaults on the fortifications. Even when specific assaults failed, they helped in wearing down the garrison and its resolve. The crusaders' maintenance of this position around Nicea was contingent on two factors: security from Turkish field forces and the provision of food supplies.

The crusaders and their blockade were challenged by the approach of a large force under Kilij Arslan on 16 May. Since Nicea's southern approaches were still unguarded, the Sultan attempted to slip a relieving force into the city before dealing with the crusaders. However, this force was intercepted and driven off by the timely arrival of the southern French contingents under Raymond of St Gilles and Adhemar of Le Puy. Even with the emplacement of these reinforcements before Nicea's southern walls, the crusaders were still caught between a garrison and a field army. They responded by dividing their forces against both groups of opponents. While the southern French and eventually Flemish contingents fought Kilij Arslan's army on the hills around Nicea, the rest of the crusader army maintained the security of their emplacements around Nicea. Although the crusaders won no decisive victory, the Sultan departed the field, leaving Nicea to its fate.[17]

It is worth noting that the crusaders were caught between garrisons and field armies on several other occasions during the Crusade, and that they responded in a similar fashion. Crusader ability or good fortune in overcoming relieving armies preserved the expedition as well as maintained a particular blockade. As this chapter is not a comprehensive military history of the Crusade, it cannot analyse these field battles or their relations to siege warfare except to state that the crusaders' victories over large relieving forces were essential in the successful siege warfare of the First Crusade.[18]

Adequate provisioning initially presented as serious a challenge to the siege's success as Kilij Arslan's field army. Acquiring foodstuffs on the march was usually not difficult, because all elements of the Crusade could forage adequately in virgin territory. However, this was not the case when the army concentrated in siege. Once foragers had picked the surrounding countryside clean, and supply became dependent on

---

[17] AA ii. 27, pp. 319–20; WT iii. 4, 5, pp. 200–2; Raymond of Aguilers, *Lib.* 43; *Gesta Franc.* ii. 14–15.
[18] For a discussion of this and other major battles, see Smail, *Warfare*.

merchants, poorer crusaders could encounter difficulties in paying for the cost of food. An acute shortage could force prices to such a level that a crisis resulted, even if there were food available in the camps.[19] It appears that there was such a supply crisis in the early stages of the siege, as the *Gesta Francorum* describes a sharp rise in bread prices because of a food shortage.[20] However, crusader supply difficulties were alleviated when Byzantine sailors and merchants began bringing supplies from Civetot and selling them in markets near the camps. The *Gesta Francorum* attributes this arrangement to Bohemond's negotiations with the Byzantines.[21] While Albert of Aachen and William of Tyre describe Byzantine logistical assistance, they do not mention Bohemond's agency but present the arrangement of market facilities as one of the rewards sent to the crusaders by Emperor Alexius after their victory over Kilij Arslan.[22]

Crucial as this Byzantine assistance was, it may not have solved food-supply problems for all crusaders. In summarizing the siege of Nicea, the *Gesta Francorum* mentions that many of the poorer crusaders starved to death during the siege. While it is unclear whether this occurred only before the establishment of markets or afterwards also, it is possible that the destitute could not afford food supplies during much of the siege.[23] While there is no accurate estimate of the number of destitute crusaders, survivors from Peter the Hermit's Crusade and humble followers of the Count of St Gilles were important constituents of this group.[24] The sheer cost of crusading may have impoverished the non-military personnel of all contingents. Yet, however difficult it may have been for the poor to obtain food at Nicea, the logistical problems of that siege, unlike those of the siege of Antioch, never developed into major challenges to the crusaders' blockade or cohesion.

The crusaders' victory over Kilij Arslan and Byzantine logistical assistance enabled the crusaders to blockade Nicea's land approaches. However, because they lacked naval forces and were unable to patrol Lake Ascanius, Nicea's western approaches were unguarded and the defenders received communications and some assistance. Conse-

---

[19] For the problems of logistics and siege warfare in modern times, see M. Van Creveld, *Supplying War: Logistics from Wallenstein to Patton* (Cambridge, 1977), 9, 41–2; for an excellent discussion of logistics in an earlier period, see D. W. Engels, *Alexander the Great and the Logistics of the Macedonian Army* (Berkeley, Calif., 1978).
[20] *Gesta Franc.* ii. 14.   [21] Ibid.
[22] AA ii. 28, pp. 320–1; WT iii. 5, pp. 200–2.   [23] *Gesta Franc.* ii. 17.
[24] For a discussion of the numerical strength of crusading forces, see Runciman, *His. Crus.* i. 336–41.

quently the crusaders and Alexius arranged for Byzantine boats to be dragged by oxen from Civetot to Lake Ascanius and to be manned by Byzantine military personnel, under the principal imperial official at Nicea, Manuel Boutomites.[25] At the same time, Alexius dispatched infantry and archers under Tatikios to assist the crusaders in their assaults, which had been going on intermittently since their arrival.

Crusader Assaults on Nicea

Soon after arriving at Nicea, the Lotharingians launched a probing attack which failed to penetrate Nicea's defences, demonstrating a need for siege machinery if further assaults were to be made. Although considerable desultory fighting between individuals and small groups occurred during the siege, the sources describe four major machine-supported attacks against Nicea. Crusaders undermined a section of the walls between 14 May and 16 May, when Kilij Arslan's approach compelled them to cease their efforts without breaching the walls.[26] Later in the siege two Lotharingian lords, Henry of Aische and Count Hartmund, financed and manned an armoured roof which attacked Nicea's northern wall. However, either because the roof was improperly handled or because it was inadequately designed against the garrison's defences, it collapsed, killing twenty of the financiers' knights.[27] Sometime before 3 June Raymond of St Gilles and Adhemar of Le Puy employed men from their household troops to topple a prominent tower—called the Gonatas Tower by the Byzantines—along the southern wall. Although a breach ensued, the sap was mistimed, allowing the garrison to build a makeshift retaining wall during the night.[28]

Lotharingian crusaders communally financed an armoured roof in the final stages of the siege, designed by a Lombard engineer who came forward offering his services after the leaders had despaired of forcing Nicea's fortifications. According to the sources which describe this event, this roof was successful where the earlier device of the two Lotharingian lords had not been, because the sides of its roof were better sloped, and the city surrendered.[29]

[25] *Gesta Franc.* ii. 16–17; AC iii, xi. 2, pp. 11–12; Raymond of Aguilers, *Lib.* 44. Albert and William underplay the Byzantine role in this operation (AA ii. 32, pp. 323–4; WT iii. 6, 7, pp. 203–4).
[26] *Gesta Franc.* ii. 14.   [27] AA ii. 30, p. 322; WT iii. 7, pp. 203–4.
[28] 'quidam de familia episcopi et comitis' (Raymond of Aguilers, *Lib.* 43–4; *Gesta Franc.* ii. 15; AC xi. 1, pp. 9–10; AA ii. 31, pp. 322–3; WT iii. 9, pp. 205–6).
[29] AA ii. 35–6, pp. 325–7; WT iii. 11, pp. 208–9.

There has been much debate over the veracity of this story, and considerable confusion has been introduced into secondary accounts by attempts to understand this story as another version of the attack of Raymond's and Adhemar's armoured roof mentioned above. Because of the almost supernatural appearance of this engineer and the public financing of his machine, as well as the lack of confirmation from other sources, nineteenth-century positivist historians ascribed the story to Albert of Aachen's desire to illustrate the unseen hand of God in the Crusade.[30]

Hagenmeyer complicated the debate by shifting the date of the attack of Raymond's and Adhemar's armoured roof to 10 June, and arguing that Raymond of Aguilers, Albert of Aachen, and William of Tyre all give different versions of the same event. He based this on Raymond of Aguilers's ambiguous chronology, rejecting the *Gesta Francorum*'s clear statement that the attack occurred before the arrival of Robert of Normandy and Stephen of Blois on 3 June.[31] Yet it is difficult to believe that Albert of Aachen and William of Tyre have conflated their descriptions of events, because they clearly refer to the attack of the count's and bishop's armoured roof before they tell of the Lombard and his machine.[32] Moreover, Raymond of Aguilers's narrative sequence places the count of St Gilles's attack before the arrival of the counts of Normandy and Blois.[33] Thus it seems most appropriate to accept the *Gesta Francorum*'s dating of this attack before 3 June, and accept that Albert of Aachen's and William of Tyre's accounts of the Lombard and his machine are not different versions of the story of the attack of the count of St Gilles's and Adhemar's armoured roof. Assuming that the story applies only to the Lotharingian contingent rather than to the whole of the crusading army, there is no reason for dismissing the story completely, and, as we shall see, there is some independent confirmation of Latin assaults in the final phases of the siege.[34] Most importantly we should realize that those authors who include this story of

[30] Krebs's reply to von Sybel is representative of this debate and discusses the problem of the Lombard engineer (F. Krebs, *Zur Kritik Alberts von Aachen* (Munster, 1881), 59).
[31] 'sic pro nichilo ebdomadibus quinque pugnatum est. tandem per Dei voluntatem quidam de familis episcopi et comitis...' But does Raymond of Aguilers mean five weeks from the siege's commencement on 6 May, or from the arrival of the count of St Gilles on 15/16 May? While Hagenmeyer chose the latter interpretation, he noted that the matter was not clearly resolved (Raymond of Aguilers, *Lib.* 43–4; *Gesta Franc.* ii. 15; Hagenmeyer, 'Chron. Prem. Crois.', 6 (1898), 288–9).
[32] AA ii. 31, 35–6, pp. 322–3, 325–7; WT iii. 9, 11, pp. 205–6, 208–9.
[33] Raymond of Aguilers, *Lib.* 44.
[34] See below, pp. 23–4.

the Lombard magister and his machine hoped that their readers would understand that the design and financing of the Lombard's device, as well as its success, were different from other efforts at Nicea.

The pattern of siege-machinery production and assault illustrated in Latin sources emphasizes the role and resources of individual commanders working within the major contingent divisions of the Crusade. The degree to which crusading lords may have designed the machinery they financed is unclear. However, the importance of their financial and manpower resources to the production and application of siege machinery is manifest.

The production of siege machinery required at least three basic ingredients: timber, the nails and ropes to fasten wood and hide components, and personnel with the engineering talent and craftsmanship to design and assemble machinery. The first requirement was found in the forests around Nicea, which provided the crusaders with an easily available supply of timber.[35]

How the crusaders obtained their second basic requirement, fastenings, is not clear. While they presumably acquired them from Byzantine sources, there is no evidence illustrating how this was done. The most plausible assumption is that the crusaders purchased nails and ropes from Byzantine merchants, either in Constantinople or at Nicea.

Who the engineers and craftsmen responsible for machinery were, and how they came to be in the army, are central questions for the origins of crusader siegecraft. However, the limited and occasionally contradictory evidence precludes comprehensive answers. Nevertheless, hypotheses may be developed to form a clearer understanding of the problems involved in the development of crusader siegecraft.

In the first place, large-scale imperial Byzantine assistance in the actual construction of siege machinery appears unlikely. There is no reference in Greek or Latin sources to a unit of Byzantine engineers serving with a crusading leader or with Tatikios's archers and infantrymen positioned near Raymond of St Gilles's camp.[36] While Anna Comnena records that her father, the Emperor Alexius I, sent the crusaders siege machines, some of which he had specially designed himself, it is unlikely that such machinery ever reached Nicea, if it was ever constructed. No such devices appear in any of the Latin sources, some of which do mention the emperor's role in arranging the blockade of Lake Ascanius.[37] Moreover, Anna's terminology is vague, and these

[35] WT iii. 6, p. 202. [36] AC xi. 2, p. 12.
[37] Ibid. 10–11; see below, pp. 23–4.

devices do not figure subsequently in her narrative. Thus her story of imperial siege machinery at Nicea appears primarily intended to extol her father's military knowledge and concern for the crusaders. While it seems clear that there was no official Byzantine assistance, this does not preclude the possibility that individual Byzantines took service with Crusade leaders, either through the recommendation of Alexius Comnenus or via their own contacts.

Whatever their ultimate origins, the siege-warfare specialists evident at Nicea were men who served either in a seigniorial household or as unattached participants. The familial troops of Raymond of St Gilles and Adhemar of Le Puy employed in the attack of their armoured roof are examples of the former, and the Lombard siege engineer who appears in the works of Albert of Aachen and William of Tyre exemplifies independent experts and craftsmen. While some lords undoubtedly brought their own experts, those who wished machines to be constructed found the personnel to build them within the Crusade. The expedition clearly attracted engineers from throughout the Latin west, as well as knights and pilgrims. Perhaps the most important factor in the organization of machinery production at Nicea was that this was the first siege of the armed pilgrimage.

Siege Machinery

Undermining and breaching walls under the protection of armoured roofs was the principal means of assault at Nicea. Raymond of St Gilles and Adhemar of Le Puy, Lotharingians Henry of Aische and Count Hartmund, and the entire contingent of Lotharingian crusaders had three large roofs built which protected their crews with varying degrees of success. The *Gesta Francorum* refers to machines built between the crusaders' initial arrival and the coming of Kilij Arslan (6–16 May) which enabled a section of Nicea's walls to be undermined but not breached when the Sultan's approach compelled the crusaders to cease their efforts. Although the *Gesta Francorum* mentions wooden towers and other wooden devices, the short period of construction involved and the description of the attack make it very doubtful that wall-dominating siege towers were employed. Rather, armoured shelters and roofs, and probably artillery, were employed in this assault. This does not preclude the possibility that the construction of siege towers was undertaken but not completed by the time of the city's surrender.[38]

[38] *Gesta Franc.* ii. 14.

Artillery, archery, and crossbow fire supported the crusaders in their attacks on Nicea. While the impression given by the sources suggests considerable artillery activity, there are few specific references to pieces or their effects. Raymond of St Gilles employed several pieces in his attack on the Gonatas Tower apparently without causing significant damage.[39]

What was to become a characteristic tool of crusader siege warfare, the mobile wall-dominating siege tower, is missing from most accounts of the siege of Nicea. While the *Gesta Francorum* refers to *turres ligneas*, it is not clear from this reference that these were siege towers, and such devices do not appear in other substantial accounts. In describing the final phases of the siege to his wife, Stephen of Blois mentions the assembly of lofty wooden, armoured towers.[40] If such machines were under construction, they were never brought to bear against Nicea's defences. However, Stephen of Blois attributes the Nicean decision to surrender to a fear of these devices.[41]

The Capture of Nicea

While there is a general agreement that Nicea surrendered to Manuel Boutimites and his imperial troops on 19 June, the sources offer differing assessments of the crusaders' immediate role in forcing the city's capitulation. Although several Latin writers present the dragging of boats on to Lake Ascanius as the last important act of the siege, soon after which the city surrendered, others give amplified or slightly different versions of events.[42] Stephen of Blois's story, in which the prospect of an assault by crusader siege towers is emphasized, is mentioned above. Albert of Aachen and William of Tyre concentrate on the attack of a communally financed armoured roof designed by a hitherto unknown Lombard expert in their explanations of the city's surrender. While these accounts may under-emphasize the Byzantine role, they

---

[39] Although Albert of Aachen and William of Tyre mention artillery damage in this assault, Raymond of Aguilers states that all of the count of Toulouse's attacks prior to that of the armoured roof were in vain (AA ii. 31, pp. 325–6; WT iii. 9, pp. 205–6; Raymond of Aguilers, *Lib.* 43–4). For a discussion of siege artillery, see Appendix II.

[40] Stephen of Blois, 'Epistula ad Adelam', i, in Hagenmeyer (ed.), *Kreuzzugsbriefe*, 139.

[41] 'quod turci adspicientes timore subacti, urbem imperatori per nuntios reddidere' (ibid. 139).

[42] Accounts in which the completion of the blockade is the siege's last important act are: Raymond of Aguilers, *Lib.* 44–5; *Gesta Franc.* ii. 16–17; FC i. 10, pp. 187–8.

indicate that Nicea underwent determined crusader assaults until its occupation by Byzantine soldiers.

Anna Comnena substantiates this view in her account of Nicea's surrender to Boutimites. Although a surrender to the Byzantines had been discussed since the beginning of the siege, Alexius issued Tatikios with last-minute instructions for his troops and the crusaders he was advising in order to ensure that the city passed safely into Byzantine authority without a Latin escalade or possible claim on the city. Tatikios was to ensure that the Latin assaults made during the actual surrender and Byzantine take-over were directed away from the main gates and other easy points of entry until the city was secured by Boutimites. Thus, as the crusaders pressed their attacks from one side of the city, Boutimites took possession from the other as though in an assault rather than a negotiated surrender.[43] While Anna's account may not confirm the details of the stories of Albert of Aachen, William of Tyre, or Stephen of Blois, she does show that the crusaders did not simply wait for the blockade to have its effect, but also kept up a steady pressure of assault throughout the siege.

One effect of the city's passing into Byzantine hands in the manner that it did was that the crusaders were denied the revenge and booty that a successful assault and concomitant sack would have facilitated. The inability to take plunder was a potential impediment to the Crusade's continuation. Waging siege warfare was expensive, as provisions had to be purchased and machinery financed. This latter expenditure fell especially heavily upon crusading leaders. Thus crusaders had financial outlays to recoup in the plunder of a besieged city, and to, a degree, the sack of one city helped pay for the siege of the next. Alexius realized the importance of the lack of a sacking at Nicea to the Crusade's continuation when he sent crusading leaders money gifts after the city's surrender.[44]

In summarizing the siege of Nicea, it is appropriate to note the nature and degree of Byzantine assistance evident from the sources and to comment on the crusaders' technical difficulties. Byzantine logistical aid was crucial in the siege's success. Their supplies of foodstuffs maintained the crusaders in a close blockade of the city, and the fastenings for crusader siege machinery probably came from Byzantine sources also. Byzantine troops and especially boats were useful, if not

[43] AC xi. 2, pp. 12–13.
[44] *Gesta Franc.* iii. 18; Raymond of Aguilers, *Lib.* 44; WT iii. 13, pp. 211–12; AC xi. 3, p. 16.

essential, in isolating the city. However, there is no clear evidence indicating that the Byzantines provided the crusaders with siege technology or specific machines. It is apparent that crusader operations against Nicea were organized and conducted within the expedition, and it seems that the siege engines involved in these efforts were built and operated by crusaders. While the crusader strategy of close blockade and rapid assault is clear, some of their attacks lacked a degree of finesse necessary to their success. The armoured roof of Henry of Aische and Count Hartmund collapsed outside Nicea's walls because of either improper handling or inadequate design. Raymond of St Gilles's and Adhemar of Le Puy's household troops mistimed the igniting of their sap under the Gonatas Tower. While a breach ensued, it occurred during the night when the crusaders could not exploit their success with an assault, allowing the defenders time to build a retaining wall which withstood crusader assaults at daylight. These failures suggest that the Crusade's siege engineers, whoever they were and however they came to serve in the army, were learning aspects of their art during the Crusade's first siege.

Antioch, 21 October 1097–28 June 1098

The capture of Antioch was one of the great sieges of the age and occupies a central place in the history of the First Crusade. The methods by which the city was reduced illustrate a different style of siege warfare from that waged at Nicea. Antioch was not assaulted until the city was betrayed after a blockade of seven months. The crusaders' primary tool was the counter-fort. However, the attack on Antioch was not a classic blockade. The size and structure of the crusader army, the strength of Antioch's fortifications and supporting field armies, as well as crusader logistical difficulties often rendered the besiegers more beleaguered than the besieged.

The Decision to Besiege Antioch

As the crusaders approached the Orontes valley in the autumn of 1097, leaders held a major council of war to decide when Antioch would be attacked.[45] The long march across Anatolia had weakened the army's

[45] Raymond of Aguilers, *Lib.* 46–7; J. France, 'The Departure of Tatikios from the Crusader Army', *Bulletin of the Institute of Historical Research*, 44 (1971), 144–5.

health as well as horse- and manpower, and the lure of lordships had drawn crusaders from the main force. These factors argued that the attack be postponed until the arrival of Byzantine and European reinforcements in the spring. Such a delay would alleviate supply difficulties, since the army could break up into smaller groups. By wintering along Antioch's lines of communication the crusaders could also begin the city's isolation.

A long-distance blockade would have been consonant with Byzantine experience and practice, as imperial troops had conducted such a siege of Antioch from the hill fortress of Baghras twelve miles north of the city in 968.[46] The imperial Byzantine adviser to the crusaders, Tatikios, recommended such a strategy in October 1097, as he did also in February 1098.[47]

Yet other views prevailed. Raymond of Aguilers reports that the count of Toulouse counselled that the crusaders ought to continue trusting in the Lord, who had already delivered Nicea.[48] The primary arguments for an immediate advance centred on the political and military benefits of keeping the army together. The focus of a close siege and major operations might prevent further departures, and the best protection against a counter-attack lay in keeping the army united.

In any case, the crusaders decided not to break up for the winter, nor to conduct a long-distance blockade, but to march directly on Antioch. Adhemar of Le Puy and Robert of Flanders led storming parties across the fortified 'Iron Bridge' which guarded the Orontes, crossing twelve miles upstream from Antioch on 20 October 1097. The next day Bohemond led an advance guard down the south bank of the river to an encampment opposite the north-eastern corner of the city's walls. The army followed along to positions opposite the northern walls, where it remained for seven months.

Antioch's Defences

Antioch's fortifications and topography determined the peculiar course of the siege. Antioch's walls, buttressed by four hundred towers, encompassed more than three and a half square miles of city, hillside, gardens, pasturage, wells, and cisterns.[49] Its citadel, located 1,000 feet

---
[46] G. Schlumberger, *Nichéphore Phocas: Un empereur byzantin an dixième siècle* (Paris, 1890), 707–11, 719–23.
[47] France, 'Departure of Tatikios', 145.
[48] Raymond of Aguilers, *Lib.* 47; see below, p. 30.
[49] See Map 1.

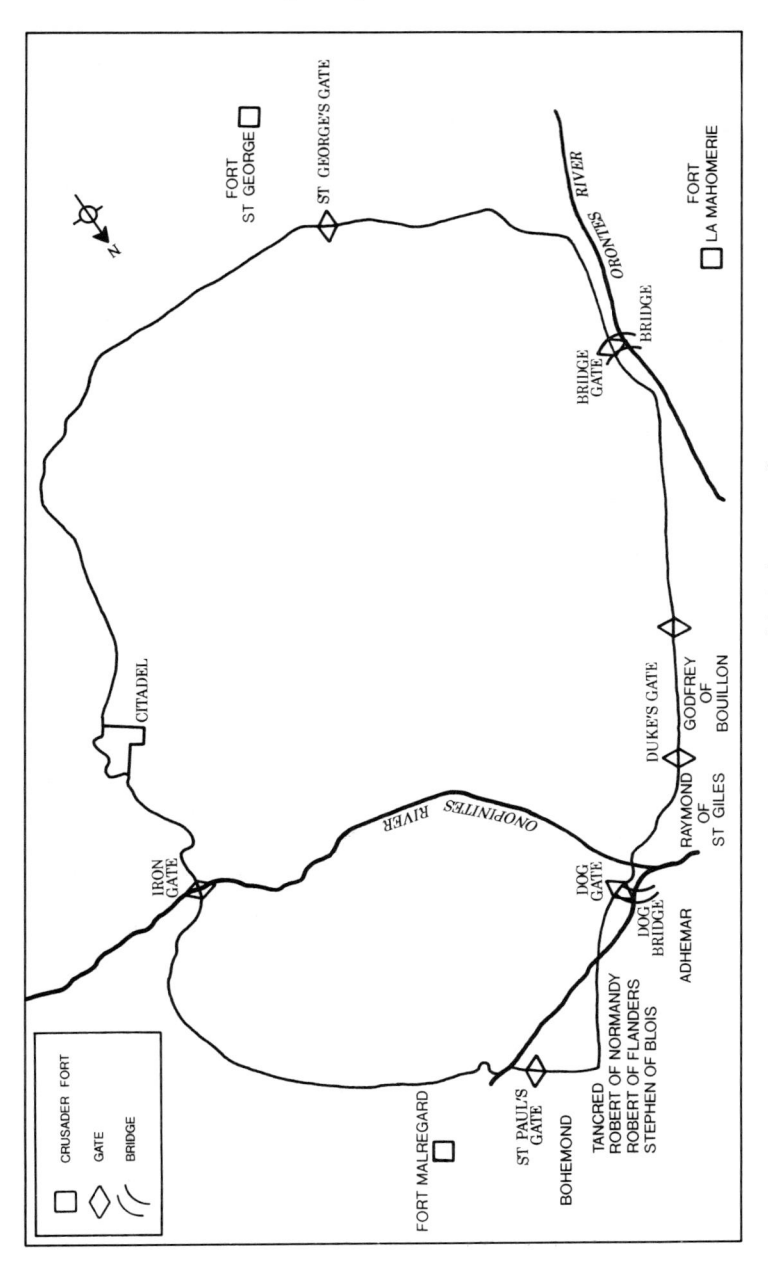

Map 1. The siege of Antioch, 1097-1098

above the valley floor atop Mt Silpius, dominated the river plain. Six gates, one on each of the eastern (St Paul), western (St George), and southern (Iron) walls, and three along the northern (Dog, Duke, and Fortified Bridge), permitted easy egress.

Choosing a site on the slopes of the Orontes valley, the city's builders had taken advantage of hillside, river, and plain in designing Antioch's fortifications. All of the southern and much of the eastern and western walls ran along a hill range, making approaches for large numbers of troops difficult. Although the northern wall, stretching along the river plain, was more accessible, the area's watercourses complicated an attacker's approach. The Orontes meandered along the northern wall, eventually touching the north-west corner. This river served less as a moat than as a barrier, separating besiegers operating immediately outside the northern wall from the hinterland. Two small streams flowed from Mt Silpius, combining into the Onopinites just outside the Dog Gate before joining the Orontes. This area was also marshy and difficult to traverse.

Two bridges further complicated besieging operations. The Dog Gate opened on to a bridge spanning the Onopinites, allowing easy movements across the marsh and so facilitating sorties. A second, fortified bridge, crossing the Orontes near the north-west corner from the Bridge Gate, was of central importance, as it allowed the defenders access to the area across the Orontes. This was the only Orontes bridge from the Iron Bridge twelve miles upstream, and so troops operating on the southern (city) bank of the river were denied an easy crossing. The Fortified Bridge also allowed garrison cavalry to harass lines of supply from the port of St Symeon, thirteen miles west of Antioch.

Although Antioch had been taken twice since Justinian I's remodelling of the fortifications which Choroses II had captured in 538, on both occasions the city's fall had involved treachery.[50] While the Turkish commander of Antioch, Yaghi Siyan, could not guarantee the loyalty of all his citizens, he had taken every precaution in garrisoning and provisioning the city. He commanded a large garrison, including a vigorous and well-supplied cavalry force. Yaghi Siyan had also solicited assistance from Muslim rulers and overlords in northern Syria.[51]

[50] Schlumberger, *Nichéphore Phocas*, 719–23; Runciman, *His. Crus.* i. 217.
[51] An understanding of Antioch's defences in 1097 must be put together from contemporary sources as well as from modern works concerned with the Selucid and medieval periods: G. Downey, *A History of Antioch in Syria* (Princeton, NJ, 1961), 597–679;

As they arrived, the crusader leaders grouped their contingents outside the gates along the eastern and northern walls of the city. Bohemond established his camps opposite St Paul's Gate, with Tancred, Robert of Normandy, Robert of Flanders, Stephen of Blois, and Hugh the Great on his right. The southern French under Adhemar of Le Puy and Raymond of St Gilles held the area opposite the Dog Gate. Godfrey of Bouillon and the Rhineland crusaders held the area between the Duke's Gate and the Fortified Bridge, although they did not control this bridge.[52]

While the initial phase of operations at Antioch involving a close blockade paralleled those undertaken at Nicea, the crusaders launched n attaks on Antioch's fortifications. They doubtless considered waging a full-scale assault, as those factors which had argued generally for an advance on Antioch also argued for a rapid resolution of the siege.[53] However, the need to rest and recuperate delayed an immediate attack. An unsuccessful storming may have involved heavy casualties and weakened the expedition's unity. Most importantly, the strength of Antioch's fortifications precluded many of the assaults by which Nicea was pressured.

That Antioch was considered too well fortified to be assaulted is stressed in several accounts. The count of Toulouse's chaplain presents a clear formulation of this opinion: 'It is so well fortified that it need not fear attack by machinery nor the assault of man, even if all mankind came together against it.'[54] One high-ranking military participant who did not complete the Crusade, Stephen of Blois, shared this opinion, noting the city's size and strong fortifications.[55] This reflects

G. Elderkin (ed.), *Antioch on the Orontes* (Princeton, NJ, 1934); E. G. Rey, *Étude sur les monuments de l'architecture militaire des croisés en Syrie et dans l'île de Chypre* (Paris, 1871), 183–204.

[52] There is no consensus among contemporary sources about the disposition of the crusaders before Antioch. While all concur that Bohemond's camps faced St Paul's Gate, they disagree about the placement of the southern French. I have followed William of Tyre and Albert of Aachen rather than Ralph of Caen or the *Chanson d'Antioche*. Neither *Gesta Francorum* nor Peter Tudebode nor Raymond of Aguilers gives a full report of the army's disposition (WT iv. 13, pp. 251–3; AA iii. 38–9, pp. 365–6; Ralph of Caen, *Gesta Tanc.* 49, p. 642; Richard the Pilgrim, *La Chanson d'Antioche*, ed. S. Duparc-Quioc (Paris, 1976), vv. 2865–964, pp. 162–8).

[53] Runciman has identified an 'assault party' within the Crusade's leadership led by the count of Toulouse and opposed by Bohemond. There is no evidence for this assertion, and J. France refutes it forcefully (Runciman, *His. Crus.* i. 217–18; France, 'Departure of Tatikios', 138).

[54] Raymond of Aguilers, *Lib.* 47–8.

[55] Stephen of Blois, 'Epistula ad Adelam', ii, in Hagenmeyer (ed.), *Kreuzzugsbriefe*, 150.

the consensus of crusader leadership, which prompted it to take a different course of action from the one that had been pursued at Nicea. Fulcher of Chartres records that, once the leaders perceived how difficult it would be to take Antioch, they swore to maintain themselves in blockade until it fell.[56]

While the precise chronology of strategic decision-making remains obscure, the outlines of crusader policy are clear. Having decided to blockade Antioch, the crusaders emplaced themselves in exposed positions, from which an assault was possible. However, no direct attack was made until a defender surrendered a portion of the walls on 2 June 1098. Nevertheless, the crusaders maintained their original positions on the south bank of the Orontes, from which they waged a close blockade. Although this positioning isolated the besiegers from the hinterland and exposed them to sallies and snipers, it kept open the possibilities of large-scale assaults and so added pressure on the defence. While the pursuit of such an aggressive and opportunistic policy in other circumstances would require little commentary, Muslim strength, the fact that the crusaders operated in hostile territory far removed from friendly bases, and supply problems were to make this strategy perilous as well as bold.

The Blockade of Antioch

Although blockading operations commenced with the crusaders' arrival, Antioch was not completely isolated until April 1098. The blockade developed step by step from the crusaders' base camps along the northern walls. As at Nicea, the basic requirements for maintaining a blockade were security and provisioning. Not only did the crusaders face three relieving armies, but they were continually harassed by the garrison. This harassment exacerbated the supply problems that provisioning an army during the Syrian winter without large-scale imperial Byzantine logistical assistance entailed. Thus the central struggle of the siege was in keeping the garrison contained. The crusaders' primary tool in this operation and in the extension of their blockade was the counter-fort. Blockade from small fortifications was a basic feature of medieval siege technique before and after the First Crusade, and in this instance the crusaders applied their experience to the problems of the siege of Antioch. However, the role of counter-

[56] FC i. 15, p. 218.

forts at Antioch involved a novel employment of these basic tools of siegecraft, which usually provided security for detachments of attackers harassing an already isolated defensive position. Although counter-forts performed this role at the siege of Antioch, they were also bulwarks protecting attackers operating in hostile territory and concentrated in vulnerable positions outside one of Syria's great cities. Crusader ability to construct and hold isolated positions was contingent on the availability of materials and serviceable horses. Because of a scarcity of these resources, the winter season, and Antioch's strength, the early and middle stages of the blockade involved desperate short-term defensive measures.

Establishing positions. Having established their camps and battle positions, the army rested and recuperated. The garrison's disinclination to sally out in the first two weeks of the siege and the easy foraging of the autumn season allowed the army a respite in its camps. While Raymond of Aguilers criticized this inactivity, the army's condition doubtless dictated it.[57]

Once garrison cavalry became active in early November, it began harassing foragers in the trans-Orontes. Consequently, the crusaders built a pontoon bridge out of captured boats near the southern French camps, which allowed them a rapid river crossing.[58] As the area around Antioch was picked clean, foragers ventured farther afield, where they were ambushed by the garrisons of small fortresses and castles. While Bohemond ambushed the garrison of one such fortress, Harenc, in mid-November, this did not alleviate the more basic problem of provisioning.[59]

At Christmas, Bohemond and Robert of Flanders led a long-distance raid into territory near Aleppo to collect flocks and booty. This force fortuitously encountered a substantial force under Duquaq of Damascus intended for Antioch. While the Damascenes were defeated, the raiding party lost its flocks and returned without supplies. During the time when this detachment was absent from the camps the garrison

[57] Raymond of Aguilers, *Lib.* 48–9; *Gesta Franc.* v. 28.
[58] While the exact place of this bridge's construction in the sequence of events at Antioch is unclear, it must have been built before mid-November 1097 (AA ii. 42, p. 368; WT iv. 14, pp. 253–4; Raymond of Aguilers, *Lib.* 48–9; Hagenmeyer, 'Chron. Prem. Crois.', 6 (1898), 517.
[59] *Gesta Franc.* v. 29.

attacked in strength and caused significant casualties, especially among the knights of Raymond of St Gilles's contingent.[60]

Just as security in the hinterland became more precarious in November, so did the safety of the encampments. Garrison cavalry ambushed small parties of foragers using the crossing afforded by the Fortified Bridge.

Albert of Aachen and William of Tyre relate that the defenders also used the smaller Dog Bridge to harass southern French positions. These two writers also describe attempts to destroy this bridge which involved the only machine-supported attacks made against Antioch's fortifications.[61]

Adhemar's and Raymond's troops initially attempted to break up the bridge with hammers and picks, but found garrison missile fire and the strength of the bridge's masonry impregnable. A mobile armoured shelter termed a mole (*talpa*) was built and pushed to the centre of the bridge under garrison artillery fire. Presumably the machine was intended to protect crews and their defenders who would break down the bridge. However, a garrison sally drove off the mole's defenders and it was burnt. Three artillery pieces were then set up against the gate and curtain wall which opened into the bridge. While this barrage kept the defenders occupied, it did not damage the fortifications. Finally the bridge was obstructed with heaps of timber, stones, earth, and other bulk. These events have seldom been incorporated into secondary histories of the siege, partially because of source problems, and perhaps because these assaults were so limited in scope and success.

While these events are related very clearly in Albert of Aachen and William of Tyre, there are no references in the *Gesta Francorum* or Raymond of Aguilers. That a follower of Bohemond should not mention these events is less surprising than their omission by the count of Toulouse's chaplain. However, Raymond of Aguilers's account of this phase of the siege, and in fact of military events during the siege, is so telescoped and abbreviated that it is difficult to follow. Peter Tudebode describes the emplacement and destruction of the mole, but he dates it to March 1098, during the blockade of the Fortified Bridge.[62] Tudebode's version lacks other elements of the story of the attack and a certain degree of logic, as the mole is used to

---

[60] Hagenmeyer, 'Chron. Prem. Crois., 6 (1898), 524–6; *Gesta Franc.* v. 31–2.
[61] AA iii. 39–41, pp. 366–8; WT iv. 15, pp. 254–6.
[62] Peter Tudebode, *His. Hiero. itin.* (1977), viii. 50–1.

obstruct the bridge after it had been blockaded by the construction of a counter-fort.[63] While the *Chanson d'Antioche* mentions the attacks by pick, artillery, and by an engine, the bridge attacked seems to be the Fortified Bridge rather than the Dog Bridge.[64] Confusing as these events were to even contemporary historians, their primarily defensive purpose illustrates the crusaders' problems. The failure of the three attacks also illustrates the strength of Antioch's defences.

Bohemond's followers were also harassed during this period, primarily by snipers who operated atop a hill overlooking Bohemond's camps. This prompted the construction of the crusaders' first counter-fort, Malregard, on this hill. Malregard seems to have been an earthen rampart fortification. This position was built not so much to extend the blockade, as Bohemond's camps already covered St Paul's Gate, but to protect these encampments.[65]

The construction of this fortress has been linked to the arrival of a Genoese flotilla of thirteen ships at St Symeon in late November.[66] However, there is no evidence that these Genoese sailors and combatants were involved in Malregard's construction. The *Gesta Francorum* does not mention the Genoese, and their historian, Caffaro, does not refer to the building of Malregard. He stresses the privations suffered by the Genoese along with the other crusaders in keeping the garrison contained. His account also relates that part of the Genoese force was massacred on its way to Antioch, emphasizing the vulnerablity of the supply lines to St Symeon.[67] Whatever the Genoese contribution, its importance illustrates how valuable combatants of any kind were in the winter of 1097–8.

Crisis of supply. Food supply became an acute problem during December and January. The same factors which had resulted in logistical difficulties at Nicea before the arrangement of Byzantine markets contributed to a major crisis at Antioch. While the expeditionary

---

[63] See below, pp. 35–7. While the Hills have cited this as evidence for Tudebode's originality, his lack of detail and apparent confusion over which bridge was attacked argues the opposite (Peter Tudebode, *His. Hiero. itin.* (1974), 57 n. 18; Peter Tudebode, *His. Hiero. itin.* (1977), viii. 79).

[64] The greater amount of detail and consistency in the accounts of Albert of Aachen and William of Tyre run contrary to the argument of Duparc-Quioc, who believes that Albert of Aachen used a primitive version of the *Chanson d'Antioche* in writing his history (Richard the Pilgrim, *Chanson d'Antioche*, 177–83).

[65] *Gesta Franc.* v. 29–30; Anselm of Ribemonte, 'Epistula ad Manassem', ii, in Hagenmeyer (ed.), *Kreuzzugsbriefe*, 157.

[66] Mayer, *Crus.* 54.   [67] Caffaro, *De liberatione civitatum orientis*, in *AI* 102–4.

force under Bohemond and Robert of Flanders defeated Duquaq of Damascus, it did not bring back the supplies for which it had been dispatched. Almost all of the sources date a major famine to the period immediately after the return of this force around 30 December.[68] Scarcity, lack of forage, soaring food prices, and poverty produced starvation, desertion, and death. Crusaders of all stations withdrew for an easier winter and perhaps to abandon the expedition. Peter the Hermit and Walter of Melun were dragged back by Tancred after departing, and Bohemond made conspicuous examples of them.[69] Robert of Normandy spent much of this period in Lattakiah on the coast.[70] In marshalling the visionary experiences which supported the Holy Lance of Antioch, Raymond of Aguilers reports the flight of lowly crusaders to Edessa, and the discoverer of the Lance, Peter Bartholomew, and his master, William Peter, tried to reach Cyprus during this period.[71]

It is difficult to ascertain how many crusaders and of what station died during this famine. Its effects must have been most acute among the non-military personnel. Matthew of Edessa, writing a generation after the Crusade, estimated that one out of five crusaders died of famine or related diseases during this period.[72]

However, there were food supplies available for those who could afford them. The Orthodox Patriarch of Jerusalem then established in Cyprus dispatched supplies. Matthew of Edessa records the role of the monks of the Black Mountains in collecting and sending provisions to the crusaders. While Latin sources record the presence of Armenian and Syrian merchants in their camps, prices were far beyond what the destitute could pay.[73] Fighting men and their leaders may have had the credit or wealth to obtain provisions, and it should be noted that Stephen of Blois was able to double the funds his wife had originally given him during this period.[74]

The supply crisis affected another area of crusader military strength—horses. The lack of adequate pasturage and fodder weak-

---

[68] Hagenmeyer, 'Chron. Prem. Crois.', 6 (1898), 529.
[69] *Gesta Franc.* vi. 33–4.
[70] C. W. David, *Robert Curthose, Duke of Normandy* (Cambridge, Mass., 1920), 232–4.
[71] Raymond of Aguilers, *Lib.* 70–2, 117.
[72] Different manuscript traditions give one out of seven (Matthew of Edessa, *Chronicle*, ed. and trans. E. Doustourian, Ph.D. (Princeton, NJ, 1978), ii. 300).
[73] Matthew of Edessa, *Chronicle*, ii. 300–1; *Gesta Franc.* vi. 33.
[74] Stephen of Blois, 'Epistula ad Adelam', ii, in Hagenmeyer (ed.), *Kreuzzugsbriefe*, 149.

ened the number of war horses and the condition of those that survived. The *Gesta Francorum* estimated that there were only one thousand knights who had managed to keep their horses in good condition. Anselm of Ribemont wrote that there were only about seven hundred horses in the entire army.[75]

This crisis provoked several responses. Following an earthquake on 30 December, Adhemar of Le Puy organized a three-day ritual of fasting, expiation, and alms-giving to improve the spiritual and material conditions of the poor. While some crusaders continued to leave, these activities seem to have maintained cohesion among the poor, and at least improved the morale of the clergy.[76]

Tatikios, the imperial Byzantine representative, revived the debate over close- versus long-distance blockading and advised the latter. The count of Toulouse argued against a change in strategy, offering to organize and finance a confraternity which would underwrite the cost of remounts.[77] Fear for the Crusade's unity and Bohemond's ambitions were other reasons for continuing the close blockade. Inertia doubtless also played a role, and, in any event, the crusaders maintained their positions before Antioch. For reasons which have remained obscure, Tatikios left the Crusade.[78]

Arduous as the winter siege was, the crusaders' situation improved in February. The advent of spring facilitated foraging and pasturage. The crusaders defeated the large relieving force of Ridwan of Aleppo in the celebrated battle of the Lake of Antioch and successfully defended their camps against Yaghi Siyan on 9 February 1098. Included in the rich spoils of war were much-needed horses.[79]

Extending the blockade. The arrival of materials and reinforcements and the destruction of a garrison cavalry force enabled the crusaders to alter the course of the siege. On 4 March a number of ships, including English vessels, arrived at St Symeon. The origins of this fleet are obscure, and, while Runciman has repeated Orderic Vitalis' story of English Varangians under the command of Edgar Atheling, there is no evidence to

---

[75] *Gesta Franc.* vi. 34; Anselm of Ribemonte, 'Epistula ad Manassem', ii, in Hagenmeyer (ed.), *Kreuzzugsbriefe*, 157.
[76] W. Porges, 'The Clergy, the Poor, and the Non-Combatants on the First Crusade', *Speculum*, 21 (1946), 1–20.
[77] Raymond of Aguilers, *Lib.* 54–5; France, 'Departure of Tatikios', 144–5.
[78] Ibid. 144–5.
[79] *Gesta Franc.* vi. 36–8.

substantiate this legend.[80] It is clear from the one participant account of this fleet that it called at Italy before reaching St Symeon.[81]

Whatever its origins, the squadron included workmen, tools, and construction materials. Bohemond and Raymond of St Gilles commanded an escort, which first collected more timber in the area around St Symeon before journeying toward Antioch.[82] After the escort's departure from the camps, a garrison cavalry force slipped across the Fortified Bridge and ambushed the convoy on its way to Antioch. While it is reported to have lost one hundred footmen and its materials, the convoy was not annihilated.[83] Bohemond returned to the main camps with news of the defeat, and a counter-attack was organized. The crusaders intercepted the sally party laden down with spoils as it tried to reach the Fortified Bridge, and massacred it. It is difficult to assess the importance of this victory on the crusader military situation and garrison morale. The Latin sources mention between 1,200 and 1,500 Muslim casualties, and the *Gesta Francorum* emphasizes that the crusader victory undermined garrison morale.[84] While this may reflect spirits within the crusader camps more than in Antioch, the action resulted in the first large-scale loss of life that the besiegers inflicted on the defenders.

Work on a counter-fort atop a hillock on the north bank of the Orontes overlooking the Fortified Bridge commenced soon after the victory. It is difficult to explain why the besiegers waited until March 1098 to fortify this position. Perhaps the arrival of materials and workmen was vital, and perhaps they simply did not feel strong enough to do so until then. The decision may reflect a growing importance of seaborne supplies and reinforcements in supporting the army. In any case, the construction and maintenance of this counter-fort, called La Mahomerie after the mosque located atop the hillock, was undertaken communally by the whole army. Anselm of Ribemont gives a splendid depiction of these activities:

Oh, with what effort and with what risks we fortified that place: one section of our forces held positions to the east, another guarded the camps, while the rest built the fortress. Crossbowmen and archers from the latter group watched the gate,

---

[80] David has shown that this fleet could not have been commanded by Edgar Atheling (David, *Robert Curthose*, 234–6).

[81] 'Epistula cleri et populi Luccensis ad omnes fideles', in Hagenmeyer (ed.), *Kreuzzugsbriefe*, 165–7.

[82] Ibid. 166.

[83] Stephen of Blois, 'Epistula ad Adelam', ii, in Hagenmeyer (ed.), *Kreuzzugsbriefe*, 151.

[84] Ibid. 151; *Gesta Franc.* vi. 41.

while the remainder, including our leaders, never ceased in the work of raising the mound, carrying rocks, or building the wall.[85]

The fortification consisted of two towers—presumably wooden—a curtain wall and mound and ditch. The fortress included Muslim tombstones looted from the graves near the mosque.[86] The position was placed under the command of Raymond of St Gilles, who was initially reluctant to assume this responsibility. His followers protested his acceptance, but eventually acquiesced and manned the position in turns. Perhaps this reluctance was due to the exposed nature of the position, isolated more than a mile from the main camps. The position was almost overrun at one stage and only saved by the timely arrival of a rescue force from the main camps.[87]

La Mahomerie and its garrison extended the blockade to include the crucial Fortified Bridge. Foragers were now more secure, as were the supply lines to St Symeon. Thus, four and a half months after the siege's commencement the besiegers achieved their first major advance in the city's isolation. This accomplishment required the whole expedition to concentrate its efforts in building a counter-fort and protecting these operations. This illustrates not only the difficulty of the enterprise, but also the crusaders' organizational response to their challenges.

The blockade was completed in April, when Tancred extended it to the monastery of St George opposite St George's Gate on the western side of the city. Tancred accepted four hundred marks to hold the position, although he was not satisfied with the size of the payment. The position was an isolated one and Tancred was instructed to employ only his own men in this operation, presumably to prevent drawing off booty-hungry knights from other contingents.[88]

Tancred's risk was rewarded on the first day of this operation, as he captured a large Armenian convoy attempting to enter the city.[89] With the establishment of the counter-fort of St George in early April 1098, the blockade was complete. While small units could still traverse the area outside the southern walls, the terrain hindered the movement of large groups and supply convoys. In April southern French patrols

---

[85] Anselm of Ribemonte, 'Epistula ad Manassem', ii, in Hagenmeyer (ed.) *Kreuzzugsbriefe*, 158–9.
[86] *Gesta Franc.* vii. 42.
[87] Raymond of Aguilers, *Lib.* 61–3. Peter Tudebode mentions this incident, and that the count of Toulouse hired knights to help in the position's defence (Peter Tudebode, *His. Hiero. itin.* (1977), viii. 1–2, pp. 79–80).
[88] *Gesta Franc.* viii. 43.   [89] Ibid. 43–4.

captured a large number of garrison horses and mules pasturing in a hidden valley on the southern circuit.[90]

The Capture of Antioch and Kherbogha's Siege

As the besiegers' supply situation improved with spring, that of the defenders deteriorated with the blockade's completion. Twelfth-century writers offered a number of explanations for the betrayal of Antioch to Bohemond by an Armenian tower commander, including bribery, divine visitation, and adultery.[91] An increasing shortage of food in the city must be included among the reasons for the Armenian's decision.

For whatever reasons, a section of the walls was betrayed and a storming party ascended them on the night of 2 June 1098. While the first ladder placed on the walls broke and there was considerable initial confusion, a second ladder was placed on the walls and an ordered ascent continued.[92] A general assault followed the storming party's entry into the city, and personnel from all contingents commenced sacking Antioch. Although crusaders under Bohemond were unable to capture the citadel, the city was at last taken.

After the sack and massacre of many inhabitants was concluded, the crusaders discovered that there was little food in the city.[93] While such a discovery would have been nothing but a satisfying confirmation of the blockade's success in other circumstances, it proved nearly disastrous for the crusaders who were besieged by Kherbogha of Mosul's relieving army on 5 June 1098.

Kherbogha's powerful force compelled the crusaders to defend the fortifications which they had so recently overcome. Although the crusaders attempted to hold La Mahomerie, this came under Kherbogha's control after three days of fighting.[94] The crusaders protected themselves against the reinforced citadel garrison by constructing a makeshift retaining wall at the foot of Mt Silpius which was defended by men and artillery.[95] Although their plight persuaded some to depart from Antioch and abandon the pilgrimage, morale was eventually restored after a series of visions and the discovery of what was believed to be the Lance of Longinus inside the city. This relic was used as a focus for mass propitiation and ceremonies which restored crusader

[90] Raymond of Aguilers, *Lib.* 63.
[91] Hagenmeyer, 'Chron. Prem. Crois', 6 (1898), 284.
[92] *Gesta Franc.* vii. 46–8. [93] AA iv. 25, p. 406; WT v. 23, pp. 302–3.
[94] Raymond of Aguilers, *Lib.* 50; *Gesta Franc.* ix. 50.
[95] *Gesta Franc.* ix. 61–2.

confidence and prepared them to meet Kherbogha's forces in open battle.[96] Whatever the origins and veracity of the Holy Lance of Antioch, the crusaders defeated Kherbogha on 28 June, ensuring the capture of Antioch and the Crusade's continuation. Although the surrender to Raymond of St Gilles by the citadel commander enabled Raymond to dispute Bohemond's claim to the city, Antioch was securely in crusader possession.[97]

While the siege of Antioch illustrates little of the crusaders' abilities in utilizing siege machinery, it demonstrates another facet of their siege technique. Methods of blockade and isolation based on counter-forts which were widespread in eleventh-century Europe were well suited to the strategic predicament which faced the crusaders in the autumn of 1097. Simply surviving the winter of 1097–8 as a cohesive force was a notable result of the decision to concentrate outside Antioch. In this regard crusaders drew upon their experience as much to protect the expedition as to reduce a major urban fortification. The demands of the siege also forced the crusaders to reorganize their military resources. The communal efforts of all contingents in building La Mahomerie stand out in this regard. Moreover, it is also clear that money or at least its promise was also a notable factor in deploying combatants to where they were needed. Whatever the methods employed, the capture of one of Syria's greatest fortifications confirmed the crusaders to contemporaries as successful and terrifying siege warriors.

Marrat-an-Noman, 28 November–12 December 1098

While the capture of Marrat may be secondary to the main affairs of the First Crusade, it is important in an assessment of crusader siege technology. Marrat was stormed with the aid of the first wall-dominating mobile siege tower employed by the crusaders. While the tower's success prefigures the dominant role this device was to have in crusading warfare, it was also the simplest such machine employed. Consequently the siege, the tower, and its degree of technical sophistication merit discussion.

[96] C. Morris, 'Policy and Visions: The Case of the Holy Lance at Antioch', in J. B. Gillingham and J. C. Holt (eds.), *War and Government in the Middle Ages: Essays in Honour of J. O. Prestwich* (Cambridge, 1984); Runciman, *His. Crus.* i. 241–6.
[97] Raymond of Aguilers, *Lib.* 82–3; *Gesta Franc.* ix. 70–1.

On 27 November 1098 the contingents of Robert of Flanders and Raymond of St Gilles, along with the mass of indigent crusaders which had followed Raymond since the fall of Antioch, arrived before the southern wall of Marrat-an-Noman. The city, situated in rolling countryside forty miles south-east of Antioch, lacked strong natural defences and was less well fortified than Nicea. The city's defences included a circuit wall and ditch guarding its southern side, and a citadel. Whatever the details of the strength of its masonry, or number of towers, it is clear that the fortification was significantly less formidable than Nicea, let alone Antioch.[98]

The crusaders launched an assault on the day after their arrival. This attack on Marrat's southern walls utilized two scaling ladders and failed. Raymond claimed that, had the attackers possessed four more ladders, they would have succeeded.[99]

Bohemond's troops arrived on 28 November and took up positions facing the northern walls. This prompted a second assault during which Raymond's troops tried to fill the southern ditch while Bohemond's attacked the northern wall. This attack failed completely, and Raymond of St Gilles decided to build a siege tower.[100]

Once again the crusaders experienced provisioning difficulties. The season and garrisons of nearby Muslim fortresses hindered foraging, and a lack of funds impeded purchase. As at Nicea and Antioch, these difficulties affected poorer crusaders most severely.[101]

By this stage of the Crusade many of the unattached poorer crusaders had coalesced into a group. While somewhat amorphous, the group's principal representative was Peter Bartholomew, the discoverer of the Holy Lance at Antioch. This group's leadership may have had connections with southern French clergymen who advised Raymond of St Gilles. While these poor followed Raymond, they occasionally coerced him to abandon policies for a rapid advance on Jerusalem.[102]

This group, or elements of it, correspond to those savage crusaders known as the Tafurs, ferocious fighters renowned for cannibalism and other atrocities. Whatever their effects on morale, they were valuable in

---

[98] At present, the city is located on a low mound with a pronounced depression on its southern side only. Cahen claims that a citadel built by Zengi replaced an earlier one (Cahen, *Syrie Nord*, 162).
[99] Raymond of Aguilers, *Lib.* 94–5.   [100] Ibid. 94–5; *Gesta Franc.* x. 75–9.
[101] Raymond of Aguilers, *Lib.* 95–6; Peter Tudebode, *His. Hiero. itin* (1977), xiii. 121.
[102] Raymond of Aguilers may have been a spokesman for this group (Raymond of Aguilers, *Lib.* 68–75, 84–5; J. H. Hill and L. L. Hill, *Raymond IV of St Gilles, Count of Toulouse* (Ithaca, NY, 1962), 89–90, 118–24).

performing the manual labour necessary in siege warfare. Not only did they gather materials, fill ditches, and carry rocks; they were also useful in pressing home the assault.[103]

Whatever the components and functions of these poor crusaders, their condition outside Marrat probably motivated Raymond of St Gilles to reduce the city as rapidly as possible.

Machinery and Assaults

Twelve days after the second unsuccessful assault of 29 November Raymond's contingent had completed its preparations for another assault. While Bohemond's troops attempted to scale the northern walls during the next assault, they did not assist in the construction or employment of siege machinery. There is no clear explanation for the southern French predominance in machinery production. While Raymond may have possessed more engineers and craftsmen than other lords, this may reflect his financial resources.[104]

Raymond's troops concentrated on neutralizing the defences of the city's southern walls by both undermining and seizing control of them. Sapping is listed as one of the means of attack agreed upon in a southern French council of war on 28 November, and played a part in enabling the crusaders to exploit their seizure of the walls.[105] Yet the centre-piece of the crusader assault was a wall-dominating mobile siege tower. Although the tower involved an armoured platform elevated above the height of the walls and made mobile by four wheels at its base, it was less sophisticated than subsequent towers. There was no bridging mechanism which would have allowed access from the tower to the walls. Thus the tower could not be used to protect and channel assault troops directly up on the walls. The tower was pushed by one hundred knights. While knights often fought in a tower's top storey, as was the case with this tower, the performance of such manual labour was unusual, especially when labouring troops were available. The tower's lack of certain features may explain the employment of pushers protected by body armour. If the tower's armour did not extend from the top platform down to ground level, those pushing would be exposed to missile fire and might find body armour necessary.

It is also possible that the quadrilateral formed by the tower's four

---

[103] Guibert of Nogent, *Gesta Dei Franc.* vii. 23, pp. 241–2; L. Sumberg, 'The Tafurs and the First Crusade', *Medieval Studies*, 21 (1959).
[104] Raymond's crews had been active if not successful at Nicea. See above, pp. 19–23.
[105] Raymond of Aguilers, *Lib.* 97; see below, pp. 212–13.

corner posts was not sufficiently large to accommodate all of the machine's pushers, requiring some men to shove from the outside. Were this so, it would suggest that the overall size of the tower was smaller than would later be usual. Either or both of these explanations would explain why armoured personnel were employed to propel the tower.[106]

This tower's lack of apparatus such as a bridge and complete armouring or generally small size could have been the result of a scarcity of suitable timber. While Albert of Aachen and the *Chanson d'Antioche* mention that Raymond of St Gilles was forced to raid a nearby castle to procure wood for his tower, neither of these sources is particularly reliable for the siege of Marrat.[106] However, it is clear that Raymond's forces possessed sufficient tools and fastenings for their task. The latter materials were presumably obtained from Antioch.

A second and more interesting explanation is that Raymond's engineers lacked the necessary expertise. While this may explain the lack of a bridge, the principal difficulty in siege-tower construction was the erection of a secure platform overlooking the walls which this device achieved. The most likely explanation is a decision to build swiftly. This tower was fabricated in eleven days, a record of crusading construction. Famine, and a belief that Marrat could be taken relatively easily, may have prompted Raymond of St Gilles to build a simple machine as rapidly as possible. Whatever the reason for the design of the siege tower at Marrat, the crusaders' hopes that the city would be taken were realized.

The crusaders commenced their attack on 11 December. Either on the eleventh or during the tower's construction they filled a portion of the ditch surrounding the southern wall and levelled a pathway for their tower.

The crusader tower was pushed to within a few feet of a civic tower, and a battle for control of that section of the walls and ramparts developed. The crusaders atop the towers, led by William VI of Montpellier, wielded pikes, hooks, and rocks in support of ladder attacks on the walls. One Everard the Hunter is reported to have blown his horn repeatedly while in the tower's top storey, and these blasts may have been signals to the tower's crew or assault troops. The defenders re-

---

[106] *Gesta Franc.* x. 79; Raymond of Aguilers, *Lib.* 97; Peter Tudebode, *His. Hiero. itin.* (1977), xiii. 122.
[107] The Lotharingians, about whom Albert of Aachen's account is most reliable, were not present, and the section of the *Chanson* describing Marrat's capture has an even more disputed authorship than the poem's earlier section (AA v. 30, pp. 450–1; Richard the Pilgrim, *Chanson d'Antioche*, vii. 476).

sponded with pikes and with an artillery piece which caused casualties amidst the knights fighting in the top storey. The defenders attempted to burn the siege tower with incendiaries but failed. At the same time Bohemond's troops tried to ascend the northern walls but were thwarted. Near dusk, troops on the southern side led by Geoffrey of La Tours captured a section of the ramparts. Despite support from the siege tower, the defenders contained the crusaders on the walls with hand-to-hand counter-attacks. This continued until sappers operating beneath the siege tower undermined the wall. When this was perceived by the defenders, they abandoned the ramparts and so allowed the crusaders to seize control of the walls and move into the city.[108] A breakthrough occurred at nightfall and Marrat's capture was completed the next day. Yet attempts to negotiate a surrender, the inchoate state of crusader military law, and the conflicting aims of different groups added a final chapter to the story of Marrat's fall.

According to the *Gesta Francorum*, the escalade of Raymond's troops effectively ended resistance and a wholesale sack commenced. Meanwhile, Bohemond obtained the wealthiest citizens as prisoners by sending word into the city that all those who collected themselves with their riches in the central mosque would be guaranteed their lives.[109]

Raymond of Aguilers gives a somewhat different version in that with nightfall the crusaders ceased their offensive activities and prepared to defend their gains against counter-attack. In the meantime poorer crusaders, emboldened by desperation and hunger, carried on the assault and concomitant massacre and sacking throughout the night and thereby gained a greater share of the city's plunder.[110] While Arabic sources give a different version from either of these two, their information helps clarify Latin accounts.

With the loss of the southern walls at nightfall, the defenders fled pell-mell into their houses. A truce was agreed and negotiations developed regarding a fixed sum to be paid per house in return for the safety of those inside. However, these negotiations broke down and the city was plundered, with the male population massacred, beginning at dawn on the twelfth.[111]

[108] *Gesta Franc.* x. 78–9; Raymond of Aguilers, *Lib.* 97; Peter Tudebode, *His. Hiero. itin.* (1977), xiii. 123–4.
[109] *Gesta Franc.* x. 79–80.   [110] Raymond of Aguilers, *Lib.* 97–8.
[111] The Damascus chronicle refers to disagreements between the defenders over whether to surrender; the Aleppo account simply refers to Frankish treachery (Ibn al-Qalanisi, *The Damascus Chronicle of the Crusades*, ed. and trans. H. A. R. Gibb (London, 1932; repr. 1967), 46–7; Kemal ed-din, *Extraits de la Chronique d'Alep*. (RHC Orient, 3), 586–7).

That the defenders attempted to negotiate terms is supported by Ralph of Caen, who describes the city's inhabitants prostrate before their conquerors, themselves undecided as to what course to follow. Ralph adds that those who wished to continue to sack and massacre prevailed.[112] It is clear that a surrender was attempted, and thus Bohemond's coup at collecting the valuable captives in one place did not occur in a vacuum but was one of a confused set of negotiations.

Moreover, given such negotiations, the activities of the poor in forcing the issue by continuing the assault becomes clearer, since they must have feared losing Marrat's wealth to the army's leaders. That the division of Marrat's booty concerned the poor is illustrated by Peter Tudebode's account of St Andrew's injunctions to the army through Peter Bartholomew. During a vision experienced before the assault St Andrew directed that a tenth of all Marrat's wealth should be reserved for the work of the Lord, and should be divided evenly between the bishops, church, clergy, and poor. Tudebode adds that this was agreed upon in general council—presumably a council of Raymond's followers.[113]

Raymond of Aguilers, despite his statement that the poor obtained a significant portion of plunder while the knights were guarding their positions, refers to southern French resentment at the amount of loot and fortifications obtained by Bohemond's troops.[114] It is clear that conflicting interests and ambiguous practices of booty distribution prevented a negotiated surrender of Marrat. While Marrat's capture may not have provided the poor with the wealth they desired, it did enhance the crusaders' reputation in siege warfare and for bloodthirstiness.

The crusaders' successes in northern Syria paid dividends during the march to Jerusalem. Local Muslim rulers gave the crusaders considerable support in January and February of 1099. Uncontested passage, access to markets, scouts, and gifts of money and horses were provided to facilitate the march. Although the crusaders met with resistance at Hisn-al-Akrad (the future Krak de Chevaliers), the position was abandoned to them on their second assault.[115] This easy progress must be seen partially as the fruit of earlier victories. A crusader distinction for success in sieges and savagery was important in determining the policies of Muslim rulers. Raymond of Aguilers

[112] Ralph of Caen, *Gesta Tanc.* 104, p. 679.
[113] Raymond refers to St Andrew's statement that the crusaders should pay their tithes, but gives no specific injunction (Peter Tudebode, *His. Hiero. itin.* (1977), xii. 122; Raymond of Aguilers, *Lib.* 95–7).
[114] Raymond of Aguilers, *Lib.* 98.
[115] *Gesta Franc.* x. 81–3; Raymond of Aguilers, *Lib.* 102–7.

speculated that stories of crusader cannibalism after the capture of Marrat augmented the crusaders' reputation for ferocity.[116] Paradoxically, this support was received in the absence of large-scale fighting, thus indicating a respect for crusader abilities at fortress-taking.

This phase of the Crusade also involved one of the expedition's military failures, the siege of Arqua. Because of a money payment by the emir of Tripoli and the Crusade's eventual success, the significance of this abortive eight-week siege has seldom been discussed. This reflects difficulties with the sources, as this stage of the expedition is generally not well chronicled and some primary accounts have particular problems of omission and bias.[117] Moreover, the siege's history has been dominated by debates concerning Raymond of St Gilles's bid for overall leadership, the motives of wrangling leaders, and the significance of events surrounding the ordeal of Peter Bartholomew.[118]

An examination of the military problems involved in the siege of Arqua may provide a background for these questions in the Crusade's history. Moreover, such an examination may illustrate some of the limitations of crusader siege technique.

Arqua, 14 February–13 May 1099

The contingents of Robert of Normandy, Tancred, Raymond of St Gilles, and the poor began the siege on 14 February 1099, and were joined in mid-March by Godfrey and Robert of Flanders. Whatever the internal political reasons for Raymond's decision to attack, Arqua proved a more difficult objective than Marrat.[119] Arqua, a dependency of the emirate of Tripoli, was a small city perched on the slopes and crest of a hill at the foot of Mt. Lebanon. It lay at the southern end of the plain of Buquaq, about ten miles from the sea. It was protected by at least one circuit wall, and a citadel dominated the summit. While Latin sources attest to the strength of its fortifications and vigour of its garrison, this may be seen as an explanation for failure.[120]

[116] Raymond of Aguilers, *Lib.* 101.
[117] J. France, 'The Crisis of the First Crusade: From the Defeat of Kerbogha to the Departure from Arqa', *Byzantion*, 40 (1970), 277–80.
[118] Ibid. 300–4; Hill and Hill, *Raymond IV*, 117–24.
[119] For the wider context of the siege, see Hill and Hill, *Raymond IV*, 117–26, and France, 'Crisis', 299–304. France also claims that Arqua must have appeared no more formidable than Marrat (ibid. 300).
[120] *Gesta Franc.* x. 83; Raymond of Aguilers, *Lib.* 106; Peter Tudebode, *His. Hiero. itin.* (1977), xiii. 128; G. LeStrange, *Palestine under the Moslems* (London, 1890), 398.

Whatever the state of Arqua's walls, its natural defences were formidable, especially against the kind of assault which had been made against Marrat. Mt. Lebanon guarded Arqua's southern approaches, so it could be approached by large numbers of troops only from the north and could not be assaulted from opposite sides, as Marrat had been. Arqua's location, even on a relatively small hill, complicated mass assaults and made the employment of a siege tower difficult, if not virtually impossible. Thus the centre-piece of the assault on Marrat could not be employed at Arqua.

While some troops attacked the territory of the emir of Tripoli, others in the service of Raymond of St Gilles encamped near Arqua and undertook siege operations. These operations have been obscured because the eyewitness accounts only mention casualties among the besiegers without giving details.[121] However, other writers, who did not draw upon one another, give broadly similar accounts of attacks, although the chronology and sequence of events is not clear.

The crusaders dug under Arqua's fortifications and attempted to sap its walls. Guibert of Nogent describes tunnelling and ceaseless effort, even on feast days. While they met with limited success, the garrison thwarted the sap, and mining operations ended.[122]

The crusaders also erected artillery against the city, and three pieces, under the command of Robert of Normandy, Tancred, and Raymond of St Gilles, bombarded Arqua. The defenders responded with their own rock-throwers and an artillery duel commenced. It was sometime during this action that Anselm of Ribemonte was killed by a garrison-hurled rock.[123]

These efforts were ineffective, and Arqua's defences were not seriously debilitated by the crusaders. No large-scale assault was launched, although Peter Bartholomew's last great vision and the reaction to it may have involved the question of such action. In this vision Peter was informed that true and false crusaders could be distinguished by their positioning in the ranks for a major assault.[124]

It is worth noting that the crusaders were well provisioned by land and sea at Arqua.[125] This alleviated the pressure of famine which had influenced the aggressive posture of the army at Marrat and may have

---

[121] *Gesta Franc.* x. 85; Raymond of Aguilers, *Lib.* 107.
[122] AA v. 31, pp. 451–2; Guibert of Nogent, *Gesta Dei Franc.* vi. 23, pp. 218–19.
[123] Ralph of Caen, *Gesta. Tanc.*, cv. 630; AA v. 31, pp. 451–2, Anselm's death is widely reported (*Gesta Franc.* x. 85; Raymond of Aguilers, *Lib.* 108–9).
[124] France, 'Crisis', 304; Morris, 'Visions', 43, Raymond of Aguilers, *Lib.* 112–24.
[125] Raymond of Aguilers, *Lib.* 108; *Gesta Franc.* x. 85.

been another factor in the decision not to assault Arqua. However, there were other reasons for not pressing an assault, nor maintaining the siege. Finally even Raymond of St Gilles became convinced, accepting a money payment and horses from the emir of Tripoli. These funds and horses were important in the Crusade's continuation, and with them crusaders moved south on 13 May.

There were a number of reasons for the siege's failure. A lack of motivation and unity of purpose plagued operations. The relative proximity of Jerusalem, wrangling between leaders, and a basic disagreement about the value of Arqua's capture contributed to this lack of unity. Added to these reasons must be the relationship between Arqua's defences and crusader siege technology. Despite Peter Bartholomew's urging, the crusaders were clearly disinclined to attempt to storm the city with straightforward personnel assaults. Arqua's defences and topography precluded the use of siege towers, at least without considerable preparation of approaches. While mining and artillery were employed, they did not neutralize Arqua's defences sufficiently to encourage a general assault. Some of the limitations of crusader siege methods are evident during the First Crusade itself. Moreover, it is clear that the capture of well-defended positions required the commitment and unity of purpose of the expedition.

Jerusalem, 7 June–15 July 1099

The capture of Jerusalem presented military and logistical challenges whose solutions were the measure of crusader siege technique. The siege involved the most concerted military and logistical efforts of the expedition, as well as the greatest number and types of machinery. The city's seizure illustrates not only experience gained in two years of campaigning, but also some of the salient features of crusader siege warfare in the Mediterranean.

The City's Defences and Preliminary Operations

The crusaders arrived at Jerusalem on 7 June 1099 with little option but to reduce the city as rapidly as possible. The approach of summer heat and the threat of a Fatimid relieving army precluded a long-term blockade.

The city's defences were based on a strong curtain wall reinforced by projecting mural towers. The 'Tower of David' Citadel midway along the western wall dominated that sector, as did the 'Quadrilateral' or 'Tancred's' tower at the north-western corner, making these areas difficult to attack. These fortifications took advantage of the area's topography, with ravines and elevations protecting the eastern, much of the southern, and part of the western walls. Northern and south-eastern approaches were enclosed by an outer wall. Prawer believes that portions of this fortification curved back joining the curtain wall to form oblong defensive enclosures.[126] Yet crusader narrative sources depict this *antemurale* primarily as a barrier to approaching the main walls, rather than as an independent redoubt. Whatever the structure of these defences, their primary aim was to obstruct access to the curtain wall over relatively negotiable terrain. This objective was furthered by a deep ditch which protected Jerusalem's walls from the Citadel, all along the northern section to the north-eastern corner. Another ditch masked south-western approaches opposite Mt. Zion. Thus sections of Jerusalem's walls which were not protected by natural features were guarded by a ditch and outer wall. Despite these extensive fortifications, Jerusalem was susceptible to large-scale assault.

As the Fatimids had captured Jerusalem from the Seljuk Turks in 1098, the Egyptian commander took every precaution. Nearby timber suitable for siege machinery was brought into the city, as were flocks and herds. Wells were poisoned, and a portion of the native Christian population driven outside.[127]

The crusaders deployed in a half-circle from the Church of St Stephen to the foot of Mt. Zion. Lords and contingents encamped from north to south as follows: Robert of Flanders, Robert of Normandy, Tancred, Godfrey, Raymond of St Gilles. Although Godfrey and Raymond initially encamped opposite the western wall, they both moved their base camps during the siege. Although the timing of Godfrey's shift is obscure and best dated after 15 June, Raymond moved his main camp on Mt. Zion soon after arriving to take advantage of more negotiable terrain.[128]

On 13 June the north-eastern sector was assaulted. The attackers, protected by their shields, broke through the barbican with picks and hammers and rushed the main wall. While they mounted one scaling ladder, the crusaders were unable to seize the ramparts. A crucial factor

[126] J. Prawer, 'The Jerusalem the Crusaders Captured: A Contribution to the Medieval Topography of the City', *CS* 2–3.
[127] WT vii. 23, pp. 374–5.
[128] *Gesta Franc.* x. 87; AA v. 46, pp. 463–4; Raymond of Aguilers, *Lib.* 137–8; Ralph of Caen, *Gesta Tanc.* 116, p. 687; WT viii. 5, p. 391, Prawer places Robert of Flanders opposite the south-western corner. (Prawer, 'Jerusalem Captured', 5.)

in this failure was a lack of sufficient scaling ladders, due to a scarcity of suitable timber. However, this attack served as a probe of Jerusalem's defences, even though Raymond of Aguilers and Ralph of Caen stress the injunctions of a hermit dwelling atop the Mt. of Olives in motivating what he believed would be a successful assault. Regardless of its aims, the attack demonstrated that, whatever the difficulties in obtaining materials, considerable siege machinery would be necessary in storming the city.[129] A council of war on 15 June decided to build such machinery, and began organizing the Crusade to facilitate production.[130]

Machinery Production and Preparations

In the four weeks between the council of war and a second assault the crusaders built numerous siege machines. At the least, two siege towers, a ram, armoured roofs, large portable shields, artillery, and scaling ladders were built.[131] The scale of production surpassed that of any other First Crusade siege, and compares favourably with later endeavours, particularly considering the difficulties encountered in obtaining materials. Whatever the degree of technical sophistication involved in this machinery, the amount produced must be seen as one of the necessary ingredients of crusader victory.

Three central problems confronted the crusaders in these endeavours: materials, technical expertise, and mobilizing manpower for building and supply operations. The expedition received important assistance with the first two problems from local inhabitants and European reinforcements. A native Syrian Christian guided crusaders to timber supplies some four to seven miles from Jerusalem.[132] While there is no mention of local craftsmen serving the crusaders, this possibility cannot be ruled out. Moreover, they may have provided some of the fastenings needed in machinery construction. Albert of Aachen attributes crusader employment of vinegar against Greek fire to information provided by local Christians.[133] In any case, assistance in locating timber was crucial to crusader success.

On 17 June six Christian ships arrived at Jaffa and requested an

[129] Raymond of Aguilers, *Lib.* 139; Ralph of Caen, *Gesta Tanc.* 118, 119, pp. 688–9; *Gesta Franc.* x. 88; AA vi. 1, p. 467; FC i. 27, pp. 293–4; WT viii. 4, p. 392.
[130] AA vi. 2, pp. 467–8; *Gesta Franc.* x. 88–90; Hagenmeyer, 'Chron. Prem. Crois', 7 (1900), 468.
[131] A discussion of this machinery and the evidence for it follows; see below, pp. 53–6.
[132] AA vi. 2, pp. 467–8; WT viii. 7, pp. 393–5.
[133] AA vi. 19, p. 476; for the properties of Greek fire, see below, p. 76.

escort to Jerusalem. Five of these ships were trapped in Jaffa by a Fatimid squadron and scuttled, with their crews joining the Crusade. Two of these ships were Genoese under the command of William Embriaco.[134] His sailors included expert carpenters, joiners, and craftsmen, who brought their hammers, axes, and adzes as well as nails, ropes, and fittings from their ships. These tools and abilities were invaluable in the rapid production of machinery at Jerusalem.[135]

Caffaro claimed that these Genoese also provided timber for siege machinery from their scuttled ships.[136] While it is possible that Raymond of St Gilles, who employed Embriaco, may also have obtained wood, especially the long beams involved in siege towers, from the Genoese, Caffaro's statement is not directly supported by other writers. However, there are accounts of Saracen captives transporting massive beams for the count of St Gilles's machinery, and they may refer to this timber.[137]

Whether or not they supplied timber, these Genoese provided expertise vital in the rapid production of machinery. Their experience in shipbuilding and repair was easily transferred to the heavy timber construction of siege machinery, and they may have had experience in siege-tower construction also. William of Tyre summarized the effect of these men as follows: 'What before their coming seemed difficult if not impossible was easily accomplished.'[138]

However important the assistance of native Christians and European reinforcements, it was not central to the crusaders' building achievement. The primary factor in this was the organization of the whole expedition to facilitate production. At some stage after the unsuccessful assault of 7 June, and probably during the council of war of 15 June, the army divided itself into two operational units. One division consisted of those lords encamped on Jerusalem's northern side, Robert of Normandy, Robert of Flanders, and Tancred, under the command of Godfrey of Bouillon (Northerners). Raymond of St Gilles's followers,

[134] Raymond of Aguilers, *Lib.* 141–2; *Gesta Franc.* x. 88; Caffaro, *Lib. civ. orient.* 110; WT viii. 9, p. 397; Hagenmeyer, 'Chron. Prem. Crois.', 7 (1900), 468–9; F. Cardini, 'Profilo di un crociato, Guglielmo Embriaco', *Archivo storico italiano*, 136 (1978), 421.

[135] 'Sed extraxerunt de navibus suis cordas et malleos ferri atque clavos et ascias atque dolabra et secures que permaxime nobis necessarie fuerunt' (Raymond of Aguilers, *Lib.* 147; WT viii. 9–10, pp. 397–400).

[136] 'galeas destruxerunt, et totum lignamen galearum, quod necessarium erat ad machina capiende civitatis, and Iherusalem portare fecerunt' (Caffaro, *Lib. civ. orient.* 110, 111).

[137] See below, pp. 51, 54.

[138] For a discussion of Genoese crusading and 'pre-crusading', see below, pp. 209–12; WT viii. 9, p. 397.

now based atop Mt. Zion on the city's southern side, formed the second division, under Raymond's command (Southerners).[139]

Each group appointed its own personnel to supervise gathering and building operations. Robert of Normandy and Robert of Flanders directed wood-cutting and transport for the Northerners, although Tancred played a role in locating suitable timber. Peter of Narbonne, bishop of Albara, supervised gathering and transport for the Southerners, although it is not clear whence they collected their timber. The bishop of Albara also supervised the captive Saracens, who were harnessed together to transport wood.[140]

Each unit also had its own director of construction. William Embriaco supervised Raymond of St Gilles's operations, and Gaston of Béarn those of the Northern lords.[141]

Reasons for Embriaco's selection have been discussed above, and he may also have engineered Raymond's siege tower.[142] The reasons for Gaston's appointment are less clear, as he was not associated with poliorcetics before taking the cross or during the Crusade, although he was to be subsequently.[143] As Gaston is described as a man of probity who instituted a division of labour among the workmen, his talents may have been primarily those of personnel management. Moreover, Gaston's Béarnais origins and earlier crusading service under Raymond of St Gilles before his associaton with Baldwin of Boulogne and Godfrey may have promoted harmony between the two major groupings before the Holy City.[144]

The manner of financing labour and machinery at Jerusalem also contributed to the rapid production of machinery. According to Raymond of Aguilers, all worked without payment except artisans, who were paid from public donations, and Raymond of St Gilles's own workmen, who were paid from the count's treasury.[145] While those who did non-specialist labour may have received some kind of maintenance, these freely performed efforts enabled the Crusade to mobilize its full resources.

---

[139] While aspects of this organization are evident in several accounts, Raymond of Aguilers gives the clearest formulation (Raymond of Aguilers, *Lib.* 145–7).
[140] Ibid. 146; AA vi. 2, pp. 467–8; WT viii. 6, pp. 392–3; Ralph of Caen, *Gesta Tanc.* 121, 122, p. 69; Peter Tudebode, *His. Hiero. itin.* (1977), xiv. 138.
[141] Raymond of Aguilers, *Lib.* 146, 147 n. 1; WT viii. 10, pp. 399–400.
[142] WT viii. 10, pp. 339–40; Cardini, 'Profilo crociato'; see also below, p. 54.
[143] See below, p. 171.
[144] Gaston followed Raymond of St Gilles until the capture of Antioch (Raymond of Aguilers, *Lib.* 145–6; Peter Tudebode, *His. Hiero. itin* (1977), xiii. 78; AA, v. 15, 42, pp. 441–2, 460–1.
[145] Raymond of Aguilers, *Lib.* 146.

Raymond of Aguilers also explains that crusaders were instructed by a council to lend their beasts of burden and servants to those directing the transport of materials. Moreover, every two knights were to be responsible for either one ladder or a portable shield.[146]

We cannot be certain concerning the degree to which these practices were prevalent among Godfrey's labourers and artisans. It seems likely that the considerable manual labour performed was also done without recompense.

The supply of food and especially water also involved considerable effort, expense, and loss of life. While there was no shortage of food, after 17 June water supply was precarious. Crusaders supplemented the occasional flowing of the pool of Siloam with water transported in hide containers from pools as far away as six miles. As garrison patrols made this laborious job dangerous, water was expensive in the camps.[147]

Water-carrying was presumably done by non-military personnel not directly involved in transport or production. Despite high prices, the importance of this activity to the maintenance of the siege should not be overlooked.

A picture of crusader organization is completed by Albert of Aachen's account of the production of the osier, rope, and hide armour necessary to protect machinery from rocks and incendiaries. The old, young, and female were concentrated in one place, to which materials were brought and then woven or stitched together by these non-combatants. Thus even those pilgrims who would have been of little use in heavy manual labour were able to contribute.[148] It is clear that every available source of labour in the Crusade from the aged and infirm to Muslim captives was tapped and organized for the production of machinery. The organization and division of labour in material gathering, transport, and construction are evident, and should be listed among the factors of crusader success at the siege of Jerusalem.

It is appropriate to comment upon the crusaders' religious and psychological preparations for their assault. Although these were doubtless numerous, they culminated in a penitential procession from Mt. Zion to the Mt. of Olives on 8 July. The procession consisted of a large portion of the expedition organized into groups of bishops, priests, knights, and other laymen. At the Mt. of Olives the crusaders heard sermons from some of the expedition's most eloquent preachers, including Peter the Hermit and Arnulf of Chocques. Although the garrison mocked the procession and profaned Christian religious

[146] Ibid.   [147] *Gesta Franc.* x. 89; Raymond of Aguilers, *Lib.* 140.   [148] AA vi. 3, p. 468.

symbols, there was no major Muslim military challenge.[149] It should be noted that the marchers' route was protected from Jerusalem by the vale of Kidron. This procession was consonant with the crusaders' belief in the divine sanction behind their expedition and with similar activities at Antioch in December 1097 and June 1098. The procession expressed the crusaders' unity and common purpose as well as their penitence. In any case, the procession was an important factor in generating morale and enthusiasm for the coming assault.

Crusader Siege Machinery

Despite the interest of historians since the twelfth century in the capture of Jerusalem, it has been difficult to establish a consensus about the number, type, sophistication, and specific employment of crusader siege machinery.[150] This stems from difficulties in the primary accounts and an uncritical use of their information. It should be noted that there is no single eyewitness account which provides detailed descriptions of all of the principal crusader machinery. This is equally true of histories written in the twelfth century, all of which concentrate on the deeds and machinery of one of the two main units. Prawer, well aware of these problems, has recently provided an excellent discussion of crusader siege machines and their deployment in light of his detailed topographical knowledge.[151] Yet it is possible to interpret the sources differently on some points. More importantly it is necessary to examine how the crusaders organized and facilitated the production and deployment of this machinery in understanding how they conquered the Holy City.

The crusaders based their attack on two mobile siege towers, one under the command of Raymond of St Gilles, and the other under Godfrey. Although William of Tyre mentions a third tower, this is a mistake. Eyewitness sources describe only two towers, and most of the later accounts follow this.[152]

---

[149] This event is described in varying degrees of detail by all of the participant writers, and mentioned in the leaders' official letter to Paschal II in September 1099. Peter Tudebode gives a particularly detailed account of the procession and mentions his own role in it (Peter Tudebode, *His. Hiero. itin.* (1977), xiv. 135–8; Raymond of Aguilers, *Lib.* 144–6; *Gesta Franc.* x. 90; AA vi. 8, pp. 470–1; WT viii. 11, pp. 400–1; 'Epistula ad Paschalem', in Hagenmeyer (ed.), *Kreuzzugsbriefe*, 170–4).

[150] Prawer, *His roy.* i 224–30; Runciman, *His. Crus.* i. 279–88; Grousset, *His. crois.* i. 153–63.

[151] Prawer, 'Jerusalem Captured', 9–10.

[152] Both Runciman and Grousset follow William of Tyre (Ralph of Caen, *Gesta Tanc.* 121–4, pp. 691–3; WT viii. 12, p. 402; *Gesta Franc.* x. 90–1; Raymond of Aguilers, *Lib.* 148–50; FC i. 27, p. 295; Peter Tudebode, *His. Hiero. itin.* (1977), xv. 138–9; AA vi. 11, p. 472; Runciman, *His. Crus.* i. 285; Grousset, *His. crois.* i. 157).

While William of Tyre gives a lucid depiction of the crusader towers, it is difficult to confirm his description completely from contemporary sources, since the most detailed accounts centre on Godfrey's tower and its assault.[153] Moreover, it seems that, although these two towers had similar features and appearances, they were built somewhat differently. Godfrey's tower had to be put together from shorter pieces of timber than desired because of a scarcity of long beams.[154] This presumably refers to the tower's corner posts, which usually provided the frame for the entire structure. Godfrey's tower, it seems, was built in stages, which were joined and scarfed together into one structure. The bottom storey would have been the tallest, with the longest beam available composing its frame, with successive storeys added and reinforced.[155] Ralph of Caen reports that this tower experienced structural difficulties during its advance and attributes this to the uncured timber which was used in its construction.[156] It is also possible that the tower's composite frame contributed to these difficulties.

By inference, Raymond's tower was built differently and there is a description of Muslim captives transporting massive beams to Raymond's building crews, strongly suggesting that Raymond's tower was built around four massive corner posts.[157]

Whatever their internal structures, both towers were covered over with layers of osier, rope, and hide armour to ward off rocks and incendiaries. Since those who propelled the towers at ground level are not described as knights nor as possessing individual protection, it is likely that some of the towers' armouring encompassed them. It is not known whether these machines moved along wheels or rollers. Because the sources concentrate on the successful assault of Godfrey's tower, we have more details about his device than about the count of St Gilles's. Godfrey and his brother Eustace commanded troops in the tower's top storey, which overtopped the walls by a 'spear's length'. Based on his knowledge of Jerusalem's defences, Prawer estimates this

---

[153] WT viii. 12, p. 402.

[154] 'turrim unam ex lignis exiguis, quia magna materies in locis illis non habetur compergerunt' (FC i. 27, p. 295).

[155] For this and other aspects of Romanesque carpentry, I am indebted to Dr C. Corrie, of the Institute of Historical Research, London.

[156] Ralph of Caen, *Gesta Tanc.* 123, 124, p. 692; see below, pp. 58–9.

[157] 'et sarracenis, quasi servis suis opera indicebant. Quia 1 vel ix portabant suo collo trabem maximam Quam non deferrent quatuor paria boum ad machinas construendas Iherusalem' (Raymond of Aguilers, *Lib.* 146; Peter Tudebode, *His. Hiero. itin.* (1977), xv. 138).

device to have been approximately fifty feet in height.[158] Missile fire from these personnel was intended to dominate and clear a section of the city walls.

In the middle storey of Godfrey's tower, which was level with the top of Jerusalem's walls, two brothers, Lithold and Englebert of Tournai, commanded a group of knights whose task it was to protect their machine and to seize the walls at a propitious moment.[159] These assault troops were provided with the means of bridging the wall, although there are different versions of the form of this bridge. In describing how Godfrey's men got on to the walls, Raymond of Aguilers tells us that they lowered an armoured covering which had protected the tower's front, extending upwards from the middle storey.[160] This may refer to a swinging drawbridge which also served as an extra layer of armouring, as William of Tyne and the Hills believed.[161] Prawer rejects this opinion for Godfrey's tower at least, and, drawing on Albert of Aachen's statement that the Northerners got on to the walls across extended tree trunks, believes that they fabricated a 'makeshift bridge' during the assault.[162] Yet it is not at all clear that Albert refers to a 'makeshift bridge' and not to planks which were intended to be pushed out from the middle storey. It is, of course, possible that Raymond of St Gilles's tower possessed an armoured drawbridge and that Godfrey's bridge consisted of extending planks, and that Raymond of Aguilers assumed that both towers were equipped with the same apparatus. Fulcher of Chartres has a story of beams seized from the defenders and employed as the tower bridge.[163] William of Tyre tries to reconcile all accounts by describing a bridge similar to Raymond's which the crusaders reinforced with beams seized from the defenders, and this may be the best compromise.[164] In any case, it is difficult to believe that both of these towers were not equipped with some kind of bridge, since Raymond of Aguilers was present at the siege and the personnel in Godfrey's tower were clearly intended as assault troops. Moreover, there are details about the assault of the Northerners' tower discussed below which may support this interpretation.

[158] AA vi. 12, p. 473; Prawer, 'Jerusalem Captured', 10.
[159] AA vi. 11, 19, pp. 472, 477; WT viii. 18, pp. 409–10.
[160] 'solverunt cratem desursum qua muniebantur anteriora turris conducte a summo usque ad medium' (Raymond of Aguilers, *Lib.* 150; Raymond of Aguilers consistently uses the term *cratis* to describe osier and hide armoured protection).
[161] WT viii. 12, 18, pp. 402, 409; Raymond of Aguilers, *Historia Francorum qui ceperunt Iherusalem (The History of the Frankish Conquerors of Jerusalem)*, ed. and trans. J. H. Hill and L. L. Hill (Philadelphia, 1968), 127.
[162] Prawer, 'Jerusalem Captured', 10.   [163] FC i. 27, pp. 297–8.
[164] WT viii. 12, 18, pp. 402, 409.

Bridges, assault troops, and the ground-level armouring were features not found on the crusader tower at Marrat, and so were measures of greater technical sophistication of crusader machinery at Jerusalem. Although these components were secondary refinements compared to the problem of elevating a firing platform above the height of the walls common to all siege towers, they were important in allowing Godfrey's troops to seize Jerusalem's defences.

Although both units based their assaults around a siege tower, supporting devices were necessary to get the towers to the walls. Light and heavy artillery were built for both Northerners and Southerners, as were scaling ladders and portable shields.[165]

The Northerners also built a large ram covered by an armoured roof which protected labouring crews.[166] According to France's 'unknown source for the capture of Jerusalem', the Southerners also employed some kind of armoured covering for crews labouring near the wall. This defensive structure protected men who tore down masonry already damaged by artillery fire with a long pole fitted with an iron hook.[167] It may also have assisted in making and levelling a pathway for Raymond's tower. It remains, however, that this device, like the ram and artillery, was intended primarily to support the attacks of the siege towers.

The Crusader Assault

Before discussing details of the crusader assault, it is necessary to understand the defences added by the garrison during the time of crusader machinery production. Fatimid defenders raised the height of defensive towers and protected the upper works of their walls with wood and rope, and other shock absorbers against artillery in areas where they anticipated an attack. The garrison also placed artillery pieces atop their towers and prepared incendiaries.[168]

Because of the topography of Jerusalem's southern approaches, Raymond of St Gilles had little choice but to attack directly into these prepared positions. However, terrain along the northern walls permitted attacks at several places. Moreover, as the crusaders had concentrated along the western half of the northern wall, so had Fatimid

[165] Raymond of Aguilers, *Lib.* 149; AA vi. 9, p. 471; FC i. 27, pp. 295–6.
[166] AA vi. 10, p. 472; Ralph of Caen, *Gesta Tanc.* 124, pp. 691–2.
[167] J. France, 'An Unknown Account of the Capture of Jerusalem', *English Historical Review*, 87 (1972), 778–9.
[168] Raymond of Aguilers, *Lib.* 147; WT viii. 8, 12, pp. 395–6, 402; Ralph of Caen, *Gesta Tanc.* 123, p. 690; *Gesta Franc.* x. 90.

strengthening. Consequently, Northern commanders decided to shift their point of attack further east along the northern wall on the other side of St Stephen's Church. This involved the transport of all machinery before its final assembly, projectiles, and camps over a distance of approximately three-quarters of a mile. This required a prodigious effort, as it was carried out solely during the night of 9 July.[169] With this shift the crusaders were ready to put the finishing touches to their machines and propel them to the walls.

The Southerners' first problem in moving up their tower was a deep ditch within missile range of the walls. Raymond's council of war decided to fill it by having individual and small groups of crusaders cast in materials by day and night, presumably to minimize the effect of missile fire. Crusaders were offered one penny for every three stones cast in, and the operation took three full days and nights. While it is not clear how many survived to collect their reward, the ditch was filled.[170] Labouring crews, protected by armoured roofs and portable shields, must have evened out the fill and levelled a pathway for the tower. Thus, by 14 July Raymond's tower was in a position to threaten the walls.

The Northerners spent three days after their shift joining the storeys of their siege tower together and fitting it out, and they commenced their assault on 13 July.[171] They also had to fill a deep ditch, and level a pathway for their tower, as well as smash a gap in the outwork guarding the northern wall. With their artillery deployed, Godfrey's troops moved their ram up to breach the barbican. The crusaders defended their device against incendiary attacks, and protected a pathway created after a gap had been made probably by 14 July.[172] However, the ram itself was now an impediment to the tower's advance, and the crusaders burnt it once it had completed its task. Ironic as the action may have seemed to those who had defended the device against garrison attempts to burn it, the ram had served its purpose.[173] More importantly, the siege tower enjoyed restricted lateral mobility, and the terrain permitted the approach of ponderous machinery in only a

[169] *Gesta Franc.* x. 90; Raymond of Aguilers, *Lib.* 147; AA vi. 9, p. 451; Ralph of Caen, *Gesta Tanc.* 123, p. 691; FC i. 27, pp. 295–6; WT viii. 12, pp. 401–2.
[170] The dating of this operation is not clear. If the Southern siege tower could be moved near the walls on 14 July, as Raymond of Aguilers implies, then this action must have commenced on the eleventh (*Gesta Franc.* x. 91; Raymond of Aguilers, *Lib.* 149; Hagenmeyer, 'Chron. Prem. Crois.', 7 (1900), 475–8).
[171] *Gesta Franc.* x. 90; Hagenmeyer, 'Chron. Prem. Crois.', 7 (1900), 476–9.
[172] AA vi. 11, p. 472; Ralph of Caen, *Gesta Tanc.* 124, pp. 691–2.
[173] The crusaders' destruction of their own ram is another example in which Albert of Aachen is more detailed and therefore clearer than William of Tyre (AA vi. 11, p. 47).

limited number of places along that sector of the city.[174] Moreover, the sequence of events has its own logic, especially assuming that the tower had limited lateral mobility. Despite, and partially because of, the destruction of this ram, Godfrey's tower was also approaching the walls on 14 July.

As the crusader tower approached, an artillery battle developed, with the garrison concentrating on the siege towers and the crusaders on Muslim pieces. It should be noted that crusader artillery operated throughout the siege exclusively in support of the major assaults of men and machines. That is, crusader rock-throwers strove to debilitate upper works, as anti-personnel weapons, and to neutralize defending artillery. Crusader artillery did not attempt to breach the walls nor bombard the city independently while the besiegers built their primary assault devices. Moreover, during the main assault crusader artillery was outnumbered, particularly on the southern side. Raymond's tower was more heavily damaged than Godfrey's, presumably because Raymond attacked into prepared positions. Although Northern defenders hurriedly erected artillery protected by bags of grain and osiers atop a civic tower near Godfrey's siege tower, these weapons were less effective than those directed against Raymond's structure.[175] The defenders projected incendiaries of several types, including coals, fire brands, inflammable jars, and pitch-coated objects studded with nails.[176]

Despite the efforts of all concerned, 14 July ended without decisive advantage to either attackers or defenders, although the crusader towers were still intact near the walls. Besiegers and besieged spent the night under arms protecting their respective positions.

With the beginning of action on 15 July both Godfrey and Raymond attempted to close the gap between their towers and the walls. By midday Raymond's advance had been stalled and his tower damaged. As a council of war debated whether to withdraw the tower for repair, they received news that Godfrey's troops had achieved a decisive breakthrough.[177]

Having cleared the outer fortification of the embers of their ram, the Northerners pushed their tower at an angle from its original line of approach towards Jerusalem's main walls. During this approach the tower experienced structural difficulties and appeared in danger of

[174] Prawer, 'Jerusalem Captured', 9-11, Figs. 2, 3.
[175] AA vi. 15, 17, pp. 474-6; Raymond of Aguilers, *Lib.* 149.
[176] AA vi. 17, p. 475; Raymond of Aguilers, *Lib.* 148.
[177] Raymond of Aguilers, *Lib.* 149-50; *Gesta Franc.* x. 90-1; Peter Tudebode, *His. Hiero. itin.* (1977), xv. 106-7.

collapsing.[178] This may be attributed to defending artillery material and design features discussed above, the difficulties of negotiating the terrain, and perhaps the stress placed on the tower's stability by the men and equipment in it, particularly as it moved.[179]

Whatever the causes of its difficulties, the tower was brought against a section of the walls near Herod's Gate, where an intense battle for control of the walls developed. The tower halted a few feet from the walls, probably within ten feet, at the maximum length of its bridging apparatus or perhaps just beyond that distance. Although combatants on the tower and walls did not yet grapple hand to hand, missile weapons of every variety were employed at close range in the struggle for this sector of Jerusalem's defences.

The proximity of the siege tower to the city walls had several effects on the artillery battle raging in the course of the assault. Defending pieces which had been situated along the walls were moved into the city, where they had difficulty finding the tower's range, perhaps because of a lack of familiarity with indirect fire.[180]

The section attacked by the crusader tower was flanked by two mural towers and this allowed the defenders to deliver an enfilade against the crusader structure from two sides. The crusaders responded with an intensive artillery bombardment, with Robert of Normandy directing operations against the rock-throwers situated on one tower, and Tancred against the other.[181] Northern artillery was effective in neutralizing Muslim pieces, while defenders concentrated on burning the siege tower.

In approaching the walls Godfrey's tower crews warded off incendiary attacks, wich increased in scope the closer the tower advanced, culminating when the defenders dropped a massive beam saturated with Greek fire and other incendiaries into the narrow gap between the siege tower and walls. While the beam may also have been intended to batter the tower, it was hoped that the timber and incendiaries would result in a blaze of such intensity and duration that the tower would be ignited. Yet Godfrey's troops managed to extinguish the blaze with vinegar before the siege tower ignited.[182]

Moreover, at this juncture fire was assisting the attackers in clearing the walls. Among those objects placed along the wall's upper works to cushion them against artillery, particularly on the northern sector,

---

[178] Ralph of Caen, *Gesta Tanc.* 124, p. 692.   [179] See above, p. 54.
[180] AA vi. 19, p. 477.   [181] Ibid.; Ralph of Caen, *Gesta Tanc.* 125, p. 692.
[182] AA vi. 17, 18, pp. 476-8.

were inflammable materials. These were ignited by the attackers with incendiary projectiles on the northern side of the city sometime during 15 July. The smoke from this blaze increased confusion and pressure on the defenders, who abandoned sections of the ramparts, including those struggling against Godfrey's tower.[183] Albert of Aachen places a different emphasis on the factors contributing to the clearing of the wall. He describes the igniting of upper-work defences earlier in his sequence of events, and attributes the defenders' flight to crusader artillery and missile weapons from the top of the siege tower.[184] Undoubtedly all these factors should be taken into account in understanding how the crusaders cleared a vital section of the walls. Moreover, although the crusaders concentrated their attacks on opposite sides of the city, the defenders must have maintained the whole perimeter of walls facing crusader encampments. It is also worth noting that these tactics were successful only for Godfrey's tower, and not for Raymond's. However, the importance of Raymond's attack in engaging defenders should not be underestimated, as it was an essential component of crusader victory.

With defences cleared, the crusaders bridged the walls. Fulcher of Chartres reports that this bridge was made from beams originally emplaced as a defence against crusader artillery which the crusaders cut down and brought up to their tower. William of Tyre reconciles this story with other accounts by explaining that the crusaders reinforced their bridge with these beams. This story is more plausible if we accept that the tower was damaged when halted and was unable to close the gap sufficiently for the bridge to use the battlements themselves as an effective support for the weight of knights moving across it. It is also possible that captured timber simply increased the number of planks pushed out from the tower's middle section, if that was the type of bridge with which Godfrey's device was equipped, or that the crusaders fashioned a 'makeshift bridge' in the heat of close-quarters fighting.[185] It should be noted that Jerusalem was much more formidable than Marrat, and that the strength of the former's defences suggests that attackers designed some form of bridge for their tower, however it may have been augmented in action. In this regard the siege tower transported and

[183] Raymond of Aguilers, *Lib.* 150; FC i. 27, pp. 298–9; Ralph of Caen, *Gesta Tanc.* 125, p. 692.
[184] AA vi. 19, p. 477.
[185] For a discussion of the bridge, see above, p. 55. FC i. 27, pp. 297–9; WT viii. 16, pp. 407–8, 18, pp. 409–10; Ralph of Caen, *Gesta Tanc.* 125, p. 692; AA vi. 19, p. 477.

protected assault troops as well as provided an elevated platform for missile weapons. However, the most important aspect of Fulcher's story concerning crusader utilization of captured timber is not how it relates to ambiguous descriptions of the tower's bridge. Rather it illustrates that the Northerners had achieved a command of Jerusalem's defences in the area immediately around their siege tower at approximately 9.00 a.m. on 15 July 1099.

However they bridged them, assault troops in the middle storey of Godfrey's tower seized the walls and began moving along them and into the city. Comrades outside placed ladders against newly secured walls and poured into the city. Troops led by Tancred drove through the city towards the Temple, exploiting the breakthrough and inflicting heavy casualties on inhabitants gathered there. The gates of St Stephen and Jehosophat were opened and crusaders, including many non-combatants, flooded into the city and began a large-scale sack and massacre.[186] With the realization that the crusaders had entered the city, those fighting against Raymond of St Gilles's troops fled, and Raymond's men poured over the walls.[187] With this the pillage and slaughter became general. Only those defenders who reached the Citadel and surrendered to Raymond of St Gilles were spared the horrors of the sack of Jerusalem. Although some inhabitants, mostly Jewish, survived to be ransomed, most of Jerusalem's defenders and citizens were annihilated.[188] While there is no account of crusader casualties, they cannot have been inconsequential.

With the assistance of local guides, Genoese experts, and a prodigious effort and organization, the crusaders worshipped at the Holy Sepulchre as pilgrims and as conquerors.

Several aspects of the capture of Jerusalem should be noted in conclusion. The role of non-combatants was considerable, as these personnel performed much of the manual labour necessary, in gathering materials and water and in preparing siege engines and armour for attack. Their efforts were also essential in supporting crusader assaults. While these crusaders were no asset on the march or on the battlefield, they were invaluable at the siege of Jerusalem.

Crusaders at Jerusalem employed not only the most sophisticated

[186] *Gesta Franc.* x. 91; AA vi. 19, 20, pp. 477–8; WT viii. 18, pp. 409–10; Ralph of Caen, *Gesta Tanc.* 125–7, pp. 692–4.
[187] *Gesta Franc.* x. 91; Raymond of Aguilers, *Lib.* 150–1.
[188] Prawer, *His. roy.* 230–4; S. D. Goitein, 'Contemporary Letters on the Capture of Jerusalem by the Crusaders', *Journal of Jewish Studies*, 3 (1952), 162–7.

siege machinery of the expedition, but also the greatest quantity. The crusaders' ability to organize themselves to facilitate the construction and employment of their machinery stands out as the single most important factor in their success. While a number of factors contributed to this, we should note the focus for unity which Jerusalem provided and the experience gained by the crusaders in siege warfare during the course of the expedition. It is in this context that we may best appreciate the Genoese contribution at Jerusalem. The Genoese provided materials and expertise in carpentry vital to the rapid production of machinery and their presence may account for the greater complexity of siege towers employed at Jerusalem than that built at Marrat. Yet it was the rapid production of a large quantity of machinery as much as the technical sophistication of these devices which was so critical to besiegers compelled by season and a hostile relieving army to reduce their objective quickly. Although vital, the Genoese were but a component in a force which succeeded because of its organization for siege-machinery production and assault.

An overall assessment of siege warfare in the First Crusade emphasizes the range of techniques employed by crusaders. The capture of Antioch involved the most difficult of blockades, and the capture of Jerusalem a complex assault conducted with considerable machinery. Although crusaders' siegecraft was based on their European experience, operations in the Crusade contributed to the development of their techniques. It is difficult to perceive an ever-increasing level of sophistication in crusader poliorcetics, primarily because the capture of Antioch involved few siege engines. However, the scale and sophistication of crusader devices at Jerusalem far surpassed that of earlier sieges. While this was partially a necessary adaptation to circumstances, it was made possible by technical assistance from newly arrived Genoese sailors and a remarkable organization of the expedition itself. Thus it is difficult—or perhaps irrelevant—to argue that the crusaders based their siegecraft on Near-Eastern techniques. While we cannot rule out the possibility that native engineers served crusader lords, the evidence of the siege of Jerusalem argues that crusader success resulted from the organization of their own abilities and technological expertise from western Europe.

It is clear that the mobile wall-dominating siege tower was the single most decisive tool of crusader siege technique. Moreover, crusaders were willing to accept the substantial casualties that the effective employment of such devices in close proximity to prepared defences usually involved. Some of the military and logistical arrangements of this first armed pilgrimage, as well as the expedition's ideology, may have facilitated this readiness. The importance of the siege tower reflects not only the siege warfare of the First Crusade but also the pattern of subsequent Latin conquests on the coast of Outremer. Thus, in assessing the role and impact of these machines, it is appropriate to examine the campaigns of the Latin conquest of the Syro-Palestinian littoral.

2
THE CAPTURE OF THE PALESTINIAN COAST AND THE DEVELOPMENT OF CRUSADER SIEGE TECHNIQUE

WITH the crusaders' triumph over the Fatimids at the battle of Ascalon on 12 August 1099, their possession of Jerusalem was temporarily assured. Yet this did not establish Latin authority securely in the Holy Land, and much of the territory which would constitute the Latin states remained to be conquered. One crucial area in the establishment of Latin power in the eastern Mediterranean was the Syro-Palestinian coast. Control of the littoral was vital in several regards, the most immediate involving lines of communication with Europe.

Moreover, the area's large cities were important military centres. Not only were their garrisons and civic forces powerful, but they provided forward bases for the Fatimid fleet. The wealth of these urban centres made them rich prizes for their conquerors. Although plunder and captives could constitute an immediate reward, the place of these cities in the commercial networks of the region and beyond made them valuable objectives of taxation.

The reduction of these cities involved the military resources of the Latin states and their allies in protracted campaigns of siege warfare which merit close study. Topographical and military factors as well as impulses of assistance from Latin Europe influenced the rhythms of this offensive, and these strategic concerns are well illustrated in the sieges of these campaigns. Because a number of attacks are related only briefly by sources which are generally reliable, it is not appropriate to examine the coastal offensive by narrating every siege. Rather we shall discuss the main characteristics of siege warfare within the context of the coastal offensive. By taking this approach, the pattern of siege warfare and the factors which determined it, as well as the methods and machinery employed, will be illuminated. The operations undertaken in the capture of the littoral illustrate crusader siege technique and its development in the Levant. An understanding of this matter may cast light on wider questions of Latin expansion in the Mediterranean.

Sources

Although some of the Latin historians of the First Crusade are also important for the twelfth century, the problems of source criticism are not as complex. While there are fewer sources which discuss siege operations in detail, they are generally in agreement about the major points of operations and machinery. Fulcher of Chartres's account of the reigns of Baldwin I and II is a fundamental source for the period and has considerable information about siege activities. While Albert of Aachen's utility as a source for the First Crusade has been questioned in the past, his history has long been recognized as important and accurate for the Latin east until 1120.[1] Nevertheless, it should be noted that on the whole accounts of the coastal offensive are less detailed than those of the sieges of the First Crusade.

William of Tyre, who died sometime between 1184 and 1186, has a rightly deserved reputation as a major historian of the twelfth century. Whatever his limitations as a chronicler of the assault of Godfrey of Bouillon's forces at Jerusalem, his history of the Latins in the east, including their siege operations, is of cardinal importance. William's account of the siege of Tyre in 1124 is particularly lucid, and has been taken as an illustration of his insight and skill as a historian.[2] The Genoese crusader and historian Caffaro provides an important and sometimes eyewitness account of the major Genoese expeditions to the east.[3]

Our understanding of siege operations during this period is enhanced by a contemporary Arabic history written in Damascus during the first half of the twelfth century. *The Damascus Chronicle* of Ibn al-Qalanisi is an important source, as the author was interested in the siege operations and machinery not only of his co-religionists but also of their Latin foes.[4]

[1] A. Beaumont, 'Albert of Aachen and the County of Edessa', in L. Paetow (ed.), *The Crusades and Other Historical Essays Presented to D. C. Munro* (New York, 1928).
[2] Edbury and Rowe, *William of Tyre*, 50–2.
[3] For a discussion of Caffaro as a historian, see below, p. 195.
[4] *DC*.

Patterns of Conflict

The capture of well-fortified coastal cities was crucial for the nascent crusader states of southern Syria and Palestine. The initial phase of conquest had left Latins holding positions in the interior but with limited control of the coast.[5] The littoral was a frontier area dominated by Muslim commercial cities supported by Fatimid military and naval power. Edessa and Antioch were not challenged by such difficulties, as the former was landlocked, and the latter had secured the port of St Symeon before the capture of Antioch itself. However, the survival of the Latin states which developed around Jerusalem and Tripoli depended on the eventual conquest of the Mediterranean littoral.

Although vital, the coastal offensive was conducted intermittently and not completed until the capture of Ascalon in 1153. While this had many causes, three should be highlighted. In the first place the crusaders were numerically weak, especially in the early twelfth century. Whatever the exact military strength of Baldwin I's realm in 1100, conservatively estimated at two hundred knights and one thousand footmen, Latin writers emphasize the effects a lack of sufficient troops had on security and freedom of action.[6] These writers may have exaggerated the degree of crusader weakness, particularly by giving figures for particular leaders and areas and not for Latins generally.[7] Moreover, Latins became more numerous during the twelfth century. However, there is little doubt that a lack of numbers affected the pattern of warfare in the early history of the Latin states.

A second, related factor involved the several fronts on which warfare in the Latin states could be waged.[8] While garrisoning requirements decreased manpower available for offensive action, such operations in one area could be undercut by military disasters in another. Having allied with a northern European pilgrim fleet and prepared much of his siege machinery, Baldwin I was compelled to abandon a major attack on Sidon in 1107. The death in battle of Hugh of Tiberias and many of his troops necessitated royal military assistance in averting disaster in

---

[5] For the military situation, see Prawer, *His. roy.* i. 241–58.
[6] Ibid. 250; FC ii. 6, pp. 388–9; WT ix. 19, pp. 445–6.
[7] J. S. C. Riley-Smith, 'The Motives of the Earliest Crusaders and Settlement of Latin Palestine', *English Historical Review*, 98 (1983), 724.
[8] For a discussion of the development of military frontiers, see J. Prawer, *Crusader Institutions* (Oxford, 1980), 472–8.

Galilee. While Baldwin received a cash payment from Sidon, this offensive opportunity was lost.[9]

Truces between cities and Latin rulers in one area occasionally facilitated offensive activities in another. Such an agreement enabled Baldwin I to march to the siege of Tripoli in 1109 without harassment from the garrisons of Tyre, Sidon, and Beirut.[10] Since Cairo and Damascus were sometimes preoccupied with internal strife or external Muslim opponents, they were not always able to give coastal cities full support, and this was on occasion important in the outcome of a major attack. However, the difficulties of waging warfare simultaneously on several fronts contributed to the piecemeal fashion in which the coastal offensive was conducted.

The Latin states' lack of warships was a third important factor in determining the pattern of the coastal offensive. Fatimid ability to resupply and reinforce by sea, even when cities were under land attack, was crucial to civic morale and military effectiveness. Acre was closely besieged by Baldwin I and some five thousand troops in the spring of 1103. The Latins brought artillery and a siege tower against the city's defenders, who were compelled to begin negotiating for terms after protracted fighting. A Fatimid squadron of twelve ships collected from Levantine ports arrived before the surrender had taken place and stiffened the defenders' resolve. They sallied and damaged several attacking machines, killing Baldwin's archery master. The king decided that his forces could not wear down the city, and abandoned the siege after destroying much of Acre's orchards. While several factors contributed to this failure, William of Tyre noted that a lack of Latin sea power was the most decisive.[11] Despite this setback and other similar failures, including a major siege of Sidon in 1108, the Latin states did not develop their own sea power sufficiently to counterbalance Fatimid strength. While Gerard of Sidon commanded a Latin kingdom squadron of some fifteen ships at the siege of Ascalon in 1153, it failed to keep out an Egyptian relieving fleet.[12] Other means of neutralizing Fatimid naval power were necessary.

By allying with Christian shipping which periodically visited Outremer, Latin rulers obtained naval assistance. While pilgrim fleets whose members returned to Europe occasionally provided

---

[9] AA x. 3–8, pp. 632–3; DC 74–5.   [10] AA xi. 11, p. 677.
[11] AA ix. 19, 20, pp. 601–2; FC ii. 22, p. 406; WT x. 26, pp. 439–40.
[12] WT xvii. 24, pp. 784–95.

aid, naval support usually came from one of the Italian maritime republics, Genoa, Venice, or Pisa.[13] In return for their assistance, Italians received tax-free trading enclaves with their own judicial and ecclesiastical rights in captured cities, as well as a portion of the spoils.[14]

Whatever their effects on the constitutions and fiscal resources of Latin states, these arrangements remedied crusader naval weakness. We should note that war galleys did not conduct long-term blockades similar to those of the 'age of sail'. Rather, galley fleets were beached except for patrol craft during a siege, unless a relieving force attempted to sail into a beleaguered city. Christian squadrons sometimes defeated Fatimid warships before an offensive against a major city commenced, usually ensuring an untroubled siege. The Venetian triumph over a substantial Fatimid fleet at Ascalon in 1123 is the most celebrated such action, as it greatly facilitated a major attack on Tyre during the next campaigning season. Yet, as naval warfare is beyond the scope of this study, its importance to siege warfare must be summarized in the realization that land-based attacks on coastal cities were much more likely to be successful if Christian ships prevented Fatimid seaborne relief.[15]

This partnership between land-based Latin rulers and maritime forces was vital to the establishment of the Latin states as well as to the timing of attacks on particular cities.[16] Moreover, this partnership was crucial in the conduct of assaults. While the importance of naval support has long been recognized, the role of maritime personnel in siege operations has seldom been discussed in any detail.[17] Yet to appreciate this role and the conditions which made it so important, we must concentrate upon the details of crusader assaults.

---

[13] A Norwegian pilgrim fleet contributed to the capture of Sidon in 1110, and pilgrim ships were important in the capture of Ascalon (FC ii. 44, pp. 543-8; WT xvii. 23, pp. 792-3.

[14] Runciman, *His. Crus.*, ii. 16; W. Heyd, *Histoire du commerce du Levant au Moyen Âge* (2 vols; Leipzig, 1885), ii. 131-44; Mayer, *Crus.* 65, 74.

[15] Foster discusses naval warfare and the number and size of Christian fleets which assisted the Latin states; while Guilmartin's book concerns naval conflict in the twilight of the galley age, his observations about the special characteristics of galley warfare are illuminating (S. M. Foster, 'Some Aspects of Maritime Activity and the Use of Sea Power in Relation to the Crusading States, 1096-1169', D.Phil. thesis (Oxford, 1978); J. F. Guilmartin, Jun., *Gunpowder and Galleys: Changing Technology and Mediterranean Warfare at Sea in the 16th Century* (Cambridge, 1974).

[16] Prawer, *His. crois.* i. 254.

[17] Foster discusses 'Sea Power and Siege Warfare' in chapter 5 (Foster, 'Sea Power', 267-97).

Raiding and Counter-Forts

First it is appropriate to discuss another aspect of the context of crusader sieges, raiding and counter-forts. The proximity of the crusader establishment to Muslim coastal cities resulted in a large frontier zone which in the early twelfth century included many of the principal Latin lines of communication. The warfare in this area centred on the raiding, skirmishes, and ambushes characteristic of 'frontier-style' warfare. Muslim and Latin troops waged 'low-intensity' campaigns of depredation against agricultural areas, commerce, and pilgrims. While truces and tributes as part of a *modus vivendi* were not uncommon, they were never permanent.

In waging this warfare, crusaders built fortifications whose garrisons were intended to harass and contain Muslim cities and their soldiers. The positions erected by Fulk of Anjou around Ascalon are well-known examples of this policy.[18] This employment of counter-forts dates from the early history of the Latin states, as the fortresses built against Tyre demonstrate. Toron, ten miles from Tyre, was built by the prince of Galilee in 1107 and Baldwin I fortified Scandalion some five miles from the city in 1117.[19] These fortifications served more as forward bases for controlling disputed territory between areas of crusader and Muslim domination than as part of a close blockade. This harrying from counter-forts was widespread in European siege technique before and after the First Crusade, and represents one of the basic elements of medieval siegecraft.

While distinctions between offensive and defensive actions blurred in this form of warfare, some counter-forts were crucial in capturing major cities. The County of Tripoli was founded from Raymond of St Gilles's fortress of Mt. Pilgrim, in 1102 or 1103. Built with Byzantine assistance atop a foothill two miles from Tripoli and supplied through the port of Tortosa, Mt. Pilgrim was a secure base for Raymond and his successors. While exacting a heavy tribute from Tripoli, they built up their power and expanded into the interior.[20] Vital as this fortress was in operations against Tripoli, Mt. Pilgrim's crusaders were unable to take the city by blockade or harassment. A major attack involving Baldwin I and Genoa was conducted in 1109, when the city finally fell

---

[18] WT xiv. 22, pp. 659–60; xv. 24–5, pp. 706–9; Prawer, *His. crois.* i. 328–31.
[19] WT xi. 5, pp. 502–3; xi. 30, p. 543; FC ii. 52, pp. 605–6; *DC* 75–6.
[20] AA ix. 32, p. 610; WT x. 26, pp. 485–6; AC xi. 7, pp. 35–6; Hill and Hill, *Raymond IV*, 153–5; Caffaro, *Lib. civ. orient.* 119.

to Raymond's successors.[21] While Mt. Pilgrim was invaluable in the Latin offensive as well as in operations against Tripoli, the city's capture was consummated by assault.

The Assault

Crusader siege technique centred on the assault. While every position was closely invested and blockaded, crusaders were never content in playing a waiting game from secure positions. They continually pressured defenders with every attack that topography, resources, and their technology allowed.

Despite different logistical and military conditions, there was continuity in the style of siege warfare waged by the First crusaders and that of the nascent Latin states. Latin besiegers after 1099 were not hampered by the provisioning difficulties which had hindered the sieges of the First Crusade. While Muslim field armies occasionally supported beleaguered cities in the early twelfth century, those forces never posed the threat that Al-Afdal's Fatimid army of 1099 had to Jerusalem's besiegers. However vulnerable the Latin states were in the early twelfth century, and they became more stable during the first two decades of that era, they were far more securely established than the conquerors of Antioch and Jerusalem. However, combatants of the Latin states waged siege warfare in a manner similar to that conducted by the First crusaders in the later stages of the expedition. A preference for large-scale assaults, frequently supported by complex machinery, characterized crusader siege warfare on the coasts of Outremer. While a western propensity for close-quarter fighting may partially explain this, those factors discussed above which determined the pattern of the coastal offensive played a key role in the conduct of siege warfare. The local numerical superiority and naval support provided by pilgrim and Italian fleets presented crusader commanders with valuable opportunities. Moreover, many of these squadrons brought material, manpower, and technical assistance which greatly facilitated the production of machinery frequently necessary in large-scale assaults.

A policy of close assault complemented offensives intended to take advantage of the time required by large Muslim forces based outside Palestine to mobilize support for beleaguered cities. Hamblin has argued that the crusaders' positional advantage in the coastal offensive

[21] AA xi. 1–2, pp. 663–4; WT xi. 9–10, pp. 507–10; FC ii. 41, pp. 526–33; Caffaro, *Lib. civ. orient.* 123; *DC* 87–8.

usually permitted something like two months of large-scale military endeavour before a substantial Fatimid army could be expected to reach a Palestinian battlefield.[22] This was not the case for Ascalon, which was itself a forward base, and does not take into account the smaller-scale aid that forces based in coastal cities could provide, such as those which relieved Acre in 1103 and Sidon in 1108. Moreover, some fortresses were besieged for periods considerably longer than two months without the advent of a powerful Fatimid land or naval expedition. Yet the importance of the crusaders' ability to concentrate for a period against major cities is manifest. Nevertheless, it is clear that the forces of Egypt, Damascus, and even Mosul were strong incentives for a rapid resolution of crusader sieges.

Siege Towers

The mobile wall-dominating siege tower was the foremost weapon of the Latin conquest. As in the siege of Jerusalem, artillery and armoured roofs were necessary in supporting siege towers, which were nevertheless centre-pieces of assaults. While some cities were stormed by ladder assault or surrendered before a major assault was delivered, the capture of key cities was based on tower assaults. Siege towers were important in capturing Haifa, Tripoli, Beirut, Sidon, Tyre, and Ascalon, and were employed in unsuccessful attacks on some of these and other cities. At least fifteen siege towers are mentioned in the Latin conquest of the coast, highlighting the importance of this device for crusader poliorcetics.[23]

Breaching by sap or ramming was relatively unimportant in capturing positions during this stage of the crusader conquest. This may reflect a lack of trained personnel or the strength of early twelfth-century civic fortifications. However, the personnel resources of large cities facilitated resistance to such methods of attack. Civic populations provided craftsmen and labourers, who were quickly mobilized to repair breaches and build retaining walls behind damaged fortifications. This happened at Ascalon in 1153, when crusaders failed to exploit a fortuitous breach with sufficient troops, enabling defenders to re-take and restore the breach.[24] While siege towers were the ultimate

---

[22] W. J. Hamblin, 'The Fatimid Army during the Early Crusades', Ph.D. thesis (Michigan, 1984), 181–2, 225–7. I am grateful to Dr Hamblin for allowing me to consult his important dissertation.
[23] See Appendix I.   [24] WT xvii. 27, pp. 797–9.

weapon in Near-Eastern siege warfare, as we shall see below, the ability to produce and employ them was important in the Latin conquest of the seaboard of Outremer.

Machinery Production and its Organization

The production of siege machinery along the Syro-Palestinian coast involved the same basic problems of materials, technical expertise, and the organization of labour which had challenged Jerusalem's besiegers. Yet Latin rulers came to possess territories and populations which provided resources for military activities. In 1107 Baldwin I prepared siege machinery for his aborted siege of Sidon at Acre, suggesting that this key port city was already important in the kingdom's military organization.[25]

However, the procurement and transport of timber suitable for siege machinery presented difficulties for a land-based power operating in territory raided by enemies. William of Tyre noted that Beirut's besiegers in 1110 were particularly fortunate because of the timber available in close proximity to operations.[26] Consequently, Christian shipping played a crucial role in transporting and frequently providing materials, especially the large, long beams necessary for siege towers and heavy artillery. Fulcher of Chartres describes the large timbers brought by the Venetians in the great expedition of 1123 which eventually captured Tyre.[27] Ships provided vital materials even when preparations had not been made in advance. In 1100 Haifa fell to a Venetian–crusader siege tower which, along with supporting artillery, had been made from Venetian ships.[28] A siege tower erected but not completed at Caesarea in 1101 was built from the masts and oars of Genoese ships.[29] Baldwin III used masts and spars from purchased pilgrim ships for his siege tower at Ascalon in 1153.[30] The Fatimids followed a similar policy by shipping long timbers for siege artillery while a land force moved on Jaffa from Ascalon during their attack on Jaffa in 1123.[31] While Christian ships presumably also carried the nails, fastenings, ropes, and hides necessary in building and armouring machinery, the transportation of cumbersome and usually expensive timber was a principal contribution to the logistics of crusader siege warfare.

[25] AA x. 4, pp. 632–6. [26] WT xi. 13, pp. 515–16.
[27] FC ii. 14, pp. 656–7. [28] TNV 276–7.
[29] FC ii. 9, p. 401; Caffaro, in AI i. 10–12. [30] WT xvii. 24, pp. 793–4.
[31] FC iii. 17, 20, pp. 662, 672.

The same kind of technical expertise brought to the First crusaders by William Embriaco and his Genoese sailors was included in the maritime forces organized to assist in the capture of Syro-Palestinian cities. Carpenters, labourers, and engineers participated in the expeditions which left Genoa and Venice to conquer the cities of Outremer. William Embriaco commanded the force which built machinery and assisted in the storming of Caesarea in 1101, and William was closely involved in the nascent Genoese commune in the early twelfth century.[32] The Genoese flotilla of 1109 which assisted in the capture of Tripoli, Jebail, and Beirut as well as the great Venetian expedition of 1123 was a maritime force equipped and prepared for urban conquest.[33]

We cannot be certain that every group of Italian ships which assisted crusaders militarily provided the degree of assistance of these great expeditions or even that of William Embriaco's men in the siege of Jerusalem. However, it is clear that the technical contribution of Italian maritime crusaders to the conquest of Outremer's seaboard was significant.

Considering the importance of these maritime expeditions and the concessions they received, as well as the interpretations of modern writers, it is appropriate to ask if Italians enjoyed almost a technological monopoly in the production of large-scale machinery in the early twelfth century. While the evidence and the nature of military experience in the region make a definitive answer to such a question difficult, it seems that Latin rulers did not depend exclusively upon large-scale Italian maritime assistance for their siege machinery. Godfrey of Bouillon had two siege towers built during his unsuccessful siege of Arsuf in 1099 without Italian aid.[34] Baldwin I had two siege towers built at Tyre in 1111 without the assistance of an Italian maritime fleet.[35] We do not know the identities or origins of those men who designed and built these machines or those built at Acre for the siege of Sidon in 1107. While some or all may have been Italians, it is clear that

[32] Caffaro, in *AI* i. 9–14; Cardini, *Profilo crociato*, 424–31.
[33] Caffaro, in *AI* i. 14: *Lib. civ. orient.* 119–24; FC iii. 14, pp. 656–7; WT xiii. 6, pp. 593–4; see also below, p. 82.
[34] William Embriaco was returning to Genoa, and Daimbert of Pisa had not yet arrived in Jerusalem (AA vii. 1–6, pp. 507–11; Caffaro, *Lib. civ. orient.* 111; FC i. 33, pp. 326–34, ii. 8, p. 398; WT ix. 19, pp. 445–6).
[35] While Byzantine naval and land forces assisted in operations against Tyre, their support was apparently limited, and there are no references in Byzantine accounts to providing materials or expertise for Baldwin I's siege engines. For the 1111–12 siege of Tyre, see below, pp. 79–82.

Latin rulers could employ sophisticated machinery without a formal Italian alliance.

In assessing the role of maritime forces we should not overlook the manpower contributed by these forces to crusader sieges. The pilgrim vessels which provided important materials for the siege of Ascalon also brought combatants and labourers. According to William of Tyre, Baldwin III forbade this fleet's sailors and pilgrims from returning home and requested their assistance in the siege, with a promise of pay—perhaps from the plunder expected from Ascalon—for their efforts.[36] As warships were usually beached during siege operations, not only their fighting men but also their sailors and oarsmen were available for labouring and military activities. Caffaro's participant account of the storming of Caesarea describes the effectiveness of such personnel. After exhortation from Patriarch Daimbert of Pisa and their leader William Embriaco, Genoese forces decided to seize the city without waiting for machinery under construction to be completed. This was accomplished by a mass escalade which overpowered the presumably startled defenders.[37] The strategic numerical weakness of the Latin states in the early twelfth century discussed above underscores the importance of these men in blockading and assaults.

The contributions of maritime forces to the waging of crusader siege warfare are manifest. By transporting and frequently providing timber and construction materials, Christian shipping facilitated the employment of siege machinery along the coast of Outremer. Moreover, expeditions organized for conquest concentrated technical and labouring personnel who built and operated such devices. These naval expeditions also included combatants who increased the ranks of besieging forces. Thus, quite apart from thwarting Fatimid relieving squadrons, maritime forces played an important role in the assaults of crusader siege warfare.

Crusader Assaults

Assaults conducted after the First Crusade show continuity with methods employed in the siege of Jerusalem. Unlike the Holy City, some coastal cities were compelled to surrender or were stormed by ladder and artillery assaults. In 1101 Arsuf surrendered to Baldwin I and his

[36] WT xvii. 23, 24, pp. 792–3.   [37] Caffaro, in *AI* i. 10–12.

The Capture of the Palestinian Coast

Genoese allies three days after their arrival, and Caesarea was stormed before a siege tower was completed.[38] The key port of Acre surrendered to Baldwin I and a large Genoese fleet in 1104 after twenty days of intensive artillery and personnel attacks. While the lack of Fatimid relief, and the strength of the besiegers, were factors in this capitulation, the constant pressure of attack, a protracted siege during the previous year, and the threat of massacre were also important in Acre's surrender. It may be noted that, despite terms of surrender, the inhabitants were set upon as they marched out of Acre by Italians seeking booty.[39]

However, other cities were subjected to siege-tower assaults and some on more than one occasion. Although every assault was unique, the same elements of terrain preparation, artillery support, and battle for control of a segment of walls were involved in every siege tower's attack. Venetian and crusader besiegers of Haifa in 1100 based their attack on a siege tower built from Venetian ships supported by seven artillery pieces. After fifteen days during which the tower was moved forward in the face of determined resistance, knights from the siege tower seized a mural tower as darkness fell. While offensive operations ceased, Latins defended their position and siege tower through the night against personnel and incendiary assaults. A crusader assault was launched at daybreak against the exhausted and demoralized defenders, and the city fell to sack and massacre.[40]

On some occasions defenders surrendered before a final onslaught carried the walls and introduced notoriously bloodthirsty crusaders into their city. Sidon surrendered to Baldwin I in 1110 after a siege of forty-seven days. The attackers had manœuvred a siege tower to the walls despite determined resistance, and Sidon's defenders capitulated to prevent the armed entry of Latin assault forces.[41] The Damascus Chronicle mentions that one reason for the Sidonese surrender was the fate of Beirut when that city was stormed by crusaders and Genoese earlier in 1110. Although Beirut's defenders neutralized one siege tower, two more were built and brought to the walls in a siege which lasted seventy-five days and which concluded in a fearful sack and slaughter.[42] At Sidon a crusader reputation for savagery at one siege contributed to success in the next.

[38] FC ii. 8–9, pp. 385–90; AA vii. 55–6, pp. 542–4; WT x. 14, pp. 469–70.
[39] AA ix. 27–9, pp. 606–7; FC ii. 25, pp. 462–4; WT x. 27, 28, pp. 486–7; DC 61–2.
[40] AA vi. 22–5, pp. 521–3; TNV 276–9.
[41] DC 107; AA xi. 31–4, pp. 678–9; FC ii. 43, pp. 543–8; WT xi. 14, pp. 517–19.
[42] DC 107; for Beirut: ibid. 99–100, AA xi. 15–17, pp. 669–71; FC ii. 42, pp. 534–6; WT xi. 13, pp. 515–16.

Decisive as siege towers were in the fates of Haifa, Sidon, and Beirut, they were not invincible. This is evident from the siege of Jerusalem, where Raymond of St Gilles's tower was damaged and stalled by the city's defenders. Defenders of Arsuf burnt two towers brought successively against their walls by Godfrey of Bouillon in the autumn of 1099.[43] Thus from the outset of the coastal offensive Muslim defenders were able to neutralize the primary tool of crusader poliorcetics.

Garrisons employed a variety of means in resisting tower assaults. Artillery was particularly important in harassing crews and damaging towers. Beirut's defenders in 1110 knocked out the first of three towers employed with artillery. The Latin tower attack at Tyre in 1124 was only made possible after clearing defending artillery from the walls.[44]

Sallies pressured those defending siege towers and occasionally damaged or destroyed machinery. Baldwin I's towers at Acre in 1103 and Sidon in 1108 were damaged by such actions, and one tower at Tyre in 1111 was completely destroyed.[45] Sidon's defenders in 1110 attempted to check the advance of Baldwin I's siege tower by tunnelling in front of the machine and sapping it. Baldwin's crews, however, spotted the tunnelling and moved their tower.[46] The building of counter-towers out of wood and rubble atop civic positions facing a siege tower's attack was an important defensive technique, first evident at the siege of Jerusalem. Such bastions were the basis of Tyre's protracted defence in 1111–12, and Qalanisi attributed the defenders' success to the equal height of attacking and defending towers.[47]

Combustibles, particularly those with a naphtha base, were crucial in destroying machinery. Naphtha, based on petroleum and often distilled and combined with a number of possible chemicals including saltpetre, had many of the properties of Greek fire except the ability to be projected through firing tubes.[48] As at Jerusalem, such incendiaries were employed when attacking machinery was close, since naphtha was valuable and most effectively delivered from point-blank range. While successful sallies could bring incendiaries against towers located some

[43] AA vii. 1–6, pp. 507–11; FC ii. 8, pp. 400–4.
[44] DC 99–100; for Tyre: WT xiii. 10, pp. 597–8.
[45] AA ix. 19–20, pp. 601–2; x. 46–51, pp. 652–5; xii. 6, pp. 691–2; DC 121–2; see also below, p. 80.
[46] AA xi. 33, p. 679.   [47] DC 122–4; see also below, p. 81.
[48] J. R. Partington, *A History of Greek Fire and Gunpowder* (Cambridge, 1960), 3–4, 22–5; D. R. Hill and A. Y. al-Hassan, *Islamic Technology: An Illustrated History* (Cambridge and Paris, 1986), 106–11.

distance from the walls, Muslim defenders attempted other means of igniting threatening siege towers. Ascalon's defenders piled wood and other kindling doused in oil in a gap between the city walls and a Latin siege tower, hoping that the intensity of the blaze would burn the siege tower. However, a change of wind blew the flames back on to the city walls and eventually created a breach. While rubble damaged the siege tower's lower frame, the breach was almost exploited by Latin troops.[49] During the siege of Tyre of 1111 a Tripolitanian sailor devised a mastlike swinging boom from which incendiaries were dropped on to a Latin tower from bastions built atop the city walls. This man's pyrotechnical and engineering skill resulted in the tower's destruction.[50]

In understanding crusader siege warfare it is important not to focus exclusively on the technical duels of engineers and their artifices. Central as these were, and much as they interested contemporary chroniclers, they were but part of a larger conflict. A weakening of defensive capabilities was a prerequisite to seizing walls. This was achieved through a variety of means, including damaging the upper works of fortifications with artillery, and inflicting casualties with artillery and missile weapons especially from siege towers. The length of time involved in these 'softening-up' operations varied from siege to siege, and they were not always successful. The attack on Jerusalem of July 1099 described above provides an excellent example of this type of conflict. However, Jerusalem's capture was notable in the short period of time between the assembly of siege towers and the city's storming—three days. Subsequent tower assaults involved a longer period of struggle, which in turn necessitated greater attacking organization as well as casualties.

Because the risks to machinery were greatest when it was drawn up in bridging distance of city walls, crusaders adopted a policy of moving towers into a position which dominated the walls but was not in the area of maximum vulnerability. From this position archers, crossbowmen, rock-throwers, and supporting artillery battled with defenders until a propitious moment for assault, when the tower would be moved up and the walls bridged. The distance between walls and siege towers during this phase of an assault varied, but in some cases was slight. The incendiary attacks on Latin towers at Ascalon and Tyre discussed above illustrate that, on these occasions, the distance was no more than

[49] WT xvii. 27, pp. 797–8. [50] *DC* 122–4; see also below, p. 81.

a large pile of combustibles or a swinging boom—perhaps as small as ten feet.

William of Tyre describes the grinding down of Tyre's garrison during the siege of 1124 and the strains such warfare placed on attackers.[51] Latin forces at the siege were compelled to respond to a Damascene army which attempted to draw enough Latins from the siege works to enable sallies from Tyre to destroy machinery. While this may have placed additional strains on those actually conducting the siege, William underscores the difficulties involved in manœuvring and guarding siege towers and artillery against a vigorous defence:

> Even those whose duty it was to guard the engine dared approach them only at the utmost speed, nor could they remain within except at great peril. From their positions in lofty towers, the enemy, armed with archery and crossbows, poured forth a deluge of bolts and arrows; meanwhile an unending barrage of huge rocks hurled from within the city pressed the attackers so hard that they scarcely dared put forth a hand.[52]

Despite casualties, it was essential to debilitate defenders and defences in preparation for a city's storming. At Ascalon in 1153 crusaders compelled surrender by personnel assaults, even though their siege tower had been rendered immobile short of the walls by a garrison incendiary attack.[53] As a result of this action, Ascalon's walls were breached and crusaders were narrowly prevented from carrying the gap. However, those inside restored their position with a retaining wall, and Ascalon's defences appeared secure. Although crusaders are reported to have lost confidence, they decided after a council of war to continue their efforts. They delivered repeated personnel assaults against the damaged walls, supported by dominating missile fire from their siege tower. After three days and substantial casualties on both sides, Ascalon's defenders surrendered.[54] In this case the cumulative effect of crusader attacks debilitated defenders to a point where they believed they could no longer resist, despite seaborne Fatimid relief earlier in the siege.

In assessing the successes of crusader siege technique, the importance of crusader willingness to wage a battle of attrition should not be underestimated. While intended to prepare a position for escalade, this

---

[51] WT xiii. 6–13, pp. 593–602.   [52] Ibid. xiii. 6, p. 594.
[53] For this incendiary attack, see above, p. 71.
[54] WT xvii. 24–30, pp. 793–805; *DC* 315–16.

struggle accelerated the process of grinding down garrison morale and manpower already begun by close blockade and investment. Moreover, the psychological stress that an approaching siege tower exerted should not be discounted, especially when a negotiated surrender appeared possible. In this fashion crusaders applied a strategy of close assault to a relentless conclusion, and this approach to siege warfare characterized the Latin conquest of the Syro-Palestinian littoral.

Technique

In focusing this discussion of crusader siege technique, it is appropriate to provide narrative accounts of exemplary campaigns. The two major Latin attacks on Tyre of 1111–12 and 1124 are suitable, not only because our sources describe them well, but also because they illustrate Latin siege methods and Muslim responses at their most sophisticated.

Tyre's defences were among the most formidable of the major coastal cities attacked by the crusaders. Surrounded by water and a double sea-wall on three sides, Tyre could be approached only from the east. Defences here consisted of a triple wall reinforced by many towers located closely together. Because Tyre was the remaining Muslim coastal city north of Ascalon when attacked, its defenders included refugees from cities already captured. Tyre was also able to solicit military assistance from Damascus as well as from Cairo.[55]

Tyre, 1111–1112

Baldwin I commenced his siege of Tyre on 29 November 1111, by investing the city closely along its eastern approaches. Baldwin's financial resources had been increased by the capture of a payment of 20,000 besants from Tyre to the atabeg of Damascus shortly before the siege began.[56] However, Baldwin's naval support was not substantial. Baldwin was supported by a Byzantine squadron which joined his forces after the siege had begun, and also by supply ships which

[55] WT xiii. 5, pp. 591–2.   [56] AA xii. 2–3, pp. 689–90.

reached Tyre during the course of the siege. Yet, he lacked naval forces sufficiently powerful to maintain a sea blockade of Tyre.[57]

Baldwin's troops began operations by digging in against Tyre's garrison and a Damascene army. Lines of contra- and circumvallation were dug around crusader camps, and winter quarters were built from Tyre's fruit groves.[58]

While the crusaders built their machines, the atabeg of Damascus led a field force which attempted to draw out the crusaders to facilitate a garrison sally, and harassed crusader supply lines which ran to the port of Sidon. However, the crusaders refused to be drawn and brought in supplies by sea.[59]

In mid-February 1112 two siege towers were moved against the city. These towers dominated Tyre's walls and were built from wood obtained from Christian ships.[60] While Qalanisi gives the height of these towers as greater than sixty-six and eighty-three feet respectively, it is not clear how accurate his figures are. Whatever the precise measurement of their height, they overtopped Tyre's fortifications.[61]

Tyre's defenders responded by building up certain civic towers with stone and wood until their height equalled that of the taller siege tower.[62] On 2 March 1112 defenders sallied out and piled incendiaries near the smaller tower in an area dominated by Muslim missile fire. The wind blew flames on to the smaller tower, and Muslim soldiers fought hand to hand to keep crusaders from their machine. Despite determined crusader efforts, the smaller siege tower was destroyed.[63]

Undaunted, the crusaders continued the advance of their remaining tower. Having filled three ditches, they brought their siege tower near one of the newly fortified civic towers. Defenders ignited the wooden supports of the structure, hoping that flames would spread to the crusader tower. While the crusader tower was undamaged, rubble and neighbouring defensive fortifications prevented further advance

---

[57] Qalanisi mentions that some two hundred vessels, including thirty warships, were burnt in the crusader retreat. Even if these figures are roughly accurate, they probably included very small transport craft which accumulatled during the siege (*DC* 125; AA xii. 4, p. 690.)

[58] AA xii. 5, p. 691; *DC* 121.   [59] *DC* 121.   [60] Ibid. 125

[61] The towers' measurements were forty and fifty cubits respectively. Gibb suggests that Qalanisi's figures may be approximate (*DC* 122 n. 1; AA xii. 6, pp. 691–2; FC ii. 46, pp. 559–61).

[62] While Fulcher implies that these defensive structures were taller, Qalanisi's more detailed account shows that neither side's towers enjoyed a significant advantage (*DC* 122–5; FC ii. 46, pp. 544–6; WT xi. 17, pp. 521–2).

[63] *DC* 122.

against that section of wall. Consequently, the besiegers withdrew their tower slightly and moved towards another section of Tyre's walls.[64]

Crusaders attempted to topple the newly built fortifications by battering with an iron-headed ram approximately ninety feet in length, suspended by ropes inside the siege tower. It is not clear whether the ram battered the new fortifications directly or the walls below them, but the latter seems most likely because of the ram's size and difficulties of suspending it in the siege tower's upper storey. Defenders responded by catching the ram's front with grappling irons and ropes, and then pulling until the attackers broke their own ram to preserve the tower's stability. Several rams are reported to have been employed and made useless, some in the above-mentioned manner and others by the weight of rocks which were tied together and dropped on to the ram.[65] It may be noted that this is one of the few twelfth-century examples of a battering ram's suspension and operation from a siege tower. It is likely that this was a makeshift response to defensive counter-towers, and was not an attempt to breach the walls directly. The difficulties encountered in utilizing the ram and the danger its operation presented the siege tower may explain why this technique was rarely used.

Eventually the crusaders moved their tower closer to the walls, even though they had not destroyed opposing fortifications. A sailor from Tripoli who had been in charge of operations against the rams then built a large mast-and-yard-like swinging boom which enabled him to drop materials on to the siege tower's summit. As the boom dropped incendiaries on to the tower, pots of oil and other combustibles were also thrown and the resultant blaze overwhelmed the crusaders' firefighting abilities. Once the tower's top was ablaze, flames eventually engulfed the structure and destroyed it. While Anna Comnena mentions a lapse in crusader vigilance during Lent in her account, Albert of Aachen and Qalanisi make it clear that the incendiaries dropped from the boom were central in the siege tower's destruction.[66] The tower's burning and the threat of a larger Damascene army persuaded Baldwin to abandon the siege on 10 April 1112.

Despite its failure, Baldwin I's siege of Tyre was an expression of crusader military capabilities after an initial period of establishment. That the Latins maintained themselves in a winter siege of a well-fortified and -supported city is itself notable. Smail has cited this siege of Tyre as an example of an action in which a field force thwarted siege

[64] Ibid. 123.   [65] Ibid. 123–4.
[66] Ibid. 124–5; AA xii. 6, p. 692; AC xiv. 2, pp. 442–3; WT xi. 17, pp. 521–2.

operations.[67] While the Damascene army played a role in the successful defence of the city, it was not a principal cause of the siege's result. The vigour of Tyre's defenders and their skills in military engineering were central. So was the inability of the Latin kingdom to prosecute the siege with substantial naval and military support from outside Outremer.

Tyre, 1124

Although one may date the beginning of what was to be the final campaign against Tyre to the Venetian naval victory over the Fatimids at Ascalon in May of 1123, the second siege of Tyre formally began on 16 February 1124 with the arrival of a large Venetian fleet and an army from the kingdom of Jerusalem under Patriarch Gormond. The Venetians had made elaborate preparations for a siege, as has been discussed above, and the patriarch had raised money for the campaign from ornaments of the Church of Jerusalem.[68] Although Baldwin II was in captivity, the full military resources of the kingdom were mobilized by Gormond.

As in the first siege of Tyre, Latins dug in against the city and Damascene relief and began building siege machinery. The Venetian fleet, which provided construction materials, was beached, except for one patrol galley. The Latins based their attack on two siege towers, one commanded by the Venetians, and one by Gormond's troops.[69]

They met fierce resistance from artillery and missile fire, as well as from sally parties, in advancing their towers. Casualties were substantial on both sides, but the arrival of reinforcements under Pons of Tripoli added significantly to Latin strength.[70]

During this period Damascene forces concentrated across the Litani River and attempted to draw besieging forces to facilitate sorties from Tyre against siege machinery. Moreover, the garrison of Ascalon made attacks on Jerusalem during the siege.

When both a Damascene attack and Fatimid seaborne relief seemed imminent, the besiegers divided themselves into three groups. Some Venetian sailors were kept near their boats in case patrols should make contact with a Fatimid squadron. Other Venetians and crusaders were left guarding the siege machinery. A third mounted force, under Pons

[67] Smail, *Warfare*, 25 n. 1.
[68] See above, p. 68; FC iii. 27, pp. 693–4.
[69] WT xiii. 6, pp. 593–4.
[70] Ibid. xiii. 7, pp. 594–5.

of Tripoli and the kingdom's constable, kept watch on the Damascenes.[71]

Meanwhile Tyre's defenders and the Latins fought for control of the walls. Defenders erected an artillery piece which damaged the Latin towers and which the besiegers were unable to neutralize. Consequently they sent word to Antioch, requesting the services of an Armenian artillery expert named Haverdic. He arrived, was paid from public funds, and directed besieging artillery accurately enough to neutralize defending pieces.[72] It should be noted that this man's expertise was required, not for knocking down fortifications, but for directing counter-battery fire.

Although Tyre's defenders did not destroy either siege tower, the city was never stormed, despite the intensity of besieging operations. Tyre surrendered on 7 July 1124 because its provisions had been exhausted during five and a half months of blockade.[73]

Whatever the effects of the mobile-tower attacks on Tyre's garrison and its resolve, this siege illustrates that crusaders were never content with blockade. Close assaults and battles of attrition, as well as the losses they usually entailed, were seen as essential in their siege warfare.

Amalric I in Egypt

Whatever their long-term effects on the strategic position of the Latin states, Amalric I's campaigns in Egypt involved siege operations largely consonant with earlier patterns. In the wake of the inconclusive if dramatic and much-discussed battle of Babein in mid-March 1167, Amalric I pursued Shirkuh and his nephew Saladin, who had moved into Alexandria.[74] The Latin king, then protector of the Fatimid Caliphate, concentrated some eight miles from the city and brought the ships at his disposal into a long-distance blockade of Alexandria's food supplies. After a month of such isolation, Shirkuh slipped out of the city with much of his expeditionary force and was followed by Amalric I. However, the king was convinced by a prominent Alexandrian that the city could be taken because of the food shortage and doubtless antipathy to the orthodox representatives of Nur-al-din.[75] In any event,

---

[71] Ibid. xiii. 8, 9, pp. 595–7.   [72] Ibid. xiii. 10, pp. 597–8.
[73] Ibid. xiii. 13, pp. 600–2; *DC* 171–2.
[74] For the battle and some of the discussion surrounding it, see Smail, *Warfare*, 131–2, 183–5.
[75] WT xix. 26, pp. 903–5.

the Latin monarch concentrated his forces in a close siege of Alexandria at the end of June 1167.

Famine was the principal means by which the attackers debilitated Alexandria and its mercenary defenders under Shirkuh's nephew Saladin. However, Amalric I sought to accelerate the city's exhaustion with attacks based on a mobile siege tower.[76] While William of Tyre's account emphasizes the involvement of Latin kingdom personnel, and particularly that of his predecessor in the See of Tyre, Archbishop Frederick, it is clear that Pisan forces played a prominent role in the phase of the siege, especially in the construction of siege engines.[77] Despite the considerable efforts of the Latins to force the issue, Alexandria continued its resistance as Shirkuh collected his forces. Ultimately no protagonist could impose a solution, and Alexandria surrendered as part of the general armistice of 1167 after a close siege of seventy-five days.[78]

The attack on Alexandria developed not as part of a crusader–Italian maritime agreement, but from the course of Amalric I's battle for Egypt. That Pisans, then engaged in a bitter conflict with Genoa, played a prominent role in building and employing artillery and a siege tower against Alexandria in the context of a thorough blockade is noteworthy. Whatever their naval strength, these Pisans were clearly sought after as practitioners of close assault. Although Pisan ships and men supported Almeric's operations against Bilbeis and Tanis in 1168, this campaign was inconclusive and the bloody treatment of Bilbeis worsened Fatimid–crusader relations.[79]

As the Latin position in Egypt disintegrated, Amalric I strove to restore his fortunes with the aid of a formal Byzantine alliance negotiated in part by William of Tyre. The compact bore fruit in a concentration of Latin kingdom and Byzantine forces after the summer heat of 1169. As Almeric's men took the coastal route down into Egypt from Ascalon, a powerful Byzantine fleet of 150 vessels, which included horse transports and ships for carrying siege equipment, made its way to Damietta on 27 October 1169, after stopping at a number of points *en route*. Amalric I's forces had arrived two days before, and his men and

---

[76] Ibid. xix. 28, pp. 903–5.
[77] Ibid. xix. 28, pp. 903–5; BM 45; *Regesta Regni Hierosolymitani, 1097–1291*, ed. R. Röhricht (Innsbruck, 1904), no. 449, p. 117.
[78] WT xix. 26–32, pp. 901–8; R. Röhricht, *Geschichte des Königsreichs Jerusalem (1100–1291)* (Innsbruck, 1898), 327–9.
[79] WT xx. 5–9, pp. 920–5; BM 47.

the Greeks established positions north of the city between Damietta and the sea.[80]

Because of the outcome of the siege, our sources are not forthcoming on the details of operations and the reasons behind them. However, it is clear that the besiegers failed to mount a major attack soon after arriving, and thereby lost the initiative. Three days after the besiegers' establishment, substantial reinforcements, dispatched by Saladin based in Cairo, entered Damietta by boat.[81]

Damietta's defences included the well-fortified Chain Tower on the left bank of the Nile; this secured a strong chain which excluded hostile shipping from the city's main harbour and the rest of this branch of the Nile.[82] The attackers' unwillingness—or perhaps inability—to reduce this fortification prevented them from isolating Damietta at any time during the siege. While this had a number of effects, the most immediate was that attackers had little choice but to attempt to overwhelm the defences directly facing their encampments.

Although the besiegers strove to destroy Damietta's northern walls by sapping an artillery bombardment, siege towers were the principal weapons of the Christian attack. William of Tyre reports that a massive seven-storey structure was first moved against the walls. The defenders responded with their own counter-tower and neutralized the Christian device. A second tower proved ineffective because it was employed over unsuitable terrain.[83] It is not clear if these towers were financed, built, and commanded by both major contingents acting in concert. Given their fleet and transports, it is likely that at the least the Byzantines provided some materials. The difficulties encountered by these towers may reflect relative inexperience as well as a divided command.

As Damietta's garrison grew in strength and morale, the attackers' position deteriorated. The Byzantines, having eaten all that they had brought, were particularly affected by famine. They also suffered the brunt of the increasing number of sallies undertaken by the garrison. Heavy rains and the burning of several Christian vessels by fireship further undermined resolve. Although the Greeks advocated a general assault in order to overrun the defenders, the Latins continued basing their actions around their towers. While a general assault was

[80] WT xx. 13–14, pp. 926–8; F. Chalandon, *Jean II Comnène (1118–1143) et Manuel I Comnène (1143–1180)* (Paris, 1912), 538–40.

[81] WT xx. 15, pp. 929–30.

[82] This tower, well known to students of the Fifth Crusade, was clearly formidable in the time of William of Tyre (WT xx. 15, pp. 929–30).

[83] WT xx. 15–16, pp. 929–33.

organized, it was never delivered, as an armistice whose exact terms remain ambiguous was negotiated during the first week of December.[84]

A number of explanations have been offered for this failure, including a Latin disinclination to share the spoils of Egypt, a lack of consensus concerning the strategic objectives of the campaign, perfidy, bad fortune, and the will of God. As well as the two latter explanations, William of Tyre cited the failure to attack the city immediately and the Byzantines' inability to provide for a long siege, although he noted that Greek leaders and forces fought with distinction.[85] However, the besiegers' inability to isolate Damietta from receiving substantial aid was the primary failure of their offensive. Not only was Damietta resupplied and reinforced from Cairo, but defenders knew these communications to be secure so long as the Chain Tower was held. Muslim control of the Nile and the city's configuration permitted substantial assaults against only one sector of Damietta's walls. The degree to which such attacks were pressed home remains conjectural, as there are no reports of substantial casualties except during garrison sallies. The Latins and their Greek allies were unable to grind down defenders or overwhelm them, and a long siege could not be sustained by Byzantine logistical and fiscal resources. It may be noted that the events of the Fifth Crusade suggest that the military and naval power needed to wrest Damietta from Auybid control were considerable indeed.

Although Almeric I and William of Tyre journeyed to Constantinople to re-establish a Latin–Greek compact, no major attack on Egypt resulted from these deliberations. Almeric I, however, continued efforts to solicit military and naval assistance, including an alliance with Norman power in Italy, which had scarcely been involved in the business of Jesus Christ since Baldwin I had spurned Adelaide of Sicily in the early twelfth century. The Hauteville monarchy had significant naval power at its disposal, which had been involved in overseas conquest since the time of Roger II. Moreover, Norman forces had much experience in siege warfare, albeit of a very different kind from that usually seen in crusader operations. A major combined attack on Alexandria in 1174 in conjunction with a *shia* revolt was organized from this alignment.[86] Although it failed, the attack on Alexandria was a dramatic attempt to alter the strategic balance, perhaps even then shifting against the Latin states.

---

[84] Ibid. xx. 15–17, pp. 929–34; Chalandon, *Comnène*, 542–4; Röhricht, *Geschichte*, 344–7.
[85] WT xx. 15–17, pp. 929–34.   [86] For this attack, see below, pp. 120–1.

Factors in the Development of Crusader Siege Technique

Having outlined the salient features of crusader siege warfare, it is appropriate to consider the development of crusader siege technique and its relative sophistication in the eastern Mediterranean. While the strategic context of the First Crusade differed from that of the twelfth-century Latin states, responses to the problems of capturing cities were similar in that assault was paramount in siege technique. Moreover, there was clear continuity in the utilization of mobile wall-dominating siege towers in such operations. Yet it is difficult to perceive siege towers as a characteristically Latin weapon from the beginning of the First Crusade. Participants in that expedition employed a range of methods in the many attacks on fortified positions that their pilgrimage involved. While a siege tower may have been under construction when Nicea surrendered in 1097, it was not the principal focus of crusader assaults. For military and topographical reasons, Antioch was not subjected to a major assault until the city was betrayed. A siege tower was important in the capture of Marrat, but that city was markedly less formidable than others subjected to a major siege, and the tower built there was less sophisticated than subsequent structures. What became the characteristic style of crusader operations centring on assault based on siege towers and supporting machinery is clearly evident at Jerusalem in 1099. It has been made clear that military and logistical concerns lent an urgency to completing operations against Jerusalem, and the nature of the expedition itself may have contributed other incentives.

Crusader assaults in the capture of the Syro-Palestinian coast were also waged in the hopes of taking positions relatively rapidly, although for very different strategic reasons. Nevertheless, these attacks were based on siege towers similar in design to those employed at Jerusalem. Although later towers may have been taller and equipped with better bridging apparatus, devices were more similar than different. This kind of warfare frequently involved a pragmatic adaptation of tried techniques to different topographical and military conditions. Engineers may have refined building techniques and employed different means of manœuvring towers over different kinds of terrain. In this regard the experience gained in operations must have been important in developing the crusaders' ability to wage siege warfare. However, it is clear that the basic elements of crusading assaults maintained a notable continuity.

Wearing down and demoralizing defenders over a protracted period of time before closing with siege towers was a refinement of crusader techniques. Such actions may have been dictated as much by circumstances such as terrain, increasing defensive skills in dealing with such devices, the availability of naphtha, and a disinclination among Latins resident in Outremer to risk all on a decisive push forward, as by an increasing poliorcetic sophistication. Yet it remains clear that the methods which proved successful at Jerusalem were the basis of twelfth-century crusading siege warfare.

A second related question involves the degree to which crusaders enjoyed a technological superiority in the early twelfth century. It has been suggested that such a superiority may explain western successes against Islam, and the siege tower has been identified as one possible manifestation of advanced western military technology.[87]

While a conclusive answer is elusive, it seems that the focus of such a view of siege warfare on the eastern Mediterranean is too narrow. Muslim defenders neutralized and destroyed crusader siege towers almost from their first employment. More importantly, it is manifest that the kind of siege warfare adopted by the twelfth-century Latin states was well suited to the capture of coastal cities with the resources available to them. Considering the very different military problems which Muslim commanders in Cairo, Damascus, and Mosul faced, it is not surprising that they did not produce crusader-style machinery regardless of Near-Eastern technological capabilities.

The capture of the Syro-Palestinian coast owed much to a marriage of convenience between land-based Latin rulers and Italian maritime powers. If any group of twelfth-century Mediterranean conquerors might be thought to have enjoyed a technological superiority, the Genoese, Pisans, and Venetians would seem the most likely. While a fuller discussion of the Italian maritime states' contribution to Latin expansion in the Mediterranean follows in Chapter 6, it is worth noting here that it is difficult to identify an Italian technological monopoly in the eastern Mediterranean. As we have seen, crusaders employed sophisticated machinery without the assistance of Italian fleets. Genoese and Venetian squadrons facilitated the waging of siege warfare, however, and gave crusader commanders a certain flexibility in planning and conducting operations. Squadrons which had not made preparations for siege campaigns could convert ships to siege engines, and the similarities between poliorcetics and naval engineering during this

[87] White, 'Thrust', 107.

period explains much of the Italian role in building siege machinery. The Tripolitanian sailor so important in the defence of Tyre in 1111–12 illustrates another way in which naval technology could be adapted to siege warfare. His example suggests that Muslim commanders could have found the expertise which White has suggested they lacked in the shipyards of Egypt, if not elsewhere.

Important as the capture of the Syro-Palestinian coast was to the survival of the Latin states, crusaders waged considerable siege warfare in the interior. With the exceptions of Marrat and Jerusalem, such sieges were usually conducted differently from those of the coastal offensive. Banyias is the only city reported to have been attacked by a siege tower during the joint crusader–Damascene siege of that city in 1139.[88] Long- and short-distance blockade was paramount in this form of siege warfare. While a range of methods was employed in directly attacking fortifications, artillery and sapping were prominent. Harenc was besieged in 1157, where labourers attempted to excavate the hill on which the fortress was situated while an intense artillery attack was maintained. Although the mining was ineffective, the artillery attack, death of the garrison commander, and imminent crusader assault led to surrender.[89]

Logistical and geographical considerations influenced much of the pattern of this warfare. The timber for Fulk of Anjou's siege tower at Banyias came from Damascus, and obtaining such materials in the interior of Outremer may usually have been more difficult. The transport of the beams, wood, hide, and other components of siege towers may have been difficult also, especially in hostile territory. Yet it is difficult to develop a detailed picture of this kind of siege warfare in Outremer. The sources are not particularly descriptive of techniques and machinery. Moreover, the political rivalries and alliances which formed the backdrop of much of this warfare also affected the pattern of conflict, and in some cases these also are far from clear. Political division was occasionally important, not only in the conflicts between Latin and Islamic powers, but also among groups sharing the same religion. The outstanding example of the effects of such rivalries in siege warfare is the calamitous siege of Damascus in 1148, which may be seen as a fitting conclusion to the Second Crusade in the Holy Land. While rivalries among Latins resident in the crusading states and among their European co-religionists clearly affected the conduct of

[88] WT xv. 9–11, pp. 685–91.   [89] Ibid. xvii. 18–19, pp. 838–40.

this perplexing siege, the details have remained obscure, and perhaps veiled.[90]

Consequently it is appropriate to shift our focus to another area of the Mediterranean which involved some similar elements but which is more clearly depicted. Moreover, such a shift will reveal forms of siege warfare significantly different from that of the First Crusade and the crusading states.

[90] Ibid. xvii. 1–6, pp. 760–9; A. J. Forey, 'The Failure of the Siege of Damascus in 1148', *Journal of Medieval History*, 10 (1984), 13–23.

3

SIEGE OPERATIONS IN THE ESTABLISHMENT OF NORMAN AUTHORITY IN ITALY AND SICILY

THE establishment of Norman authority in southern Italy and Sicily was one of the age's great military and political achievements, rivalling the creation of the crusader states in its significance for the Mediterranean. Norman military endeavours included much siege warfare conducted against Muslim, Byzantine, and Italian fortifications and urban centres. While the region's terrain and political divisions affected patterns of siege warfare, the techniques employed were also influenced by the military resources available to Norman commanders. However different the strategic context, Norman siege operations may be compared with major undertakings in other areas of the Mediterranean. Although many sieges of the Norman establishment occurred before the First Crusade, they merit discussion in order to identify patterns of conflict. At the least such an examination will provide an example of Latin siegecraft in the Mediterranean before the twelfth century and perhaps underscore some of the distinctive characteristics of operations in the Levant. Moreover, it will set the stage for one of the twelfth century's great practitioners of siege warfare, Roger II.

If the rise of the Norman states was the region's most important political development of the later eleventh century, the subordination of Apulia and Capua to the ruler of Sicily was of equal significance for the first half of the twelfth century. Skill in siegecraft was a hallmark of Roger II's warfare and an important factor in his eventual victory. In discussing the siege operations of his campaigns, we shall take the measure of one of the twelfth century's most effective siege trains. Moreover, we shall observe how a Norman commander adapted very different military resources from those of the crusader states to the problems of reducing fortifications.

Sources

While Norman military exploits were important concerns for eleventh- and twelfth-century historians, their accounts of the politics and warfare of their rulers, cities, and religious establishments present difficulties for an understanding of Norman siege operations. Not only are sources often highly partisan, but they offer comparatively few detailed accounts of major sieges. Moreover, some of the most illuminating descriptions of techniques and machinery come from Byzantine and Arabic writers. While Latin historians and annalists were not uninterested in siege warfare, these events were generally not central to their historical concerns. Consequently, accounts of operations are often brief, simply recording that a particular position was attacked and taken. Nevertheless, some writers give important information, and Chalandon provides a discussion of the historical sources for the Norman period. His work also offers a fundamental political narrative of the Normans in Italy and Sicily.[1] Yet it is appropriate to discuss certain works of particular importance to this enquiry, and some of the problems of interpretation they present.

Writers who described the achievements of Norman conquerors of the later eleventh century provide important information about military affairs. Although highly partisan and replete with anecdotal material, these works illustrate the problems Normans faced in their campaigns and how they were met. Geoffrey Malaterra wrote such an account of the military activities of Count Roger I of Sicily in his conquest of that island and in assisting his brother's capture of Apulia and Calabria.[2] William of Apulia, another historian of Hauteville campaigns in southern Europe, wrote a verse account of Robert Guiscard's achievements, probably during the pontificate of Urban II.[3] A monk of Montecassino during the time of Abbot Desiderius known as Amatus wrote a comprehensive history of the Normans in Italy sometime in the last five years of the eleventh century which records many of the deeds of Norman warriors and leaders and particularly those of

---

[1] F. Chalandon, *Histoire de la domination normande en Italie et en Sicilie* (2 vols.; Paris, 1907).
[2] Geoffrey Malaterra, *De rebus gestis Rogerii comitis et Roberti Guiscardi ducis et fratris eius*, ed. E. Pontieri (RIS 5; Bologna, 1927).
[3] William of Apulia, *La Geste de Robert Guiscard: Gesta Roberti Wiscardi*, ed. M. Mathieu (Palermo, 1961), 17–25.

Robert Guiscard. Although existing only in a fourteenth-century French translation, the work's historical value is beyond question.[4] One of the most illuminating sources for Norman military affairs of the later eleventh and early twelfth centuries is the *Alexiad* of Anna Comnena. Although she wrote considerably after events and reflects a thoroughgoing Byzantine bias and an even stronger one towards her father Alexius I, she provides important accounts of Norman attacks on Durrazo in 1081–2 and 1107–8. Her account of Bohemond's offensive of 1107–8 is particularly lucid and describes Norman siege devices and their employment in considerable detail. This may reflect the presence of her husband and fellow historian Nicephoras Bryennius in that campaign and it is possible that he provided a relatively complete narrative as a basis for her account.[5]

The considerable siege warfare of Roger II is recorded primarily by two writers who differ considerably in their assessment of the Norman–Sicilian kingdom's first monarch. Alexander of Telese, abbot of San Salvatore and supporter of Roger II, wrote a contemporary account of the king's conflicts in southern Italy from 1135 to 1137 at the behest of Roger II's sister.[6] The chronicle of the notary Falco of Benevento recounts events in the region from 1102 to 1140, reflecting a profound hostility towards the Normans and particularly Roger II.[7] Although these two Italian writers held conflicting opinions about the Anacletan schism and Roger II's kingship, their works demonstrate the importance of the Sicilian ruler's siege technique in his southern Italian wars. The chronicle attributed to Romuald of Salerno is a complex text of much-disputed authorship which nevertheless contains useful information and observations concerning Roger II's wars.[8] Although there is agreement that this archbishop and diplomatic representative of William II wrote the final section of this world history concerning the Peace of Venice in 1177, his involvement with earlier sections of

---

[4] Amatus of Montecassino, *Ystorie de li Normant: Storia de' Normanni*, ed. V. Bartholomaeis (FSI 76; 1935), pp. viii–lxvi; Chalandon, *Normande*, i, pp. xxi–xxiv.

[5] Bryennius' conversations with Bohemond regarding the treaty of Devol in 1108 illustrate his involvement in the campaign (AC xiii. 10, p. 124).

[6] Alexander of Telese, *De rebus gestis Rogerii Siciliae regis libri quatuor*, in *Cronisti e scrittori sincroni Napoletani*, i. *Storia della monarchia*, ed. G. del Re (Naples, 1845).

[7] Falco of Benevento, *Chronicon de rebus aetate sua gestis*, in *Cronisti e scittori sincroni Napoletani*, i. *Storia della monarchia*, ed. G. del Re (Naples, 1845).

[8] Romuald of Salerno, *Chronicon sive annales*, ed. C. A. Garufi (RIS 7; Città di Castello, 1935).

this work are less clear-cut.[9] Matthew has suggested that the narrative of Roger II's campaigns during the 1130s is based on a written account.[10] While this cannot be demonstrated conclusively, the chronicle does give information pertaining to Salerno and its environs and is sympathetic to the cause of Roger II. It may be mentioned that an interest in siege warfare is notable in this section, and also evident in the chronicle's account of affairs in Lombardy during the wars of Frederick I Barbarossa in that region.[11]

It may be noted that, considered as a whole, these sources do not compare favourably with the historiography of the Crusades or northern Italy regarding military affairs and particularly siege operations. Historians of the Normans in southern Europe seldom give information about fortifications or topography in their accounts of major sieges. Moreover, these writers frequently employ vague and general terms in describing devices used in attacks on fortified positions. Despite these limitations, it is possible to develop an understanding of siege operations in Norman Italy and Sicily which illuminates broader aspects of twelfth-century Latin siege warfare.

Siege Operations in the Establishment of Norman Authority

During the early phases of the conquest the basis of Norman siege technique was blockade and the principal means the counter-fort. Although major positions were isolated in the course of protracted campaigns, unlike operations in the crusader states large-scale assaults as a consummation of such undertakings were rare. From their early days in southern Italy Normans established themselves in mountain fortresses from which they descended on travellers, inhabitants, and livestock. The great conqueror Robert Guiscard began his career as a Calabrian brigand operating from such positions.[12] This was partially a consequence of the scale and gradual pace of the conquest,

---

[9] H. Hoffmann, 'Hugo Falcandus und Romuald von Salerno', *Deutsches Archiv für Erforschung des Mittelalters*, 23 (1967), 116–70; D. J. A. Matthew, 'The Chronicle of Romuald of Salerno', in R. H. C. Davis and J. M. Wallace-Hadrill (eds.), *The Writing of History in the Middle Ages: Essays Presented to Richard William Southern* (Oxford, 1981), 239–74.
[10] Matthew, 'Romuald', 270 n. 1.
[11] For these campaigns, see below, pp. 132–51.
[12] Chalandon, *Dom. norm.* i. 118–19.

which one recent writer who has stressed these factors in the evolution of the Norman Italian states has called an 'infiltration'.[13] Whatever the reasons behind it, Normans easily developed this practice into a method of harassing and preparing objectives before attempting their capture.

Not only were fortifications important in the initial phases of a position's reduction; they were the primary tool of Norman siegecraft. Normans interdicted communications and supplies by building small fortifications—castles—in close proximity to besieged positions, often directly in front of entrances. Troops stationed in these forward positions maintained an active blockade and were often able to wear down opponents and their will to resist. Although not rapid, such means of fortress-taking were economical in manpower once counter-forts were erected. Norman mercenaries were well suited to such warfare, which to a degree made a strategic virtue of the Normans' lack of a territorial base in the early period of the conquest. The successes of this policy were, of course, also due to other factors, including Norman prowess in battle and their fearsome reputation. Yet it remains that the construction of such fortifications was a clear demonstration of Norman persistence in a fashion which conserved often scarce military resources.

This method of siege warfare was employed throughout the eleventh century. Richard of Capua brought that city to terms with a blockade based on three such counter-forts in 1057.[14] Richard of Capua and Guiscard blockaded Salerno and its citadel for eight months in 1077–8. Their forces, supported by an Amalfitan fleet, conducted operations from four castles built around the city.[15] Guiscard besieged a number of smaller positions in this fashion, including Mileto in 1062, which was held by his brother Roger I.[16] Roger I himself compelled the garrison of Squillacae to flee in 1060 by building a castle directly in front of the fortress's main entrance.[17]

As the importance of counter-forts in Norman blockading suggests, Norman manpower during this period was limited. Although a precise

[13] G. A. Loud, 'How "Norman" was the Norman Conquest of Southern Italy?', *Nottingham Medieval Studies*, 25 (1981), 16.
[14] Amatus of Montecassino, *Normant*, iv. 11, pp. 189–90; *Chronica Monasterii Casinensis* (MGH SS 7), 707.
[15] Amatus of Montecassino, *Normant*, viii. 14–24, pp. 354–65; *Chron. Mon. Casinensis*, 735; William of Apulia, *Gesta Roberti*, iii, vv. 412–64, pp. 186–8; Geoffrey Malaterra, *De rebus gestis Rogerii*, iii. 2–4, pp. 58–9.
[16] Geoffrey Malaterra, *De rebus gestis Rogerii*, ii. 23, p. 37.
[17] Ibid. i. 37, pp. 23–4.

measure of Norman numerical strength is not possible, we may note that Norman siege activities seldom involved circumvallation, let alone the mass assaults characteristic of much of crusader siege warfare. Douglas has suggested that significant numbers of locally recruited infantry must have participated in the capture of large cities and perhaps have been involved in other actions, even though they are not mentioned in Norman narrative sources.[18] While Italian forces may have been employed in attacks against major cities such as Bari and Durazzo, Norman reliance on counter-forts argues that such practices were rare if followed at all. In this context it should be noted that, while major Norman leaders usually pursued their military and political interests separately, they combined in attacks on major cities, such as the blockade of Salerno of 1077 mentioned above. This ability of the Hautevilles to unite their relatively limited forces against the region's great cities is well illustrated in operations against Bari and Palermo.

Bari, 1068–1071; Palermo, 1071–1072

Guiscard's siege of Bari, which began on 5 August 1068, marked the culmination of his long campaign to secure Apulia from Byzantine authority, as well as the largest and most complex Norman siege up to that time. Guiscard's troops invested the city closely, while one naval force raided the Balkan coast and another tried to maintain a blockade.

Not content with these actions, Guiscard attempted to force Bari's defences with siege machinery. Although the walls were attacked with battering rams and artillery, a wall-dominating siege tower was the principal focus of Norman efforts. As the tower was brought to the walls, the defenders succeeded in igniting it and then destroyed supporting devices after the tower was consumed in flames.[19]

Although Guiscard's crews and engineers were either unlucky or outclassed by their opponents, their origins merit consideration. Guiscard recruited Calabrian and Apulian sailors for his fleet, and he may have found his siege crews from the same sources.[20] Complex machinery is mentioned in connection with Apulian sieges from the mid-eleventh century. In the course of a revolt against Byzantine authority, Baresi rebels, supported by Norman horsemen, attacked the city of Trani in a month-long siege beginning in early July 1042. A

---

[18] D. Douglas, *The Norman Achievement* (London, 1969), 79–80.
[19] William of Apulia, *Gesta Roberti*, ii, vv. 478–574, pp. 158–62.
[20] Ibid. ii, vv. 485–6, p. 158.

siege tower, whose description impressed the author of the *Annals of Bari* as something not seen in the region in recent memory, was built probably by the Baresi. However, this device was never brought against Trani, as it was burnt on the orders of the Barese commander when he changed sides.[21] It should be noted that our sources do not refer to similar devices in siege operations between the sieges of Trani and Bari. As we have seen, Guiscard's crews could not overcome Bari's defenders, and they may have included some of the region's most adept engineers.

Although thwarted in his assaults, Guiscard persevered in the siege, and among other methods tried to gain entry to the city by bribery.[22] In 1070 he led elements of his forces against Brindisi doubtless to relieve tedium as well as to eliminate possible bases for a seaborne Byzantine relieving force. However, this expedition was unsuccessful and suffered substantial casualties.[23]

Despite these set-backs, Guiscard maintained the blockade by land and sea, and set an example by wintering outside the city in a hut made of branches and thatch.[24] In isolating Bari, Guiscard had chained boats together across the entrance to Bari's harbour and then connected them to each other and the mainland with wooden bridges which enabled land-based troops to support the defence of this barrier. In February 1071 a Byzantine naval force attempted to break the blockade but was thwarted by these defences and recently arrived ships under Roger I. When no further relief appeared forthcoming, the anti-Norman faction in the city lost control and Bari surrendered on 16 April 1071.[25]

After taking Bari, the Hauteville brothers moved against another position whose capture was essential in their expansion: Palermo. The attack of 1071 completed a long campaign of harassment and isolation which followed an unsuccessful siege in 1064.[26] These operations were based on small fortifications which gradually established Norman

[21] Although the complex text was compiled over a long period of time by different writers, the section dealing with the years 1035–43 was written by a contemporary. 'The *Annales Barenses* and the *Annales Lupi Protospatharii*: Critical Edition and Commentary', ed. W. J. Churchill, Ph. D. thesis (Toronto, 1979), 122; *Annales Barenses* (MGH SS 5), 56).
[22] William of Apulia, *Gesta Roberti*, ii, vv. 530–9, pp. 161–2.
[23] '*Ann. Bar.*', ed. Churchill, 146–7.
[24] William of Apulia, *Gesta Roberti*, ii, vv. 556–60, p. 162; Geoffrey Malaterra, *De rebus gestis Rogerii*, ii. 40, p. 49.
[25] William of Apulia, *Gesta Roberti*, ii, vv. 478–574, III–14, pp. 158–62; iii, vv. III–63, pp. 170–2; Geoffrey Malaterra, *De rebus gestis Rogerii*, ii. 50–3, pp. 48–9; Amatus of Montecassino, *Normant*, v. 27, pp. 248–55; '*Ann. Bar.*', ed. Churchill, 146–7.
[26] Geoffrey Malaterra, *De rebus gestis Rogerii*, ii. 36, p. 45.

control in the area between Roger I's base at Troina and Palermo.[27] Although the garrisons of these positions had weakened Palermo's defences, conquest depended upon a major attack.

With Apulia secured, Guiscard brought men and ships to assist Roger I in taking Sicily's most important urban centre in August 1071. Although the Normans blockaded Palermo by land and sea, their main encampments faced the city's northern and eastern walls. As these forces skirmished and prepared the north-eastern approaches for an assault, men under Roger's nephew Serlo harassed positions in central Sicily. Although the Norman blockade brought about famine in Palermo, the besiegers also encountered provisioning difficulties during the winter of 1071–2. On 7 January Roger I led Norman forces in a general assault against Palermo's north-eastern walls. This attack was made across a section of the city's main ditch, which had been filled and was supported by artillery. As this attack developed, Guiscard secretly marched a select force around to the southern side of the city. Three hundred men were sent up seven ladders into the outer city, which was rapidly taken by the Normans. The majority of Palermo's inhabitants fled into the better-fortified inner city and surrendered to terms on 10 January.[28]

The Conquest of Sicily

While the capture of Palermo confirmed the Normans as Sicily's pre-eminent military power, the conquest of the island was not completed until the surrender of Castrogiovanni in 1091. The cities and dependent fortresses of Trapani, Taormina, Agrigento, Syracuse, and finally Castrogiovanni were successively reduced during the intervening period. Roger I's forces played a dominant role in these operations, as Guiscard was occupied in Italy and the Balkans during much of the period.

Norman fortifications and their garrisons were Roger I's principal means of conquest. Small positions interdicted supplies and communications and assisted in pacifying populations which continued to resist after the main fortifications of an area had been taken. Taormina was blockaded by twenty-two such fortifications in the course of a

---

[27] Ibid. ii. 22, 38, pp. 36, 47–8.
[28] Ibid. ii. 55–6, pp. 53–4; William of Apulia, *Gesta Roberti*, iii, vv. 194–324, pp. 174–82; Amatus of Montecassino, *Normant*, vi. 17–19, pp. 278–82; Chalandon, *Dom. norm.* i. 206–9.

complex siege which also involved Norman sea power and extensive raiding in areas around Mt. Etna. Roger I's preparations proved adequate and Taormina surrendered during March 1079.[29] In securing mountainous territory between Palermo and Trapani after the latter's fall in 1077, Roger I fortified twelve castles from which his troops eventually subdued the area's inhabitants.[30] This careful, piecemeal policy was the result of a number of factors, including Roger I's ability to control his troops and his political acumen. Yet a relative scarcity of combatants must be seen as fundamental.

Roger I's preference for close blockade rather than personnel assault is illustrated by the capture of Syracuse in 1085. Having isolated the city, Roger I sailed into its harbour, while his son Jordan led horsemen overland against Syracuse. On 24 March Roger I's ships fought those of the emir of Syracuse, who was killed in the fighting which overcame his vessels. Geoffrey Malaterra speculated that Jordan could have stormed Syracuse at this juncture had this not been forbidden by Roger I.[31] A blockade was maintained while elements of Roger I's fleet raided the Sicilian coast. Syracuse surrendered in October 1085.[32]

The conquest of Syracuse illustrates not only Roger I's disinclination for assault, but also the importance of sea power. Warships were essential in preventing seaborne relief from Muslim cities in Sicily and Zirid in north Africa. Moreover, Roger I's ships occasionally transported attacking forces when terrain or an attempt at surprise made overland movement impractical. The count of Sicily descended on Trapani in May 1077 by sea and, after landing men and horses, established a close blockade. After an unsuccessful sortie and loss of their harbour, those citizens of Trapani who had not fled to the hinterland surrendered to terms.[33]

The importance of Roger I's sea power in the conquest of Sicily is manifest, and the contrast with Baldwin I striking. Of course military conditions in Outremer and Sicily differed considerably and Roger I's general command of the military initiative was perhaps the most significant. Moreover, the count of Sicily was seldom compelled to meet military challenges on two or more fronts, unlike the king of Jerusalem.

Roger I achieved his military objectives without the alliance of an

[29] Geoffrey Malaterra, *De rebus gestis Rogerii*, iii. 15–19, pp. 66–8.
[30] Ibid. iii. 11, p. 64.
[31] Ibid. iv. 2, p. 86.   [32] Ibid. iv. 1–5, pp. 85–6.
[33] Ibid. iii. 11, pp. 62–4.

Italian maritime state or pilgrim fleet. Although he was offered Pisan assistance in taking Palermo after his victory at Cerami in 1063, he declined and did not support Pisan forces after they had stormed Palermo's harbour in that same year.[34] Whether this decision was the result of long-term political vision or Roger I's difficulties and interests in 1063 cannot be determined. Yet, whatever the reasons, the independent sea power which Roger I developed was an important constituent of his military strength and that of his successor.

Muslims in the Service of Normans

The military needs of Norman rulers compelled them to recruit locally as the scale of their enterprises increased. As they moved into Sicily, Muslims became part of the subject populations from which Norman commanders drew their forces. Guiscard seems to have inaugurated a policy of employing Muslim troops during his siege of Salerno in 1076.[35] While the duke of Apulia may have been the first Norman leader to use such resources, the lord of Sicily benefited most from this practice.

Although the evidence is not conclusive, Roger I may have first utilized Muslim military forces in the complex and wide-ranging operations involved in the capture of Taormina in 1079.[36] His recruitment of Muslim troops was also connected to a wider policy of using military strength from one area of his dominions to suppress rebellion in another. Roger I dispatched Calabrian forces to suppress a rising by Muslim inhabitants in the area around Mt. Iato south-west of Palermo in 1079, while the bulk of his forces were occupied around Taormina.[37] In May 1090 a rapidly raised Sicilian force blockaded Manerius of Acerenza in his lordship in central Calabria after he had defied Roger I's court. Despite the considerable natural strength of Manerius' main fortification, the speed with which Roger I concentrated his forces compelled the rebel's surrender. Although Douglas argued that Roger's troops included Muslims, there is no clear reference to such personnel.[38] However, in assisting Roger Borsa in suppressing a

[34] Ibid. ii. 34, p. 45; BM 5.
[35] 'Il asembla troiz turmez de manieres de gent: de Latin, de Grex ed de Sarrazin' (Amatus of Montecassino, *Normant*, viii. 14, p. 354).
[36] Chalandon, *Dom. norm.* i. 334.
[37] Geoffrey Malaterra, *De rebus gestis Rogerii*, iii. 20, pp. 69–70.
[38] Ibid. iv. 16, p. 94; Douglas, *Achievement*, 79.

rebellion in Cosenza in 1091, Roger I mobilized a strong Sicilian contingent which included a substantial number of Muslims.[39]

Whenever Roger I may have begun recruiting Muslim troops, their most significant deployment came during the siege of Capua in 1098. In assisting Richard II of Capua in retaking that city after it had rebelled, Roger I concentrated a powerful force containing a large element of Muslims. Eadmer mentions 'many thousand' Muslim troops in Roger I's service; he also states that many of them, moved by the example of St Anselm's piety, would have become Christians had such conversions not been forbidden by Roger I.[40] While this prohibition may simply demonstrate Roger I's desire to prevent the strong-willed saint from meddling with the discipline of his troops, it may also reflect the arrangements with local Sicilian Muslim leaders by which Muslim troops were recruited.

Regardless of their religion and the conditions of their service, Roger I's forces took a leading role in operations against Capua, building a bridge across the Volturno to facilitate communications with Roger Borsa's encampment. They also constructed siege machinery, which probably included artillery. Whatever the specific engines, their appearance opposite Capua's defences and the mediation of Urban II persuaded the citizens to negotiate a surrender.[41] While the degree to which these efforts were the work of Roger I's Muslim troops is not certain, it is likely that these men performed much of the manual labour at the least.

It is important to note that Sicilian positions were compelled to submit to Norman authority much more often by blockade than by the assaults of men or machinery. While, in the first place, a consequence of the nature of Norman military resources, this pattern was also affected by the political and economic objectives of protagonists. Access to Sicily's considerable wealth during the period was a primary goal of Norman warfare on the island. The subjugation of Sicilian communities to the fiscal exactions of Roger I's lordship was a significant means to that end. In this regard, populations which surrendered to terms which guaranteed their possessions may have been more lucrative in the long run than those which were sacked and perhaps massacred. A number of Sicilian communities resisted Norman domination, on

---

[39] Geoffrey Malaterra, *De rebus gestis Rogerii*, iv. 17, p. 96.
[40] Ibid. iv. 26, p. 104; Eadmer, *Vita Anselmi*, ed. R. W. Southern (Oxford, 1962), 33, pp. 110–12.
[41] Geoffrey Malaterra, *De rebus gestis Rogerii*, iv. 26–7, pp. 104–6.

occasion with considerable tenacity. However, when Norman arms and perseverance appeared irresistible, communities sought to preserve not only limited religious and legal rights, but also their lives by negotiated surrenders. In such a context siege techniques which minimized casualties and the passions generated by loss of comrades were particularly appropriate.

Although blockade and negotiation characterize the siege warfare of Roger I, it is clear that his forces possessed significant engineering and construction capabilities by the time of the siege of Capua, if not earlier. This, as well as the importance of Muslim troops in Norman Sicilian armies, foreshadows the military strengths of Roger II.

Operations in the Balkans

Durazzo, 1081–1082

Guiscard's Balkan offensive of 1081 represented a major attempt to establish the Normans in that region and involved considerable military and naval resources.[42] The campaign initially centred on Durazzo and, while part of the invasion fleet was sunk in a storm Guiscard's forces established themselves outside the city on 17 June 1081. Although the remaining Norman ships were first defeated and then blockaded by Venetian naval allies of Alexius I, the Normans persisted in their attack.

As at Bari in 1068, Norman assaults centred on a siege tower intended to dominate the city's defences. This device was equipped with a drawbridge designed to swing down on to the walls from the tower's summit and was armoured from top to bottom, thus protecting those who propelled it. During September 1081 the tower was completed and moved to within bridging distance of the walls. The defenders responded with a counter-tower built opposite the Norman device's point of approach. This Byzantine structure was not an elevated platform for casting missiles and incendiaries, as we have encountered in the Levant, but rather a projection for men and equipment intended to thrust a beam horizontally out towards the attackers. As Guiscard's crews began lowering their bridge, the Byzantines thrust forward a large

---

[42] Norman historians give only brief accounts of this expedition. Anna Comnena is the principal source for this campaign.

beam which jammed the siege tower's bridge before it had descended to the walls. During the subsequent confusion defenders concentrated archery and incendiaries on to the siege tower, which was consumed in flames.[43]

Although Guiscard's crews began building another siege tower, its employment proved unnecessary. The Norman victory over Alexius I in the battle outside Durazzo ended defenders' hopes for relief and enabled Guiscard to negotiate a surrender during February 1082, after a portion of the city's defences had been betrayed. Despite this success, the Normans were unable to establish their power in the region. Henry IV's diplomatic and military offensive against Guiscard occupied his attention soon after Durazzo's fall. Bohemond's defeat outside Larissa in 1083 and Guiscard's death in 1085 ended this phase of the Norman offensive against Byzantium.

Durazzo, 1107–1108

Although Bohemond's attack on Durazzo, like that of 1081, was launched from Apulia, the returned First crusader sought wider international support, recruiting in northern France after his marriage to a daughter of Louis VI. Bohemond seems to have attempted to bring Italian maritime power into his endeavour, as Caffaro reports that Bohemond and his bride visited Genoa on their way south.[44] The Genoese annalist does not record any Genoese involvement in the attack on Durazzo, but this may reflect the offensive's outcome more than the participation of Genoese sailors and engineers. In any event, the Genoese government does not seem to have become immersed in the operation.

Bohemond's expedition crossed the Adriatic in October 1107 and invested Durazzo, while the hinterland was raided for supplies and winter quarters built. The Normans did not attempt a sea blockade, although their ships resupplied Bohemond in the spring of 1108. The arrival of these ships completed Norman preparations for attacking Durazzo's defences.

As attackers secured positions around Durazzo and the machinery for the main assault was being prepared, unsuccessful attempts to batter and sap Durazzo's walls were made from underneath armoured roofs. Although attacking miners managed to reach the foundations of the

[43] AC iv. 4–5, pp. 152–4; William of Apulia, *Gesta Roberti*, iv, vv. 240–52, p. 218.
[44] Caffaro, in *AI* 10.

city walls, they were driven out of their mine by Greek fire, and their tunnel was collapsed by defenders before a breach could be made.[45]

The centre-piece of Bohemond's assault was a wall-dominating siege tower as sophisticated as any built in the crusader states. It was equipped with a drawbridge and armoured at all levels, which protected those who propelled it along rollers. As the siege tower progressed towards the walls, the Byzantines built their own tower opposite the advancing device. This defensive structure was based on four beams, which elevated a roofed platform slightly above the siege tower's summit. As the attacking tower drew up to the city, it was moved off its rollers and reinforced for a battle over Durazzo's walls. Although the distance between the two towers was not great, the incendiaries projected by the defenders fell short. The besiegers clearly intended to maintain their tower in a position where maximum pressure could be maintained on defenders without risking the device. In this regard Bohemond's siege-tower assault resembles those we have already examined in Outremer. Defenders piled combustibles between their walls and the siege tower and ignited them. In contrast to a similar situation at Ascalon in 1153, where an unexpected change of wind direction blew such a fire back against defences and caused a breach, Bohemond's tower was completely destroyed.[46]

While Bohemond's poliorcetics against Durazzo were as ineffective as his father's, he was unable to repeat Guiscard's battlefield success against Alexius I. The emperor refused to commit his forces to a major confrontation, although the Normans sought such a decisive encounter throughout the campaign and particularly after the destruction of their siege tower. Alexius organized his response around a long-distance blockade, and with the approach of winter Bohemond's troops lost heart and he came to terms in the treaty of Devol.[47]

Although Guiscard and Bohemond brought complex siege machinery in large-scale assaults against well-defended cities, these methods were not successful. Perhaps Normans lacked sufficient artillery or failed to guard their devices with sufficient troops. It is possible that Bari's and Durazzo's defenders were more familiar with these kinds of devices and their assaults than the attackers. However, it is clear that, for whatever reasons, the Normans did not achieve a domi-

[45] Anna Comena gives the most detailed account of this campaign (AC xiii. 3, pp. 94–7).
[46] Ibid. xiii. 3, pp. 97–9.
[47] Ibid. xiii. 4–9, pp. 99–119, 139; OV xi. 24, pp. 100–4.

nation of crucial sections of city walls. While all three of the towers employed in these sieges were destroyed, Normans maintained their blockades and forced a surrender in two of the three sieges. Perhaps what they lacked in handling the close assault of siege machinery was compensated by their perseverance.

That Norman commanders employed siege machinery similar to those utilized in Outremer is noteworthy. It further supports an argument that the technology required in such operations was not acquired by the Latins in the course of the First Crusade. Moreover, although Bohemond and his followers played a prominent role in the First Crusade, they were not associated with such devices during the expedition.

The two great Norman offensives against Durazzo illustrate a different style of siege warfare from that observed in Italy and Sicily. This may reflect a desire to reduce Durazzo rapidly before Byzantine forces could be concentrated. The method of attack chosen may also be related to the resources of these expeditions and the hopes of participants to capture and doubtless plunder a major city at the outset of their campaign. In any case the close assaults of mobile towers were not central to Norman siege techniques of the eleventh and early twelfth centuries. Twelfth-century Norman rulers, like their predecessors, waged considerable siege warfare. However, they chose to reduce fortified positions by other means.

The Sicilian Conquest of Southern Italy

The campaigns of Roger II formed a different pattern of conquest from that of the Norman establishment. During more than a decade of intermittent warfare a centralized Sicilian power overcame and eventually amalgamated Norman states in southern Italy. During the initial phase of this conflict Roger II fought to secure his claim to the lands of Duke William of Apulia after his death in 1126. By 1128 Roger II had gained not only Apulia but, for a period, the submission of Robert of Capua. Roger II's support of Anacletus in the schism of 1130 and the challenge to imperial authority implicit in his coronation resulted in a nine-year conflict which set the king's Italian wars on to an international stage. Innocent II and his partisans, including St Bernard, actively supported resistance to Roger II. Lothar III provided more concrete support for Roger's Apulian and Capuan opponents and himself led two expeditions into Italy on Innocent II's behalf. Roger II's survival and final victory against Norman barons and southern Italian cities, as well as pope and emperor, are a measure of his achievement.[48]

This conflict included a civil war between prominent Norman families, and involved other groups which had come to differing terms with the rise of Norman lordship in the region. Although a variety of factors motivated protagonists, the growth of Roger II's power and authority on the mainland was paramount. The issues and passions which became intertwined with dynastic, religious, and ethnic rivalries resulted in a war notable in its brutality.[49] Yet the importance of Sicily's fiscal and personnel resources cannot be overestimated in the success of Roger II's military endeavours. During some stages of the conflict his support on the mainland was minimal, consisting of little more than Salerno and Amalfi. Sicily was not only a plentiful and secure base but also the source of fighting men with which he replenished his forces destroyed in battle. The island also provided the king with the sea power essential in waging warfare against opponents in southern Italy. Roger II's ships not only transported men and materials but also assisted in royal sieges.[50] In terms of Roger II's military resources, the establishment of the Norman kingdom reflected a Sicilian conquest of the mainland.

Siege operations were of major importance in the warfare which engulfed southern Italy. While a number of reasons lie behind this, the importance of the region's cities in patterns of conflict is paramount. Amalfi and particularly Salerno provided bases for Roger II's mainland operations, and on occasion refuge from pursuing foes. One of the most important elements in the conflict was Roger II's bid to establish and maintain his lordship over urban communities, and the attempts of those cities to maintain their independence. Roger II's skill in the former endeavour was fundamental in his military achievement.

[48] Although aspects of the king's wars are described in modern accounts, a detailed military history is needed; see Chalandon, *Dom. norm.*; E. Caspar, *Roger II und die Gründung der Normanisch-Sicilischen Monarchie* (Innsbruck, 1904); E. Curtis, *Roger of Sicily and the Normans in Southern Italy* (London, 1912); D. C. Douglas, *The Norman Fate* (London, 1976) E. Jamison, 'The Norman Administration of Apulia and Capua, more especially under Roger II and William I', *Papers of the British School at Rome*, 6 (1913), 211, 481.

[49] For a discussion of the effects of the conflict on Capua, see G. A. Loud, *Church and Society in the Norman Principality of Capua, 1058–1197* (Oxford, 1985), 148–62.

[50] For Norman sea power, see W. Cohn, *Die Geschichte der Normannisch-Sizilischen Flotte unter der Regierung Rogers I und Rogers II, 1060–1154* (Breslau, 1910).

The king fought only three major field battles, two of which—Nocera (24 July 1132) and Rignagno (8 October 1137)—were thorough defeats in which he barely saved his own life, leaving his men to be slaughtered.[51] Commentators since the twelfth century have noted that Roger preferred overwhelming odds or negotiation to the less predictable resolution of affairs on the field of battle.[52] Whatever his inclinations, Roger II excelled at the military and political challenges of siege warfare.

While Roger II conducted effective operations against castles and small towns, it was his mastery in forcing the submission of urban centres which marked his siege warfare. The king's political acumen and ability to play on the divisions within communities were notable in this regard. Yet his fearsome reputation for reprisal was equally important in the political context of his military operations. The capture and treatment of a defiant city was not only a military victory in its own right, but also an example to other communities. Roger II's calculated terror at the outset of his counter-offensive of 1133 netted dividends following his dramatic capture of Montepeloso. Many of the towns of central Apulia chose surrender rather than the consequences of resistance to the king's siege train.[53] It may be noted that Roger II and his forces seldom sought new lordships, nor even to plunder the positions they attacked. Rather the king demanded submission. In this context a reputation for ruthless efficiency and a calculated, terrible vengeance were positive factors in the battle of will at the core of every siege. As one victory contributed to the next, royal siege crews sometimes induced positions to surrender without actually attacking fortifications. Roger II's forces besieged St Agatha in 1135 and, after drawing up in close blockade, began constructing siege machinery. The sight of this activity is reported to have terrified defenders, who surrendered after three days.[54] Duke Sergius VII of Naples came to terms with Roger II in 1131 after the king's successful campaign against Amalfi and its dependent fortifications. As Alexander of Telese wrote, Naples surrendered at the 'mere report' of the fall of the city.[55] While a number of factors played a role in Roger II's siege warfare, his success depended ultimately on the efficacy of his siege techniques.

[51] Chalandon, *Dom. norm.* ii. 21–2, 80–1.
[52] Alexander of Telese, *De rebus gestis Rogerii*, iv. 3–4, p. 147; Hugo Falcandus, *Liber de regno Sicilie*, ed. G. B. Siragusa (FSI 22; 1897), 5.
[53] See below, pp. 110–12.
[54] Alexander of Telese, *De rebus gestis Rogerii*, iii. 41, pp. 136–7.
[55] Ibid. ii. 12, p. 106.

Patterns of Fortress-Taking

Roger II's siege campaigns differed from those of his father, although there was continuity in the importance of Muslim troops and sea power. Resistance to the king's authority was suppressed and his will imposed by an army which owed much of its effectiveness to unity and mobility. Cohesion was important for Roger II's forces, not only to concentrate offensive power, but also because they often operated in hostile territory. While the king forced Apulian towns and cities to accept his lordship, he seldom inspired their loyalty.

'Rebellious' field forces led by experienced commanders provided other reasons for the maintenance of royal cohesion. Avoiding such forces while besieging well-defended positions became a cardinal principle of Roger II's art of war after such a disaster outside Nocera.

Mobility was critical for Roger II's forces for another reason: the reduction of an area's fortifications and cities depended partially on appearing before as many positions as possible during a campaigning season. Rather than the slow, gradual operations of the Great Count's conquest of Sicily, Roger II waged rapid campaigns which one modern commentator has likened to a form of 'Blitzkrieg'.[56]

Although blockades were undertaken, they were seldom maintained for protracted periods of time. Naples was blockaded by land and sea in 1135–6 after the king had re-established his position on the mainland. While the city almost exhausted its provisions, Roger II's ships could not completely interdict seaborne communications. The king's siege crews built up a large earth mound outside the city upon which the main besieging fortification was to be located. The soil amassed for this mound was taken from the immediate area, and this, along with the dry season, resulted in the mound's collapse. Although the siege was maintained, it was abandoned when news that Lothar III's expedition was indeed on its way to southern Italy.[57] Although investment was part of Roger II's repertoire, he seldom dispersed forces into long-distance blockades based on counter-forts and harassment, as had his forebears. Roger II chose, or perhaps was forced, to take positions by other means.

Roger II's siege warfare was characterized by what in this form of conflict may be regarded as rapid movement and reduction of fortifications. In taking positions which were not easily overrun or in-

[56] K. J. Leyser, 'Some Reflections on 12th Century Kings and Kingship', in *Medieval Germany and its Neighbours* (London, 1982), 251–2.
[57] Alexander of Telese, *De rebus gestis Rogerii*, iii. 19–29, pp. 138–47; Falco of Benevento, *Chron.* 228–30.

timidated, Roger II's crews relied on methods which differed from those practised by Norman commanders in southern Italy and the Balkans. The king seems to have built only one siege tower and that was burnt by royal forces when Roger II abandoned the siege of Brindisi in 1129.[58] Several reasons may account for this, including a scarcity of armoured assault troops and perhaps a royal disinclination to risk men in close-quarters combat. A more tangible factor is the southern Italian interior, over which many of these campaigns were fought. The terrain of positions located away from coastal areas impeded the transport of the materials required for mobile siege towers. At the very least, moving the massive corner beams and components for armouring essential in these devices across Apulia would have required substantial effort and manpower. Moreover, siege towers required much preparation and considerable time to construct. Although it is difficult to provide an accurate estimate of 'usual construction time' given the sources and the variation in details of size and armouring between towers, such devices could take four to six weeks to build, although the First crusaders at Marrat demonstrated that simple forms of such structures could be erected rapidly. It must be noted that, as we have seen, pushing such devices in close proximity to prepared defences and maintaining them in such a position often involved substantial casualties. Whatever the abilities of Roger II's crews, a siege technique based on such siege machinery was not well suited to his military resources and the campaigns that conditions and opponents compelled him to wage.

Roger II's crews relied on methods and devices which required expertise but relatively few materials and labour-intensive preparations. Sapping, battering, and, above all, artillery bombardment were the means by which fortifications were damaged sufficiently to permit assault or bring about a negotiated surrender. Although we should not underestimate the difficulties of these techniques, they were significantly less than for those involved in siege-tower assaults. Moreover, the equipment and materials necessary for these operations were substantially less numerous and cumbersome than those required for siege towers. The quick pace of Roger II's campaigns owed something to the poliorcetics employed.

The attack on Amalfi in 1131 illustrates royal siege technique against an important city and its dependent fortifications. Operating

[58] An interpolator of Romuald of Salerno is our only source for this (Romuald of Salerno, *Chron.* 218).

from Salerno, the king's ships under George of Antioch blockaded Amalfi and attacked its coastal possessions. As these actions commenced, a force under Admiral John worked its way inland from Salerno into the mountains behind Amalfi. While George of Antioch captured Amalfitan positions along the coast and on Capri, John isolated the city from the interior. Admiral John brought his troops against the mountain fortress of Tramonti, held for Amalfi by John Sclavo. Royal forces erected artillery and an armoured roof which attacked an outwork of Tramonti. As artillery was brought to bear against the fortification, crewmen under the armoured roof extended a long pole topped with an iron hook and with it pulled down sections of masonry damaged by artillery. With his outwork in ruins, John Sclavo surrendered.[59]

Roger II's forces worked their way along Mt. Lattari and down the Dragone valley to Ravello, a position of considerable natural strength rising almost 1,000 feet from the coast and Amalfi's principal defence against attack from the east. The king joined his forces outside Ravello and directed artillery against the fortification's main tower. The bombardment damaged the citadel, which surrendered to the king. Ravello's fall isolated Amalfi from the interior, and George of Antioch's ships completed the blockade. With little possibility of relief from Naples and the rapid loss of well-fortified dependent positions, the city came to terms, enabling Roger II to take control of Amalfi's citadel and mural defences.[60]

Roger II's offensive of 1133 following the disaster at Nocera illustrates one of the king's most effective campaigns, as well as the efforts of his siege crews. One consequence of the royal defeat was a widespread revolt which undermined the position Roger II had established in Apulia and Capua since 1128. The victors at Nocera, under the leadership of Roger II's great rival, Rainulf of Alife, not only crushed royal forces but virtually eliminated them, including Roger II's siege crews. In replacing his losses Roger II drew primarily upon Sicilian resources, and Muslims constituted a substantial proportion of the new force.[61] Sicilian troops and their loyalty were essential in re-establishing Roger II's authority in the context of the Apulian revolt. So were their skills in siege warfare.

Whatever his shortcomings on the field of battle, Roger II knew the

[59] Alexander of Telese, *De rebus gestis Rogerii*, ii. 7–10, pp. 104–5.
[60] Ibid. ii. 11, p. 105.
[61] Falco of Benevento, *Chron.* 217.

value of the offensive. In the spring of 1133 he marched his forces across Calabria, striking first at Venosa in Basilicata. The city was quickly taken and burnt with most of its inhabitants massacred.[62] This marks the beginning of Roger II's policy of calculated brutality and terror which Chalandon, writing approximately a decade before the battle of the Marne, termed 'répression à outrance'.[63] These massacres and executions appalled observers in the region and beyond, and there is little doubt that Roger II took a personal interest in acts of revenge. Nevertheless, this policy must also be seen as part of Roger II's siege technique, as the deliberate destruction of one community provided a clear example for others. Although Falco of Benevento's comments concerning the unprecedented cruelty of the king during this campaign may be taken as partisan, it should be noted that Roger II's treatment of captured cities is unlike that of earlier Norman commanders. While doubtless generated largely by the political context of the revolt, the king's policy differed markedly from earlier conventions of warfare in the region. Whatever its motivations, the novelty of this policy may have increased its impact.[64]

After devastating the area around Venosa, Roger II's forces marched against Barletta, taking several positions around that city before receiving its surrender. The king then moved south-west, capturing Matera and nearby positions.[65] Resistance coalesced at Montepeloso, which was the centre of the lordship of one of the leading Apulian rebels, Tancred of Conversano. The city was well fortified and reinforced by forty knights of Rainulf of Alife, under the command of Roger of Plenco.

As Roger II's forces encamped in blockade, forcing defenders from outlying positions, his Muslim siege crews began filling a ditch which masked a barbican which itself protected the city's main gate. The labourers were soon protected by an armoured roof, and wood was brought into the ditch as fill upon which earth taken from a nearby mound was to be placed. Tancred's forces counter-attacked at night, attempting to burn the roof and wooden fill. However, royal forces saved their machine by diverting a watercourse into the ditch.

With the ditch filled and approaches prepared, attacks against the barbican commenced. In conjunction with artillery, Roger II's men under the armoured roof brought down damaged masonry with three

[62] Ibid. 217. Alexander of Telese omits describing this phase of the campaign.
[63] Chalandon, *Dom. norm.* ii. 27. [64] Falco of Benevento, *Chron.* 217–20.
[65] Ibid. 218–19; Alexander of Telese, *De rebus gestis Rogerii*, ii. 38, pp. 114–17.

long iron-hooked poles similar to one employed at Tramonti. As the barbican's defenders abandoned their position, the king's assault troops followed them and stormed the city's gates. After Tancred of Conversano was forced to hang Roger of Plenco publicly, Montepeloso was destroyed, and inhabitants who had not escaped during the storming of the city were killed. This included women and children.[66]

The destruction of Montepeloso and capture of several prominent leaders of the revolt constrained further resistance in Apulia. Although Acerenza had harboured refugees from Montepeloso, the former surrendered immediately upon the advent of Roger II's forces.[67] Bisceglie attempted to defend itself, but capitulated as soon as a section of walls had been undermined. The city's walls were then completely destroyed as punishment. Although Trani surrendered without resistance, it was devastated and its fortifications damaged. These actions isolated Bari from remaining rebel positions in Apulia and Capua and from the forces of Rainulf of Alife. A small force under Cardinal Crescentius attacked Benevento and its environs, particulary its vineyards. At the same time Roger II's main forces concentrated against Apulia's principal city. Consequently, Bari surrendered without resistance and its citizens recommenced the construction of a royal citadel.[68] Roger II's forces then moved to Troia, whose leaders attempted to assuage royal vengeance with a procession of the city's relics and clergy led by their bishop. Although the procession was not attacked, Roger II's demeanour was such that participants lost their nerve and fled. While the king did not unleash his men on the town, he had five prominent officials executed, and dispersed the population, much of which fled to Benevento.[69] Royal forces continued receiving the surrender of Apulian positions until the middle of October, when they ended operations and entered Salerno with much plunder and many prisoners.[70] In the vigorous campaigning of one season Roger II had restored much of the territory he had lost after Nocera. He had also established a reputation.

Artillery

While the interplay of several factors contributed to Roger II's successes, the effectiveness of siege engineers and labouring crews was

[66] Falco of Benevento, *Chron.* 219; Alexander of Telese, *De rebus gestis Rogerii*, ii. 41–5, pp. 117–19.
[67] Alexander of Telese, *De rebus gestis Rogerii*, ii. 47, p. 120.
[68] Ibid. ii. 49, pp. 120–1; Falco of Benevento, *Chron.* 219–20.
[69] Falco of Benevento, *Chron.* 219–20.   [70] Ibid. 221–2.

Siege Operations in Italy and Sicily

clearly important, and we should consider whether these forces enjoyed any decisive technological or organizational superiorities. We might also compare Norman operations with those we have already examined in the Levant.

Of foremost significance is the role of artillery in Roger II's siege operations, as royal weapons were prominent in the fall of a number of fortified positions. It should be noted that the use of artillery was not limited to Roger II's forces. Falco of Benevento reports that, in the wake of Roger II's campaign of 1133, a prominent follower of Rainulf of Alife, known in the notary's work as Hugh the Infant, changed sides and went over to the king. Although Hugh had supported Roger II in Apulia in 1128, Chalandon believed that this was but one such seigniorial abandonment of the rebellion in the autumn of 1133.[71] Perhaps to prevent further erosion, Rainulf of Alife led his followers and the Beneventan militia against Hugh's principal fortification, blockading it closely and attacking fortifications with siege engines, including artillery. After four days, and in light of preparations for an intensive bombardment, defenders surrendered and Hugh returned to his former lord's obedience.[72] While the destructive power of this artillery cannot be measured, it is clear that such weapons were available to southern Italian besiegers.

Although the Norman monarch had no monopoly on artillery, he may have employed large or particularly efficient devices or have been served by skilful crews. As we have seen, Roger II's artillery was particularly destructive when assisted by labourers who brought down damaged fortifications with hooks and poles. His rock-throwers were effective without such aid, as demonstrated at Nocera in 1134. The king's forces besieged the city, while infantry held the Sarno river crossings to prevent Rainulf of Alife from repeating the events of 1132. Although checked in initial attacks, Roger II persevered and had an artillery piece erected in a secure position whence it commenced bombarding Nocera. Although Falco of Benevento blamed the city's surrender simply on treachery, it is clear that this device was important in bringing about Nocera's capitulation.[73] Roger II's forces captured Tocco in 1138, after a siege of eight days during which artillery substantially damaged one of the position's towers.[74] Whatever the effects the prepar-

---

[71] Falco of Benevento, *Chron.* 199, 220–1; Chalandon, *Dom. norm.* ii. 31–2.
[72] Falco of Benevento, *Chron.* 220–1.
[73] Ibid. 226; Alexander of Telese, *De rebus gestis Rogerii*, ii. 56–60, pp. 123–5.
[74] Falco of Benevento, *Chron.* 242.

ation of these devices may have had on positions which chose to surrender without resistance, it is clear that artillery was significant in Roger II's siege operations.

This contrasts with artillery in Outremer, where its primary role was in neutralizing defenders atop fortifications and their rock-throwers erected against crusaders and their engines. While crusader artillery also doubtless damaged the upper works of fortifications, Roger II's weapons were more decisive in the resolution of events. Because of the comparative brevity of accounts of Roger II's operations, it is difficult to compare his artillery with that of the crusader states in terms of their destructive capabilities. Moreover, the greater efficacy of Roger II's rock-throwers may reflect a lower standard of fortification in southern Italy more than significant differences in artillery. Nevertheless, it appears that Roger II employed heavy artillery, which was notably more effective than devices employed by other Latin forces during the period. Although the royal siege train included primarily anti-personnel devices as well as larger pieces, there is no clear evidence for the utilization of counterweight artillery. While Roger II's artillery may indeed have delivered an especially heavy projectile, this may reflect the experience of his artillerymen with large traction devices rather than counterweight lever weapons. It is also plausible that Roger II's artillerymen used large counterweight artillery in their siege operations. However, the evidence does not allow us to demonstrate this conclusively. Yet, it is certain that whatever form of artillery Roger II's men operated, they developed considerable expertise in its employment.

Moreover, it is clear that artillery, however powerful, was well suited to the nature of Roger II's military operations. Lever artillery was relatively mobile and this was particularly important in the sieges of Norman Sicily's first monarch. While the men and beasts of burden who dragged the main beams across southern Italy may not have felt these weapons to be particularly mobile, they were in comparison to those required for siege towers. Moreover, lever-artillery pieces could be assembled and taken apart with little difficulty, enabling them to be used repeatedly on campaign without significant modification. This was particularly important in the fluid warfare of Roger II's campaigns in Italy. Roger II's men must also have become proficient at realizing the full potential of their devices.

Roger II's forces were adept at mining and sapping, and this diverges somewhat from the military experiences of the crusaders. This, however, is primarily the result of dissimilarity in the nature of the positions attacked rather than significant differences in technique. The fortifications of cities and castles of southern Italy in most cases do not compare with those of the Levant. Moreover, while crusaders on occasion created breaches, they were unable to exploit them in the face of determined resistance and countermeasures. The willingness of Roger II's antagonists to surrender even before his men had fired their sap and actually breached the walls is notable.

While the expertise, loyalty, and achievements of Roger II's siege engineers and crews are manifest, their origins are not. Muslim troops operated attacking devices and performed manual labour during the siege of Montepeloso in 1133 and it is likely that they continued doing so. The make-up of Roger II's crews before the campaign of 1133 is not known, although it is possible that Sicilian Muslims were part of these forces. It may be noted that the king's crews were eliminated in the course of the battles of Nocera and Rignagno. Nevertheless, similar techniques and levels of proficiency are evident in Roger II's sieges throughout his campaigns. Whatever their origins, these men were effective and above all utterly loyal.

Although his labouring crews were Muslim and his important field commanders Greek, it is unlikely that Roger II's siege technology derived directly from Muslim or Byzantine sources. Muslim Sicily had no developed traditions of sophisticated siege warfare or military technology. Roger II's methods, moreover, cannot be linked to specific Byzantine practices. While the first Norman king in southern Europe emulated earlier Norman commanders in using the resources of one area to conquer another, and particularly in employing Muslim troops, his siege techniques did not stem directly from those of Guiscard, Bohemond, or Roger I. Roger II's methods of fortress-taking derived from the military and political challenges of warfare in southern Italy and from the resources at his disposal.

Rogers II's Defence of his Kingdom

The military achievements of Roger II can be illuminated by the failure of his opponents to drive his forces from Italy despite the advent of a powerful imperial offensive in 1137. By this stage of the conflict Roger II enjoyed a dominant position in the heartland of his opponents. Not only were Bari and Apulia subject to his authority, but so were Capua and Benevento, as well as many of the possessions of Rainulf of Alife. The king had followed these successes by closely blockading his remaining foes in Naples in 1135–6. Although unable to constrict supplies sufficiently to compel surrender, Roger II clearly possessed the military initiative. The advent of Lothar III with substantial German combatants supported by Pisan maritime power caused the king to lift the siege and prepare his defences. The coalition against Roger II possessed powerful military and naval forces with the means to change the course of the war.[75]

The imperial invasion of southern Italy began in April 1137, with Lothar III's forces divided into two main contingents. One, under Henry of Bavaria, assisted Rainulf of Alife and Robert of Capua in operations around Monte Cassino and along the western coast. The emperor and the main German force proceeded along the Adriatic to Bari. Although Roger II shadowed Lothar III's contingent during its march, the Norman king avoided any large-scale military encounter with the emperor's forces throughout the invasion. Instead, Roger II relied on his garrisons in the cities of Bari and Salerno, as well as the difficulties inherent in allied operations in the region.

As troops under Henry of Bavaria and Rainulf of Alife re-established control over Capua and the area around Monte Cassino, Lothar III's forces and Bari's citizens laid siege to the royal citadel inside the city. The men of Bari took the lead in attacking this fortification and its stubborn Muslim defenders, undermining and sapping a section of its walls. Although allied operations were successful, they consumed a month of valuable campaigning. The survivors of Bari's garrison were hanged or thrown into the sea when the citadel fell at the end of June.[76]

While Bari was under siege, imperial forces led by Robert of Capua and Henry of Bavaria moved against Salerno, blockading the city on 17 July. They were soon joined by the duke of Naples and a powerful Pisan fleet. Amalfi and its dependent positions had already been captured by the Pisans in a well-organized and determined offensive against their commercial rival.[77]

With Bari's capture, Lothar III's forces turned west. The march

[75] For accounts of Lothar III's Italian expedition of 1135–7, see Chalandon, *Dom. norm.* ii. 52–78; W. Bernhardi, *Geschichte Lothar III von Supplinburg: Jahrbücher der Deutschen Geschichte* (Leipzig, 1879), 578–740.
[76] Falco of Benevento, *Chron.* 237; *Annalista Saxo* (MGH SS 6), 773–4.
[77] Falco of Benevento, *Chron.* 232; BM 11; Romuald of Salerno, *Chron.* 223. For Pisa's offensive against Amalfi, see below, pp. 198–9.

across the peninsula, and the length of time that this Italian campaign, waged since 1136, involved, kindled discontent. A substantial number of the emperor's knights demanded that military activities be completed quickly so that German troops might visit their families. Although Lothar III maintained discipline over his men, it was clear that the campaign's duration was limited.[78]

The emperor arrived at Salerno on 8 August and authorized Pisan forces to conduct an assault with the siege machines they had prepared.[79] Believing that they lacked the means of effective resistance and any chance of relief, Salerno's citizens surrendered to the emperor in return for a guarantee of their lives and property. This arrangement was made with the permission of Roger II's chancellor, Robert of Selby, who retired into the royal citadel. A conflict between Lothar III and the Pisans developed over whether Salerno, now under imperial protection, might be plundered. The situation was exacerbated when a party of Salernitans burnt a number of Pisan devices, including their siege tower. The Pisan contingent then abandoned the offensive against Roger II and sailed home.[80] The Pisan departure not only weakened the alliance's naval strength, but also deprived it of a skilful and well-provided siege train which had displayed its capabilities in operations against Amalfi.

With the Pisan withdrawal and German discontent, the imperial offensive was effectively ended. The emperor provided the newly invested Rainulf of Alife with funds with which he hired eight hundred German knights before Lothar III commenced his homeward journey in late August.[81] Roger II counter-attacked, leading a force composed largely of Muslims which devastated Nocera and positions around Capua and Monte Cassino. In conjunction with a force led by the newly reconciled duke of Naples, Roger II moved against Benevento and then Apulia. Rainulf of Alife, supported by the militia of Bari, Trani, Troia, and Melfi, confronted the royal army between Rigngano and Castelnuovo. Despite the mediation of St Bernard, a full-scale battle developed. Although Rainulf of Alife shattered the king's army at the

---

[78] Otto of Freising, *Chronica*, ed. A. Hofmeister (MGH SRG; Hanover and Leipzig, 1912), 20, pp. 337–8.
[79] For a discussion of Pisan actions in southern Italy 1135–7, including the siege of Salerno, see below, pp. 198–200.
[80] BM II; *Ann. Saxo*, 774–5; Falco of Benevento, *Chron.* 233–4; Romuald of Salerno, *Chron.* 223.
[81] *Ann. Saxo*, 775.

battle of Rignagno on 30 October, Roger II resumed his offensive during the following spring.[82]

Roger II's campaign of 1138 began with the reduction of positions in western Apulia. While Rainulf of Alife sought another battle, the king's forces consistently avoided such an encounter. Rainulf of Alife's position disintegrated during the year, as cities and fortifications throughout the area north of Salerno surrendered to Roger II and his sons.[83] The death of Anacletus II in January 1138 and Rainulf of Alife in April 1139 made possible an accommodation between Roger II and Innocent II. The capture of the pontiff in the aftermath of Roger II's only victorious field battle at Galluccio on 22 July 1139 further facilitated a settlement which ended the conflict between Innocent II and Norman Italy's first monarch.[84]

Roger II's forces continued operations against remaining rebel positions after their victory at Galluccio. Troia, which contained Rainulf of Alife's remains, was the first objective of royal forces and it surrendered without resistance. Although the king required that Rainulf of Alife's body be disinterred and taken from the city, he did not devastate Troia itself.[85] From Troia the king moved against Bari, which resisted for two months. Royal forces made several attempts to break down defences and maintained a close blockade from fortified camps. This and the absence of any substantial support in the peninsula compelled the city's surrender. Although all citizens were initially spared, a claim by one of the king's knights that he had been blinded while held captive resulted in the execution of Bari's prominent leaders. Although remaining inhabitants feared that the king would once again justify his fearsome reputation, the city was spared.[86] After almost a dozen years of conflict, the lord of Sicily was master of the southern Italian mainland.

The Military Achievement of Roger II

Roger II was one of the foremost siege warriors of his age. That he understood the techniques, psychology, and politics of southern Italian siege warfare is manifest throughout his campaigns. One illustration of

[82] Falco of Benevento, *Chron.* 235–6; Romuald of Salerno, *Chron.* 224–5; Chalandon, *Dom. norm.* ii. 80–1.
[83] The sequence of operations in this campaign is far from clear. I have followed Falco of Benevento (Falco of Benevento, *Chron.* 240–3; Chalandon, *Dom. norm.* ii. 82–5).
[84] Falco of Benevento, *Chron.* 243–4; Romuald of Salerno, *Chron.* 224–5.
[85] Falco of Benevento, *Chron.* 247–8.
[86] Ibid. 248–9.

Roger II's sagacity may be drawn from a campaign in which his opponents failed to overcome the difficulties of siege operations in southern Italian conditions. In the course of his war to secure the inheritance of William of Apulia, Roger II fought a coalition which included Robert of Capua, Rainulf of Alife, and the papal city of Benevento with Pope Honorius II as its leader. In July 1128 Honorius II led an expedition against Roger II's forces in Apulia, operating along the Bradano. With the advent of his foes, Roger II concentrated his forces at an important and defensible ford which also possessed an adequate water supply. Roger II's opponents had little choice but to besiege the position. However, the summer heat, the tedium of the siege, and a lack of finances generated discontent among the allies. Some abandoned the expedition, others requested permission to withdraw, and perhaps some were bribed. Faced with the disintegration of his forces, Honorius responded to Roger II's request for negotiations, and a settlement which involved papal recognition of Roger II's claim to Apulia resulted. In relating this affair, Alexander of Telese emphasized Roger II's respect for the Holy See and the difficulties and costs of provisioning, and Falco of Benevento the weaknesses of Robert of Capua. Romuald of Salerno noted Roger II's sagacity.[87]

Siege warfare dominated Roger II's military affairs, and his successes in this sphere of conflict to a degree compensated for his failures on the battlefield. Some indication of the scale of his siege activities can be obtained from the number of positions reported to have been assailed by his forces. More than eighty attacks are mentioned and more than fifty were successful. The king was no armchair general and was present at at least sixty of these endeavours. Although the evidence does not permit certainty, it is probable that Roger II himself took a prominent role in directing operations. These figures are not firm statistical data but rather a more general indication from the narrative sources. Chroniclers may not have reported or even known of every position attacked during the confused fighting of the period. More importantly, a considerable number of positions submitted upon the arrival of Roger II's forces or during preparations for a major assault. Doubtless the number of full-scale attacks and formal sieges is smaller. It is clear that the reputation of Roger II's forces as siege troops and that of the king's in exacting revenge were persuasive in effecting capitulation and political compromise. Yet that cities and castles

[87] Ibid. 199; Alexander of Telese, *De rebus gestis Rogerii*, i. 13–14, pp. 95–6; Falco of Benevento, *Chron.* 199; Romuald of Salerno, *Chron.* 217; Chalandon, *Dom. norm.* i. 394–5.

surrendered without being subjected to a major assault is in one way another indication of Roger II's consummate skill in siege warfare.

The Heirs of Roger II

Roger II's wars involved the last large-scale offensive siege operations of the Norman state in Italy during the twelfth century. Although Roger II established a lordship over north African cities which relied on Norman naval power, these conquests did not require notable siege warfare. However, some of the overseas expeditions of his grandson, William II, resulted in complex assaults on major cities. Examinations of the attacks on Alexandria and Thessalonica provide illustrations of Norman siege capabilities in the later twelfth century. Moreover, the operations conducted in these attacks demonstrate the high level of expertise, organization, and skill of William II's siege warfare.

Alexandria, 28 July–2 August 1174

A measure of the sophistication of Norman siege crews in the later twelfth century may be taken from this seaborne attack on Alexandria in the summer of 1174. Although unsuccessful, this onslaught was thoroughly organized and one of the most ambitious seaborne attacks of the twelfth century.

The principal narrative source for this expedition is a letter of Saladin's secretary al Quadi-al-Fadil, which is summarized by the thirteenth-century historian Abu Shamah. William of Tyre gives a brief account, noting that the Norman fleet numbered two hundred ships and that the attacking forces sustained substantial casualties during the five or six days they were before the city. The Pisan Annals of Bernard Maragone describe a fleet of 150 galleys and fifty horse transports which carried a thousand knights and considerable siege machinery. However, the account breaks off before giving details of operations, and when it resumes, in a medieval Italian translation, there is no reference to events at Alexandria.[88]

The attack was planned to coincide with a *shia* uprising in Egypt and military support from Amalric I of Jerusalem. However, the king's

[88] Abu Shamah, *Le Livre de deux jardins* (RHC Orient. 3), 164–6; WT xxi. 3, pp. 963–4; BM 61; Chalandon, *Dom. norm.* ii. 394–5.

death and Saladin's successful suppression of the revolt left the Normans unsupported.

The Norman fleet which descended on Alexandria on 28 July was powerful, consisting of between 150 and 200 galleys, thirty-six horse transports, and six ships which carried siege equipment. While Norman troops disembarked in the face of opposition, their ships seized Alexandria's harbour, as defenders scuttled their ships and sent for assistance. Having landed their forces, the Normans established camps on the city's eastern side and commenced preparing siege machinery for an assault. On the next day Norman siege crews brought two large artillery pieces, several smaller ones, three large armoured roofs, and three battering rams against Alexandria. Although artillery could be assembled in a short period of time, the rapid assembly and employment of these roofs and rams are remarkable and unique for the period. Portions or all of these devices must have been prefabricated. Perhaps sections of armouring were transported already put together and were attached to prepared frames when ships landed their cargoes. In any event, their rapid employment indicates considerable advance preparations.

Norman rams battered the walls throughout 29 July, as their artillery bombarded the city. The attackers cast specially prepared black rocks, probably intended to fragment on impact, against Alexandria's defenders. Whatever their purpose, that attackers transported their own ammunition is another indication of the considerable preparedness of this expedition. The efforts of Norman besiegers were rewarded in the damage done to a mural tower and in bringing down a section of the city walls. Although the Normans attacked in the wake of this success, they were unable to carry the breach. A sally party burnt the Norman roofs and battering rams on the morning of 30 July and this set-back, along with reports of Saladin's approach on 1 August, prompted the besiegers to abandon their enterprise. Their withdrawal was chaotic, and a number of men were left at Alexandria and captured by Saladin's forces on 2 August. Despite its failure, the attack on Alexandria illustrates sophisticated and well-organized siege crews and engineers, as well as the logistical and technical capabilities of the Norman kingdom.

Thessalonica, 6–24 August 1185

Eleven years after the expedition to Alexandria William II's forces undertook a major offensive against the Byzantine empire. Considerable preparations were made and a large force formed from royal troops, other Normans, and mercenaries. These men were supported by a fleet of over two hundred ships under Tancred of Lecce.[89]

Durrazo was rapidly taken in June 1185 and the expedition concentrated outside Thessalonica on 6 August. These besiegers were joined by Tancred and his fleet on 15 August, when the city was completely isolated. Although the defenders prepared for a long siege, the attackers commenced operations against the city's defences soon after their arrival.[90] Artillery was prominent and large devices maintained a particularly effective bombardment. Although Huuri speculated that this may have been one of the earliest manifestations of counterweighted artillery, the evidence is not conclusive.[91] Protected by this artillery and by archery delivered from the rigging of Norman ships, labouring crews undermined the outer walls on the city's western side. When it appeared that a section was about to be brought down, the defenders attempted to build a makeshift retaining wall behind the initial defences. However, attacking artillery thwarted these efforts. The successes of the besiegers and the likelihood of a general assault prompted a group of German mercenaries serving the Byzantines to betray the city. Thessalonica's western gates were opened and Norman sailors, soldiers, and mercenaries rushed through the city, seizing its citadels. Thessalonica was plundered and many inhabitants killed during a sack as thorough as any conducted in the Levant.[92] Despite the effectiveness of their siege operations, the Norman offensive resulted in no permanent gains for William II. Although Tancred of Lecce controlled the Bosporus with his fleet, the land force was conclusively defeated by the Byzantines while trying to cross the Striman north-east of Thessalonica on 7 September 1185.

Although William II's overseas ventures never resulted in a durable extension of Norman authority, the complex siege operations undertaken in these endeavours are a fitting conclusion for this discussion of Norman siege warfare. The poliorcetics evident in the attacks of Alexandria and Thessalonica reflect the high level of expertise developed in the Norman kingdom during the twelfth century. While

---

[89] Chalandon, *Dom. norm.* 404–15.
[90] The principal source for this siege is the eyewitness account of Bishop Eustathios of Thessalonica, whose work reflects considerable animosity against the Normans and the garrison's commander, David Comnenus (Eustathios of Thessalonica, *Narratio de capta Thessalonica*, ed. B. G. Niebur (Corpus Scriptorum Historicae Byzantinae; Bonn, 1842)).
[91] Ibid. 455; K. K. Huuri, 'Zur Geschichte des mittelalterlichen Geschützwesens aus Orientalischen Quellen', *Studia orientalia*, 9 (1941), 92; see also Appendix II.
[92] Eustathios, *Narratio*, 451–76.

almost all of the significant Norman leaders seem to have been proficient in siege warfare, the skill and political cunning of Roger II was supreme. While the rise of regular siege trains and their engineers and crews is usually associated with the Capetian–Plantagenet wars of the later twelfth century, the first Norman monarch profited from such resources in the decade of the 1130s, if not earlier. Whenever their precise formation, the crews and engineers who served Norman kings were among the most organized and technically proficient in western Europe during the twelfth century. While these men employed a range of techniques and devices, they are particularly associated with mining and heavy artillery. This stems not only from the kinds of objectives they were asked to capture, but also the size of their units and available resources. In any case, their skill and the materials at their disposal also reflect the fundamental importance of effective siege operations in the foundation and centralization of Norman rule.

4
CITIES AND SIEGE WARFARE: LOMBARDY IN THE TWELFTH CENTURY

So far in this study urban communities have been the objectives of besiegers and have seldom engaged in offensive operations, except in seaborne expeditions along the shores of the Mediterranean. An examination of warfare in twelfth-century Lombardy permits us to view siege warfare waged by inland cities against one another. The forces which generated the urban growth of the later eleventh and twelfth centuries also fuelled conflicts between the nascent city-states of the Po valley. When these conflicts became armed, they involved considerable siege operations against cities and their outlying fortifications as well as raiding and harassment.

Although Milan achieved a predominance in the area by the mid-twelfth century, the pattern of politics and warfare was profoundly altered by the attempt of Emperor Frederick I Barbarossa to exercise lordship in Lombardy. The resultant conflict engulfed northern Italy and dominated papal and imperial political relations for twenty years. In campaigns waged intermittently by Barbarossa's German forces and Lombard allies, major sieges and large-scale devastation were significant. The resources of Lombard cities provided imperial forces with the manpower and materials necessary for conducting large-scale siege operations. Lombard experts and men attracted to major sieges from other areas enabled imperial forces to wage complex assaults based around sophisticated machinery which invite comparison with operations in the Mediterranean. The sieges of Barbarossa's Lombard wars may also be compared with those of Roger II, facilitating a better understanding of siege warfare in the military and to a degree political endeavours of both these monarchs.

An examination of siege warfare in twelfth-century Lombardy also sheds light on a topic of considerable interest for this enquiry: the role

Cities and Siege Warfare

of military engineers. Because of the importance of engineers in the military capabilities of Lombard cities, and perhaps as a reflection of their own milieu, Lombard historians refer to such men and their contributions. While these writers do not provide much biographical detail, they enable us to observe something of the functions and importance of these men in siege warfare.

Sources

Lombard historians of the twelfth century were interested in sieges. This, if nothing else, illustrates the importance of siege operations in the region's warfare and politics. While not every annalist concentrates upon such military affairs, there are a number of contemporary and often eyewitness accounts of the blockades, skirmishes, and assaults so important in the fate of cities. Although Italian writers concentrate on events important to their civic communities and usually reflect their biases in their works, their accounts are nevertheless valuable. One such work is the anonymous *De bello et excidio urbis Comensis*, a two-thousand-line history of the war between Como and Milan in 1118–27.[1] Although surviving only in an eighteenth-century edition, the poem has been dated by internal evidence to a period between Como's destruction in 1127 and reconstitution in 1159.[2] The work was undoubtedly written by a partisan of Como, and the author may have been associated with Bishop Guido of Como, who died in 1125.[3] While its verse-form differentiates it from most of our sources, it provides an extremely detailed and illuminating account of this protracted conflict, which culminated in Como's defeat and destruction.

The deeds of Frederick Barbarossa inspired much historical writing in the twelfth century, including several first-hand accounts of imperial siege operations. While some were written by inhabitants of

---

[1] *De bello et excidio urbis Comensis*, ed. J. M. Stampa (Rerum Italicarum Scriptores, 5; Milan, 1724).
[2] W. Wattenbach, R. Holtzman, and F. J. Schmale, *Deutschlands Geschichtsquellen im Mittelalter* (3 vols.; Cologne, 1971), iii. 922.
[3] The bishop figures prominently in the work and his death is noted in some detail (*Bell. Com.*, vv. 1247–74, p. 439).

Barbarossa's civic allies and adversaries, and consequently maintain a bias towards their particular city, they nevertheless describe operations in considerable detail. Foremost among these is the *Historia Frederici I* of Otto of Morena, together with his son and continuator, Acerbus.[4] Citizens of Lodi, and firm imperial supporters, these men give an extremely important account of Barbarossa's military operations in Lombardy from 1155 to 1164. While an anonymous writer continued this history down to 1168, it is the work of Otto and Acerbus which is especially useful. Otto of Morena's section of the history covers the period 1153–60 and gives a very detailed description of the siege operations of which he was sometimes an eyewitness observer. His account of the siege of Crema is particularly important and he interviewed defenders after the siege in compiling his history.[5]

An anonymous Milanese wrote a history of Barbarossa's Lombard campaigns which provides a balance to pro-imperial sources although it is less comprehensive than the work of the Morenas. The *Gesta Federici I Imperatoris in Lombardia*, which covers the period 1154–77, was written by one of the men elected in 1161 to monitor the distribution of provisions and money in Milan during Barbarossa's campaign of devastation.[6] A canon of the cathedral church of Tortona provides another anti-imperial account of Barbarossa's siege operations in a history entitled *De ruina civitatis Terdonae*.[7] Although limited in its scope, the work provides a valuable narrative of the emperor's first Lombard siege.

Two verse accounts of Barbarossa's campaigns in northern Italy give useful information about imperial siege operations. Godfrey of Viterbo served both Frederick I and his son Henry VI and wrote a poetic account of their military activities in Italy.[8] An anonymous citizen of Bergamo provides another highly partisan account of Barbarossa's campaigns in Italy, which includes useful information

---

[4] Otto of Morena and Acerbus, *Historia Frederici I: Das Geschichtswerk des Otto Morena und seiner Fortsetzer über die Taten Friedrichs I in der Lombardei*, ed. F. Guterbock (MGH SRG NS 7; Berlin, 1930; repr. 1964), pp. ix–xvii.
[5] Ibid. 73.
[6] *Gesta Federici I Imperatoris in Lombardia, auctore cive Mediolanensi*, ed. O. Holder-Egger (MGH SRG; Hanover, 1892), 4–6, 48.
[7] *De ruina civitatis Terdonae*, ed. A. Hofmeister (Neues Archiv für altere deutsche Geschichtskunde, xliv (1920), 142–3.
[8] Godfrey of Viterbo, *Gesta Friderici I et Heinrici VI*, ed. G. Pertz (MGH SRG; Hanover, 1870), pp. v–x.

about siege warfare, particularly at Crema in 1159–60.[9] Although both writers drew from Otto of Freising's history of Barbarossa's early campaigns in Lombardy, their accounts of later operations contain important information.

While a considerable number of non-Italian writers discuss Barbarossa's military ventures, two works should be noted here. One of the most important works of Hohenstaufen historiography is the *Gesta Friderici I Imperatoris* of Bishop Otto of Freising and his continuator, Rahewin.[10] Uncle of the emperor and a distinguished historian, Otto of Freising provides an invaluable account of the early period of Barbarossa's reign. However, Otto's portion of the *Gesta Friderici* describes only one major imperial siege in Lombardy, the attack on Tortona of 1155. The assaults on Milan of 1158 and Crema of 1159–60 are recounted by Rahewin, a canon of the church at Freising and a writer who preferred the phrases of ancient Latin writers and Josephus to his own prose.[11] While Otto occasionally used phrases from Roman historiography and literature, Rahewin's very considerable reliance on earlier writers complicates his descriptions of Barbarossa's siege warfare. Although his account of the siege of Crema is put together from various sources far removed from twelfth-century Lombardy, it cannot be dismissed completely, as Rahewin selected those passages from classical works which he felt best expressed the course of events at Crema. His statement that the city was well supplied with grain is taken directly from Josephus' *Jewish Wars*, and there is no reason to doubt its accuracy, especially since no contemporary source suggests that Crema's defenders were threatened by a lack of provisions.[12] However, Rahewin's descriptions of assaults and particularly machinery, also borrowed from earlier sources, cannot be taken as equally accurate. His account of imperial devices at Crema, also taken directly from Josephus, places attacking siege towers atop man-made mounds, which are not referred to by other sources. Rahewin describes defensive

[9] While there is a more recent edition of this work, its text presents difficulties; I have used the following: *Gesta di Federico in Italia*, ed. E. Monaci (FSI 1; 1887): *Carmen de gestis Frederici Imperatoris in Lombardia*, ed. I. Schmale-Ott (MGH SRG; 1965); J. B. Hall, 'The "Carmen de gestis Frederici Imperatoris in Lombardia"', *Studi medievali*, 26 (1985), 969–76.

[10] Otto of Freising and Rahewin, *Gesta Friderici I Imperatoris*, ed. B. von Simson (MGH SRG; Hanover and Leipzig, 1912).

[11] Rahewin's editors and translators have identified many of his borrowings (ibid.; Otto of Freising and Rahewin, *The Deeds of Frederick Barbarossa*, trans. C. Mierow (New York, 1953), 8–9).

[12] Otto of Freising and Rahewin, *Gesta Frid. Imp.* iv. 47, p. 312.

countermeasures which not only are not confirmed by other accounts of the siege of Crema, but which are also absent from twelfth-century narratives of siege operations.[13] Thus Rahewin's uses as a historian of siege warfare are limited primarily to anecdotes and references to specific individuals.

In contrast with Rahewin, the Bohemian Vincent of Prague provides a detailed and very original account of siege operations and machinery. Chaplain to Bishop Daniel of Prague, Vincent accompanied his master on his Italian expeditions and was present at the sieges of Milan in 1158 and Crema in 1159–60, and recorded many of the events and devices which interested him in an informative and individual style. Although his continuator, Gerlach, recounted the siege of Alessandria, his work is less important.[14] While Barbarossa's Italian campaigns were described in varying degrees of detail and accuracy by a number of twelfth-century writers, those mentioned here are particularly important for an understanding of siege warfare in northern Italy.

Patterns of Conflict

The city-states of Lombardy developed powerful military capabilities along with civic political institutions during the first half of the twelfth century. Their military strength was based on well-organized and disciplined communal infantry forces whose cohesion stemmed from civic institutions and bonds of kinship and community.[15] These forces provided the manpower required in the raids, counter-attacks, battles, and assaults on fortified positions which made up much of Lombard warfare. Although each city was independent, well-established patterns of enmity and alliance usually determined the direction of conflict. Milan's rise to predominance was a central factor in the region's political and military development, and cities oriented their alliances from Lombardy's most powerful urban community. Milan's chief allies were Piacenza, Brescia, Crema, and, later, Tortona, which had traditional enmities with Milan's opponents Lodi, Como, Cremona,

---

[13] Ibid. iv. 48, pp. 313–14; see also below, pp. 139–40.
[14] Vincent of Prague and Gerlach, *Annales* (MGH SS 17).
[15] For a discussion of twelfth-century Italian military institutions, see J. F. Verbruggen, *The Art of Warfare in Western Europe during the Middle Ages*, trans. S. Willard and S. C. M. Southern (Amsterdam, 1977), 125–7; J. Beeler, *Warfare in Feudal Europe, 730–1200* (Ithaca, NY, and London, 1971), 197–8, 206.

and Pavia. While it would be an exaggeration to say that every city was surrounded by its enemies, the proximity of adversaries and the relatively negotiable terrain of the Lombard plain resulted in a situation in which a concentration of forces in one area could leave a city or its hinterland vulnerable.

The rivers and lakes of the region affected the pattern of warfare and particularly attacks on cities and their dependent territories. Control of river crossings was especially important, as it provided security for a city's agricultural areas and facilitated raiding of an opponent's territory. Moreover, the ability to utilize or interdict river and lake communications was occasionally critical in the outcome of major offensives against urban centres. This is well illustrated by operations conducted near Lodi in 1160. During that year Lodi's offensive actions concentrated on neutralizing positions from which Milan and its allies attacked Lodese territory. In May Barbarossa led contingents from Lodi and Cremona against the town of Pontirolo, which based its defence around a fortified church. With the support of a Lodese artillery piece, imperial forces captured the position and then moved against other fortifications on the south bank of the Adda. While Milanese forces observed from the opposite bank, they did not interfere with imperial operations. They were, however, successful in enticing impetuous imperial combatants into attempting to cross the Adda to give battle where a number drowned. Nevertheless, Lodi's defences in the area remained secure.[16] In August 1160, after his defeat at Carcano, Barbarossa attacked Piacenzan pontoon bridges over the Po with Cremonesi, Pavesi, and Lodese forces, supported by two rock-throwers from Lodi. The rapid deployment of artillery after an engagement had begun compelled the Piacenzi to abandon their bridge, which they destroyed, while saving the boats over which it had been placed. The Piacenzi soon built another bridge, which the emperor and his allies attempted to destroy in October. Although Pavesi boats assisted the expedition, the Piacenzi were once again able to break their own bridge while preserving their boats.[17] These actions were of a small scale, especially in comparison with the major campaigns of 1159 and 1161 against Crema and Milan. However, they were important in maintaining Lodi's security against Milanese attacks of 1160.

However effective raiding and depredation were in weakening a city-state, these communities were taken by blockade or by assaults conducted by major concentrations of forces. Lodi was compelled to

[16] Otto of Morena, *His. Fred.*, 105–6.   [17] Ibid. 125–7.

surrender to Milan and her allies after a campaign of devastation and close blockade in IIII. Although their surrender guaranteed their lives, Lodi was destroyed and its population dispersed.[18] Como was captured in 1127 after a major Milanese offensive which culminated in an assault based on considerable siege machinery. As this campaign is well described by a contemporary writer, we should examine it in some detail in order to illustrate siege warfare in Lombardy before the coming of Barbarossa.[19]

Como, 1118–1127

The war commenced in 1118 when Milanese and Comasche forces met at Baradello some two miles south of Como, where they skirmished until nightfall. Although Milanese troops slipped into undefended Como during the night, they scattered inside the city seeking plunder and were driven out by returning forces with significant casualties.[20] This conflict continued intermittently for the next nine years and involved many of the communities along the shores of Lakes Lecco and Como, as control of these waterways became vital. The importance of naval warfare in the patterns of conflict which developed should be noted. Boats were used in raiding, in transporting men and materials for siege operations, in relieving beleaguered positions, and in neutralizing opposing naval forces.[21]

Although Milan and its allies besieged Como in 1119 and 1120, the city's defences and ability to reinforce and resupply by boat enabled the Comaschi to thwart these attacks.[22] While combatants on both sides employed rams, armoured roofs, and artillery in their attacks on fortified positions, these operations were conducted on a small scale. Como's offensive actions centred on its lakeside adversaries. Menagio was attacked in 1124 by a naval and land expedition which commenced siege operations after surrounding defenders in the town's main defensive tower. Comaschi forces attacked this fortification with missile weapons and a battering-ram whose crew was protected by an armoured

[18] Landulf of St Paul, *Historia Mediolanensis* (MGH SS 20), 24–8, pp. 31–2.
[19] An anonymous Comaschi historian was the only contemporary to describe this war in detail. For a modern and well-illustrated account, based primarily on this writer's work, see *Storia di Milano*, iii (Fondazione Treccani degli Alfieri per la Storia di Milano; Milan, 1953), 325–45.
[20] *Bello Com.*, vv. 35–119, pp. 414–15.
[21] Ibid., vv. 145–97, pp. 416–18; vv. 235–6, p. 418; vv. 334–43, p. 420; vv. 580–626, pp. 425–6.
[22] Ibid., vv. 197–270, p. 418; vv. 293–313, pp. 418–19; vv. 1090–112, p. 434.

roof. Attacking crews laboured until the tower's wooden supports were exposed, and these were then ignited. The resultant blaze damaged Menagio's fortifications sufficiently to compel its defenders to flee.[23]

While Como parried repeated Milanese attacks, Milan and her allies achieved a commanding position in the area in 1127. In that year a major offensive drawing contingents from Milanese allies including some that had not participated so far in the conflict was organized. As a powerful force concentrated outside Como, timber for siege machinery was transported by boat from Lecco. In their assault the besiegers employed four siege towers and two large armoured roofs, which were constructed by Genoese and Pisan experts especially recruited by Milan. While these towers were similar in form and armouring to those used in the eastern Mediterranean, their height is not known. Nevertheless, when they were moved to the walls, crossbowmen and archers firing from these structures dominated Como's defences, while crews under the armoured roofs battered and sapped the walls. After failing to damage these machines in a sally, the defenders fled by boat and eventually surrendered. As the escape of Como's inhabitants indicates, the city was not completely isolated during the siege and possessed an ability to resupply by boat. Como was overwhelmed by the magnitude of the assault. After the surrender the city was levelled and its population dispersed.[24]

Although this conflict involved much siege warfare, the scale of the final offensive and its supporting machinery far surpassed earlier efforts. Milan and her allies clearly marshalled substantial personnel, material, and technical resources. More importantly, the assault of 1127 was similar to those conducted against urban centres in the eastern Mediterranean which we have already examined, and the role of Genoese and Pisan expertise is significant. It is, of course, possible to argue that this reference to both Genoese and Pisan technical assistance is primarily a stylistic device intended to emphasize the magnitude and power of the offensive against Como. However, the author's general attention to detail and accuracy makes this appear unlikely. Even if the reference was an invention of the author, that itself indicates something about Genoese and Pisan reputations. Thus it seems that experts from at least one of the maritime states built sophisticated machinery essential in Como's capture and perhaps organized the attack. This account not only illustrates Lombard siege warfare in the early twelfth century but also suggests that Lombard communities

[23] Ibid., vv. 884–911, pp. 432–3. [24] Ibid., vv. 1815–1961, pp. 452–6.

developed a style of siege warfare which resembled that conducted in Outremer. It also highlights the role of engineers in facilitating large-scale and complex assaults.

Siege Warfare in Barbarossa's Campaigns in Lombardy, 1155–1175

Frederick Barbarossa's Lombard sieges were but one aspect of his considerable military activities in the region.[25] Nevertheless, the reduction of powerful urban opponents was crucial in Barbarossa's overall attempt to exercise imperial lordship in Lombardy. The aims of Barbarossa's Lombard policies, and their relation to wider imperial programmes, have generated much disagreement among modern scholars.[26] Whatever the precise nature of Hohenstaufen lordship, and Frederick's and his advisers' conception of it, mastery over Lombard cities was one measure of its effectiveness. Siege operations had political as well as military significance in the conflicts and alliances of the region. They were important in establishing and maintaining the coalitions which conducted imperial warfare. Moreover, formal sieges and their outcomes influenced perceptions of imperial authority and prestige.

Barbarossa's Lombard sieges resulted from the congruent interests of the emperor in subjugating recalcitrant communities, and Lombard cities in eliminating traditional rivals. Ancona was the only major city not attacked in conjunction with a Lombard city-state. On that occasion Christian of Mainz's forces were supported by Venice, which hoped to undermine the growth of a rival in Adriatic and Byzantine

[25] Giesebrecht's narrative of events and critical commentary on the sources remain important (W. von Giesebrecht, *Geschichte der Deutschen Kaiserzeit* (6 vols.; Brunswick and Leipzig, 1881–93), v. For the period until 1158, see H. Simonsfeld, *Jahrbücher des Deutschen Reichs unter Friedrich I, Jahrbücher der Deutschen Geschichte* (Leipzig, 1908).

[26] H. Appelt, 'Friedrich Barbarossa und die Italienischen Kommunen', and G. Fasoli, 'Friedrich Barbarossa und die Lombardischen Städte', in G. Wolf (ed.), *Friedrich Barbarossa* (Darmstadt, 1975). In setting out his interpretation of Barbarossa's political vision and abilities, Munz discusses imperial policies in Lombardy at length; however, his work must be used in conjunction with Gillingham's appraisal and the exchanges it generated (P. Munz, *Frederick Barbarossa: A Study in Medieval Politics* (London, 1969); J. B. Gillingham, 'Why did Rahewin Stop Writing the *Gesta Frederici*?', *EHR* 83 (1968), 294–303; Munz, 'Why did Rahewin Stop Writing the *Gesta Frederici*? A Further Consideration', *EHR* 84 (1969), 771–9; Gillingham, 'Frederick Barbarossa: A Secret Revolutionary?', *EHR* 86 (1971), 73–8.

trade.[27] This reliance on Lombard allies reflects the importance of the communal forces, bases, and supplies that these cities provided. Critical as this support was in the strategic balance between imperial forces and their antagonists, it was of special significance in siege warfare. The material, personnel, and technical resources of Lombard cities were essential in the siege operations conducted by Barbarossa. Moreover, these capabilities significantly affected the timing and nature of the operations undertaken.

In looking at Lombard siege warfare in terms of the region's coalitions and traditional patterns of conflict, we should not underestimate the role played by Barbarossa's ultramontane forces. Major German and Bohemian territorial lords supported the emperor's campaigns with powerful contingents composed of varied kinds of combatants. These troops were prominent in the battles, escort duties, and depredations which made up much of the military activity of the emperor's Lombard wars. These same forces, moreover, were important in siege operations. Crucial in the emperor's military strength were the *ministeriales*, who were the basis of his own contingent.[28] These men were of great value in the close-quarters fighting of assault, machinery protection, and skirmishing which Barbarossa's sieges involved.

German military personnel of a lower status also assisted the emperor in his siege operations. Otto of Freising's story about a soldier described by the ambiguous term *strator* at the siege of Tortona illustrates some of the qualities of these men and the German forces who served the emperor in Italy. During attacks against the city's principal defensive position, one of Frederick's followers, bored by the siege and wishing to set an example, scaled the city's Red Tower by chopping steps into it with an axe. After overcoming an opponent atop the tower, this man returned to the imperial camps to be richly rewarded by the emperor, after declining Barbarossa's offer of the status of knighthood.[29] Whatever the social ambitions of such men, their efforts were important in the large-scale offensives waged against the well-fortified and vigorously defended cities of northern Italy.

[27] This siege of 1173 is not well described in the sources (BM 59; Romuald of Salerno, *Chron.* 265; Boncompagno, *Liber de obsidione Ancone*, ed. G. Zimolo (RIS 6; Bologna, 1937), 10–17.
[28] On *ministeriales*, see K. Bosl, *Die Reichsministerialität der Salier und Staufer* (MGH Schriften, 10, Stuttgart, 1950–1), especially ii. 587–90; B. Arnold, *German Knighthood* (Oxford, 1985), 212–19; for their military role, see Contamine, *War*, 37–8.
[29] Otto of Freising and Rahewin, *Gesta Frid. Imp.* ii. 23, p. 126.

Tortona, 14 February–18 April 1155

The attack on Tortona was Barbarossa's first major siege operation in Italy and was undertaken in the course of his journey to Rome for the imperial coronation. Tortona's association with Milan was one reason for the attack, and the desire of its imperial ally Pavia to strike a decisive blow in its long-standing conflict with Tortona was another. Moreover, Tortona doubtless appeared a less formidable challenge than the region's largest city, Milan.

Barbarossa's forces, with those of Pavia and the marquis of Montferrat, drew up in close blockade on 14 February, after a German force completed a probe of Tortona's defences during the previous week. After establishing their camps, German forces, led by Henry the Lion and his Saxon troops, stormed Tortona's walled suburb but did not carry the old city.[30]

Perched on a steep and rocky hillside, the inner city was approachable only from the west, where it was protected by a massive brick tower which dominated the area. The city was well defended and supported by one hundred Milanese knights and two hundred archers. While adequately provisioned with foodstuffs, Tortona had lost its principal water supply with the capture of the suburb.[31] The attackers ringed their encampments closely around the inner city with the Pavesi, who were later supported by the marquis of Montferrat around the northern and eastern sides. Henry the Lion established his forces in the suburb his men had stormed, which lay along the southern slope of Tortona's approaches. Barbarossa's men encamped on the western side, where attacks against Tortona's fortifications were principally conducted.[32]

Operations centred on destroying the brick tower along the western walls. Rock-throwers were set up against it, which damaged its upper works but did not render it indefensible. The besiegers attempted to sap the tower, first having tunnelled into the hillside, perhaps under the protection of an armoured roof. While the sappers reached the tower's foundations, the defenders located the tunnels and sank countermines, suffocating the attackers.[33]

During these operations defenders harassed the besiegers' camps,

---

[30] Otto of Morena and Acerbus, *His. Fred.* 22–3; Otto of Freising, *Gesta Frid. Imp.* ii. 20–1, pp. 122–4.
[31] Otto of Freising and Rahewin, *Gesta Frid. Imp.* ii. 20–1, pp. 122–4; *Gesta Fed. Lom.* 16.
[32] Otto of Freising and Rahewin, *Gesta Frid. Imp.* ii. 21, p. 123.
[33] Ibid. ii. 21–3, pp. 123–6.

and there was considerable skirmishing. Tortona's only water supply was a spring located between the city and the Pavese camp, which defenders continued to draw from, despite Pavese attempts to prevent this. Because the attackers could not completely isolate the spring from the defenders, it was polluted first with animal corpses and later with sulphur and pitch.[34] Their lack of water, the close blockade, the impossibility of Milanese relief, and a fear for their survival should the city fall to a full-scale assault prompted the defenders to seek terms during an Easter truce which began on 24 March.[35] Barbarossa, however, did not initially agree to a negotiated surrender and military operations recommenced at the end of the Easter festival. During the truce the defenders constructed an artillery piece which they brought to bear against a particularly effective attacking rock-thrower and damaged it, presumably by destroying the frame on which the main beam rotated. However, the besiegers' artillery weapon was soon repaired and continued its bombardment.[36] Nevertheless, the attackers' lack of progress and Barbarossa's desire to reach Rome persuaded him to grant terms on 18 April. Although Tortona was abandoned to the destructiveness of imperial forces, its inhabitants marched to Milan with what they could carry.[37]

While it is not clear which groups were responsible for the attackers' poliorcetics, Barbarossa's own followers were prominent in the siege and operations were conducted from his troops' encampments. However, the artillery and mines directed against Tortona were largely ineffective in debilitating the city's defences. The loss of the suburb which contained the main water supply was the decisive event of the siege. While a number of factors doubtless prompted the emperor to accept a negotiated surrender, one which cannot be discounted is the significant casualties his forces would have suffered in storming the city. Despite Tortona's defeat, the main centres of anti-imperialist power in Lombardy remained untouched by this expedition.

Crema, 2 July 1159–27 January 1160

The siege of Crema in 1159–60 was one of the region's great sieges of the Hohenstaufen period, involving many assaults and considerable siege machinery. The capture of this key Milanese ally, located

[34] Ibid. ii. 21, pp. 124–5.   [35] Ibid. ii. 24, p. 126; *Ruina Terd.* 153–4.
[36] Otto of Freising, *Gesta Frid. Imp.* ii. 26, p. 131.
[37] Ibid. ii. 24–6, pp. 126–32; *Ruina Terd.* 155–6; Otto of Morena and Acerbus, *His. Fred.* 23.

between Milan's adversaries of Cremona and the recently refounded Lodi, was critical in the imperial offensive against Milan. Although the difficulties encountered in reducing this small-sized Lombard city illustrate the strengths of twelfth-century civic fortifications when manned by highly motivated defenders, the eventual success of Barbarossa and his Lombard allies is the measure of their poliorcetic techniques as well as their determination.

The siege of Crema began on 2 July 1159, when forces from Cremona encamped outside the city. Traditional enmities as well as imperial strategic interests lay behind the decision to attack Crema. In 1159 the fragile peace between the emperor and Milan which had resulted from his campaign of 1158 disintegrated and Lombardy was engulfed in warfare.[38] Barbarossa's allies included Cremona and Bergamo, as well as the recently reconstituted cities of Lodi and Como. Milan depended upon Brescia, Piacenza, and Crema for support. Cremona and Crema were long-standing rivals, and the recommencement of regional conflict offered the former an opportunity to strike decisively. Although Barbarossa's siege of Milan in 1158 had forced an armistice, the city had not been defeated. In renewing his offensive the emperor chose not to attack Milan directly but rather to eliminate a smaller adversary in the region.[39] Cremona's ambitions and the emperor's aim of reducing one of Milan's staunch allies located in a strategic position led to seven months of arduous siege warfare.

Barbarossa's forces arrived on 9 July and included contingents from Swabia, the Rhineland, and Bohemia. While the emperor led knights and horsemen from Lodi and Pavia in raids on Milanese territory, imperial forces continued concentrating at Crema. Barbarossa's followers established their camps on the east side of the city between the walls and the Serio River. The Cremonesi had encamped along the southern and south-western approaches. The emperor's half-brother Count Conrad, his cousin Frederick of Swabia, and other German lords established their contingents opposite the western and northern walls. Pavian troops occupied the north-eastern approaches. On 21 July Henry the Lion arrived and led his powerful force to positions outside the northern gate, completing the ring of besiegers around Crema. Although more reinforcements, including a contingent of Bavarians under Welf VI, continued to arrive, Crema was completely

[38] For a discussion of the 1158 siege of Milan, see below, pp. 143–4.
[39] For a discussion of this phase of Barbarossa's Lombard warfare, see Delbrück, *Kriegskunst*, iii. 349–55.

isolated from this point. Welf VI took up Barbarossa's original position when the emperor moved his headquarters into Cremonese positions in order to be near the siege tower his Italian allies were building and to direct its assault.[40]

Crema's defenders were particularly constricted by a group not hitherto involved in Lombard siege warfare: destitute and presumably itinerant personnel who prowled the area immediately outside Crema's defences attacking anyone on whom they could lay their hands. Derisorily termed the 'children of Arnold'—presumably after Arnold of Brescia, executed by Barbarossa in 1155—this group harassed Crema's defenders whenever they ventured from the city, despite the casualties inflicted on them.[41] While we know little about the composition of this group or its means of supply, it seems likely that it was made up from the marginal populations of Lombard towns hostile to Milan and her allies. Although armed only with knives and stones, they seem to have terrified Crema's defenders. If nothing else, the story about them illustrates how major sieges served to draw to them personnel of every sort, who came in the hopes of military and perhaps economic opportunity.

Crema, situated in a marshy area near the Serio River, was protected by a strong double circuit of towered walls approximately one quarter-mile in circumference. A massive wet moat connected to the river encompassed the city. Four gates—the Serio (eastern), Rivolta (southern), Umbriano (western), and Planengo (northern)—permitted easy egress for sally parties. The city's defenders had been reinforced by four hundred Milanese knights and foot soldiers and by men from Brescia. Crema's inhabitants were adequately supplied with provisions and water and it was clear that the city's capture depended on the besiegers' success in overcoming Crema's defences.[42]

Although the precise chronology of the besieging operations is not recorded, the sequence of events is clear. In the first place attackers maintained a close blockade, while machines to support an assault were prepared. Barbarossa, German lords, and Italian contingents all built or had machinery made, including a considerable number of artillery pieces and small armoured shelters.[43] Artillery was directed against Crema's fortifications while more complex devices were under construction, and then supported those devices during their attacks. Some

---

[40] Otto of Morena and Acerbus, *His. Fred.* 69–75; Giesebrecht, *Kaiserzeit*, v. 202–3.
[41] Otto of Morena and Acerbus, *His. Fred.* 73.
[42] Ibid. 74–5; *Gesta Fed. Lom.* 36–7; Giesebrecht, *Kaiserzeit*, v. 198–201.
[43] Otto of Morena and Acerbus, *His. Fred.* 74, 87.

of these pieces were situated in close proximity to Crema's walls and on at least one occasion were the objective of a defensive sally. However, German troops, including Count Conrad and Otto of Wittelsbach, succeeded in saving these devices.[44] This illustrates not only that artillery was continually employed if for no other reason than to maintain morale, but also the importance of German forces in maintaining the security of machinery deployed in close proximity to active defenders. However, the Cremonesi took the lead in machinery production by building several artillery pieces, including three particularly large ones, armoured roofs, and, above all, a siege tower reported to have been more than one hundred feet in height.[45]

As this siege tower was the basis of the first major assault, the emperor moved his camp amidst the Cremonesi opposite the city's southwestern walls in order to direct the attack or at least be involved in it. Vincent of Prague describes this tower in one of the most detailed accounts of such a device.[46] The oak tower consisted of six storeys, with archers and crossbowmen located at the summit. The bottom storey came up to the height of Crema's walls and contained a bridging apparatus. Each intervening stage was smaller than the one below it and contained crewmen and combatants. The tower was initially armoured with one layer of osiers and hides on three sides. Vincent also mentions that the tower's horizontal frame was laid out like a carriage.[47] Presumably portions of the tower's frame extended behind the vertical structure and armouring to aid in the tower's movement by enabling more men to propel the device and facilitate steering. The tower was moved along rollers with a crew of five hundred men consisting of Germans and Italians. While no other writer describes this tower in such detail, some were clearly impressed by the structure. Otto of Morena, an experienced observer of siege warfare, noted that the tower's size and sophistication was greater than anything which had been seen in northern Italy.[48]

Vincent of Prague also has a story about the engineer who designed this machine. A certain man from the kingdom of Jerusalem came to Barbarossa during the early stages of the siege, claiming to have taken many Saracen fortifications with his devices and offering to construct a siege tower which would enable the attackers to place an assault force

---

[44] Ibid. 76–8.
[45] Seventy *brachiis* (I have used twenty inches as a measurement for one *bracchium*).
[46] Vincent of Prague, *Ann.* 677–8.   [47] 'in modum rhede' (ibid. 677).
[48] Otto of Morena and Acerbus, *His. Fred.* 73.

on to the walls. The man's services were accepted, and the Cremonesi offered to provide the funds, materials, and craftsmen to build the device.[49]

While elements of this story may be discounted, as this engineer may well have exaggerated his earlier achievements, Vincent of Prague's account is very specific and not contradicted by other historians. It is clear that this account is not the result of a confusion with another engineer, Marchesius of Crema.[50] At the very least, Vincent of Prague's account illustrates the kind of stories about such men and the knowledge they spread which circulated in the mid-twelfth century. Yet it should be noted that Cremonese finance, materials, and workmen were essential. While an itinerant expert in siege warfare and technology may have enabled a larger and more complex device to be built, his work was based on the labour, skills, and resources of a Lombard city-state.

In supporting the siege tower's attack, Barbarossa had built a large armoured roof termed both a *gattus* and a *testudo*.[51] While large armoured roofs often preceded siege towers in order to fill ditches and level pathways, this device also protected men from defensive missile fire as they laid down rollers for the tower's movement. This shelter was also important in protecting the labourers whose task it was to fill Crema's deep moat. However, these men lacked sufficient earth, wood, and other materials to accomplish this. Consequently, Barbarossa turned to Lodi's citizens for assistance. They donated two hundred empty barrels, which were filled with earth and cast into the moat. They also brought two thousand cart-loads of earth and wood for the same purpose. With such assistance labourers were able to fill the ditch and create a pathway sufficiently stable to bear the siege tower.[52] This illustrates not only the importance of large-scale manual labour in operations, but also one aspect of Barbarossa's reliance on Lombard resources in his siege warfare.

Although artillery attacks continued while these machines were under construction, the completion of the tower and its pathway sometime after September 1159 heralded the beginning of a major assault against Crema's south-western defences. As the tower moved forwards immediately behind the armoured roof and over the ditch, defensive

[49] Vincent of Prague, *Ann.* 677.
[50] See below, pp. 140–1; Giesebrecht, *Kaiserzeit*, v. 211–12.
[51] Otto of Morena and Acerbus, *His. Fred.* 78; *Gesta Fed. Lom.* 37.
[52] Otto of Morena and Acerbus, *His. Fred.* 78.

artillery pieces mounted on Crema's fortifications damaged the tower substantially. In the hopes that he could dissuade the Cremasci from continuing their bombardment, Barbarossa suspended Cremasce and Milanese captives on the outside of the siege tower. Unfortunately for the emperor and the captives, the defenders continued projecting rocks, and Barbarossa was compelled to withdraw the tower.[53] This tower was further armoured with another layer of osiers and hides and then with woollen sacks before a second attack was undertaken.[54] It is clear from the tower's re-armouring that the besiegers were unable to neutralize Crema's rock-throwers with their own artillery.

During this period a ram operating under the armoured roof had battered a breach of some thirty feet in Crema's outer wall.[55] The defenders responded by building an earthen and timber retaining wall and wooden fighting platform. This not only blocked the breach but also enabled men to continue fighting from this section of civic fortifications.[56] Defensive forces made repeated sallies against the armoured roof and dug tunnels from the city to the device in order to burn it with incendiaries. Incendiaries were also dropped on to the roof from a swinging boom manœuvred from the fighting platform. However, determined efforts by imperial forces guarding the device and from archers atop the tower preserved the roof.[57] It is worth noting that, while Cremasce incendiaries included oil, sulphur, and pitch, there is no reference to Greek fire or any naphtha-based substance.

Even with the intensity and duration of the assault and the resistance it generated, neither side enjoyed a decisive advantage by 6 January 1160. Despite the tower's dominant position near the walls and the casualties inflicted on defenders by that structure's archery and crossbowmen, the attackers could not gain sufficient control of the walls to permit an escalade. According to Otto of Morena, the scales were tipped when one of Crema's premier military engineers changed sides.

The technical aspects of Crema's defences were directed by the city's chief engineer named Marchesius, who grew dissatisfied with his employers Crema and their Milanese allies during the course of the siege. One night in January 1160 he crept over the city walls in order to take service with the emperor. Barbarossa rewarded him immediately

---

[53] Ibid. 79–81; Otto of Freising and Rahewin, *Gesta Frid. Imp.* iv. 56, pp. 289–91; Vincent of Prague, *Ann.* 677; *Gesta Fed. Lom.* 37–8; *Gesta Fed. Ital.*, vv. 3000–30, pp. 112–13.
[54] Otto of Morena and Acerbus, *His. Fred.* 84.   [55] Ibid. 84.   [56] Ibid. 84.
[57] Ibid. 84–5.

with a horse and twelve pounds of silver.[58] Marchesius, who clearly understood Crema's defences, advised the emperor on the disposition of his attack and recommended building a bridging machine intended to increase the size of an assault force able to be placed on to the walls.

Once his plan had been accepted, Marchesius had his device built, presumably by Cremonese craftsmen, although it is not clear who financed this machine. This device was intended to place troops on to a particularly advantageous section of the walls in conjunction with those attacking from the siege tower's bridge. Since archers and crossbowmen from the tower already dominated the outer walls, there was no need to erect an elevated firing platform above Marchesius' machine. His device enabled a bridge approximately sixty feet long and nine feet across to be extended slowly on to the walls. The bridge was covered with osier and hide armour, which protected personnel moving along it. Approximately thirty feet of this armoured bridge could be projected forward.[59] Presumably this device had a vertical armoured frame and a mechanical apparatus for extending and balancing the bridge. It is not clear whether the bridge inclined sharply or was extended almost horizontally from a device whose height equalled the walls.[60] In order to bring this machine into a suitable position the armoured roof had to be removed. As the withdrawal of first the tower and then the roof would have been a time- and labour-consuming exercise, which would also have left the armoured roof vulnerable, it was burnt by the attackers. The siege tower was then moved up into the roof's former place and the bridging machine brought to the tower's side. These activities were completed by 21 January, when imperial forces were ready to deliver an assault on the walls.[61]

Count Palatine Conrad and his knights formed an assault party from the siege tower, while other German and Lombard leaders and their forces attacked from Marchesius' machine. Although the attackers

[58] Ibid. 87–8. This story is recorded in another account, although Marchesius' name is not given (*Gesta Fed. Ital.*, vv. 3059–75, pp. 114–15).

[59] Otto of Morena and Acerbus, *His. Fred.* 88–9; *Gesta Fed. Ital.*, vv. 3095–9, pp. 115–16. Giesebrecht believed this was some kind of scaffolding device (Giesebrecht, *Kaiserzeit*, v. 211–12).

[60] From Otto of Morena's account this machine resembles the ancient *sambuca* which permitted assault troops to ascend fortifications. It appears that Otto was unaware that Marchesius' machine had a Roman antecedent. For a description of the *sambuca*, see Vegetius, *Epitoma rei militaris*, ed. C. Lang (Stuttgart, 1872; repr. 1967), iv. 21, pp. 142–3.

[61] Otto of Morena and Acerbus, *His. Fred.* 88–9.

dominated a section of Crema's outer wall, the Cremasci organized a strong defence based on their inner wall and troops stationed at the foot of that wall. While the imperial assault forces which rushed over the bridges made initial progress, fierce resistance prevented a breakthrough. A Cremasce artillery piece shattered a section of the siege tower's bridge. The defenders then counter-attacked and the assault ended in confusion.[62]

Despite this set-back, attacking devices maintained their positions, while defenders suffered significant casualties, especially from archers and crossbowmen atop the siege tower. Fighting continued after the assault was abandoned and the burdens of continued battle and vigilance put considerable strain on defenders. These factors, and a fear of reprisals should the city fall to a general assault, prompted the Cremasci to ask for terms similar to those offered to Tortona, which were granted on 26 January.[63]

The Cremasci surrendered to Barbarossa on 27 January and the besiegers spent the next five days destroying Crema's fortifications. On 31 January all of the attacker's machines were burnt, and Otto of Morena noted that they had cost more than 2,000 marks of silver and much time and effort.[64]

With Crema's surrender, Barbarossa retired to Pavia and many of the German contingents returned north. As there were few German or Bohemian reinforcements during 1160, the emperor spent much of that year campaigning at the head of Lombard forces, parrying Milanese thrusts against Como and Lodi.[65] In explaining why Barbarossa did not attack Milan in 1160, some modern commentators have underlined the duration and arduousness of the siege of Crema. Thus this siege has been viewed as a costly and time-consuming affair, illustrating the stagnation frequent in medieval campaigns requiring the reduction of well-fortified and determined cities.[66] While such interpretations reflect the degree of difficulty and time required in capturing a small twelfth-century Lombard city, they should not obscure an understanding of the significance of Crema's capture.

Barbarossa's lordship in Lombardy in this period hung on the

---

[62] Ibid. 89–91; Vincent of Prague, *Ann.* 678; *Gesta Fed. Lom.* 38.
[63] Otto of Morena and Acerbus, *His. Fred.* 92–4; Vincent of Prague, *Ann.* 678; *Gesta Fed. Lom.* 38; Otto of Freising and Rahewin, *Gesta Frid. Imp.* iv. 71–2, pp. 316–18; *Gesta Fed. Ital.*, vv. 3100–49, pp. 116–17.
[64] Otto of Morena and Acerbus, *His. Fred.* 95–6.
[65] Ibid. 104–28; see above, p. 129; Delbrück, *Kriegskunst*, iii. 351–4.
[66] Beeler, *Warfare*, 210–13; Munz, *Barbarossa*, 169.

Cities and Siege Warfare 143

reduction of Milan and its allies. To a degree his support in the region depended on maintaining the course set at Roncaglia and on perceptions of imperial power and determination. That the emperor was prepared to mobilize imperial resources and wage a difficult siege during summer and winter in subduing a well-fortified and determined Milanese ally was a demonstration of resolve. Crema's Lombard foes won an important victory, and our sources stress Cremonese animosity towards Crema throughout the siege.[67] Moreover, Crema's capture and the retreat of its population into Milan were important factors in the larger offensive against that city. When the emperor received substantial German support in 1161, these forces and Lombard allies had the freedom of movement and positional advantage to wage a decisive campaign against Milan.

Milan, 7 August–7 September 1158

Before turning to Barbarossa's decisive offensive against Milan, it is appropriate to discuss his first major attack on that city. It developed during the summer of 1158 and commenced with a close blockade by the emperor's powerful German and Lombard forces of the city's main gates on 7 August. Although Milanese and imperial forces fought each other outside the city's fortifications on several occasions, neither side achieved a significant advantage. A fortification known as the Roman arch served as an outwork on Milan's south-eastern side and was the focus of much fighting. Imperial labourers attacked the tower's foundations under protection of archers and troops drawn up to prevent a sally. Fearing that their structure would be sapped and toppled, the Milanese garrison surrendered. The attackers then bombarded Milan with an artillery piece erected atop this fortification. The defenders responded with two similar devices which projected their rocks from inside the city. The imperial rock-thrower was unable to neutralize these opposing weapons and to damage Milan's fortifications substantially.[68]

The *Annales Pisani* report that the city government sent archers, crossbowmen, and builders as well as tents to assist the imperial cause. Although this source relates matters in such a fashion that the Pisan advent appears important, they arrived towards the end of the siege and seem to have taken little active part in operations.[69]

---

[67] Otto of Morena and Acerbus, *His. Fred.* 95; Vincent of Prague, *Ann.* 678.
[68] Otto of Morena and Acerbus, *His. Fred.* 54–5; *Gesta Fed. Lom.* 31–2.
[69] BM 18–19.

While no large-scale personnel assault was launched and the Milanese did not commit their forces to a major battle, there was frequent skirmishing. Imperial forces concentrated on devastating Milanese agricultural areas and maintaining a close blockade. Their successes in these operations, the lack of progress in debilitating Milan's fortifications, and the summer heat, as well as a willingness to reach a political compromise over the issues at stake, resulted in a truce and settlement on 7 September.[70] However, the war resumed in 1159.

Milan, 29 May 1161–4 March 1162

On 29 May 1161 a powerful imperial force drawn from Germany, Bohemia, and Lombardy converged around Milan in Barbarossa's second major offensive against that city. This attack was also characterized by blockade and devastation; however, these endeavours were conducted more thoroughly and on a larger scale than those of the campaign of 1158. As before, imperial troops encamped around Milan, destroying crops and trees right up to the city's main defences. Instead of attacking Milan's fortifications or enclosing the city in a close blockade, Barbarossa's men devastated the agricultural resources of Milan. By mid-June the circle of destruction had spread some fifteen miles from the city.[71]

Although contingents from the major Lombard cities returned to their homes at the end of June, many of the emperor's Germans maintained a presence to the south and south-west of Milan near Lodi and Pavia. Military activities centred on continued devastation, and the elimination of outlying Milanese positions. The area immediately around Milan was ravaged once again in August and, although the Milanese came out in strength, they were unable to prevent the destruction of the bulk of the season's harvest.[72]

During the autumn of 1161 three imperial positions along Milanese lines of communication with Brescia and Piacenza were fortified and garrisoned.[73] These fortifications and the powerful contingents in them isolated Milan from its remaining allies and sources of supply.

---

[70] Otto of Morena and Acerbus, *His. Fred.* 53–8; *Gesta Fed. Lom.* 29–32; Vincent of Prague, *Ann.* 677; Otto of Freising and Rahewin, *Gesta Frid. Imp.* iii. 36–47, pp. 209–24; Giesebrecht, *Kaiserzeit*, v. 162–5.
[71] Otto of Morena and Acerbus, *His. Fred.* 135–8.
[72] Ibid. 141–6; *Gesta Fed. Lom.* 47–8.
[73] Otto of Morena and Acerbus, *His. Fred.* 146–8.

They were also essential forward bases in the blockade Barbarossa and his followers maintained during the winter of 1161–2.

As in the previous year, Barbarossa spent much of the winter fighting against a Lombard city-state. However, the emperor did not direct machinery assaults against fortifications nor co-ordinate a close investment. Milan was debilitated by a long-distance blockade, conducted by forces situated in recently constructed fortifications, and in the Lombard communities allied to the emperor. Barbarossa established winter quarters in Lodi, selecting that city rather than Pavia, because the former location facilitated communications with imperial forces operating against Brescia as well as with those isolating Milan from Piacenza. Men from Lodi, Cremona, and Pavia also monitored territory around their cities and dependent fortifications to prevent supplies from reaching Milan. Imperial forces patrolled vigilantly and severely punished those caught. The penalty for provisioning the proscribed city was amputation of the right hand, and it was reported that on one day alone twenty-five men from Piacenza were punished in this manner.[74] Although some supplies reached the city, they were few and expensive, and unable to sustain the population. Moreover, it was evident that Barbarossa possessed the military capability and will to repeat what had been done to Milan's agriculture in 1161. Negotiations began in late February, and Milan formally surrendered on 4 March 1162.[75] The city's fortifications were then destroyed, the legal identity of the commune abolished, and Milan's former citizens dispersed to four 'villages'.

The defeat of Milan significantly altered the balance of power in Lombardy and was one of the emperor's outstanding military successes. The capitulation of 1162 culminated a long campaign which, as Delbrück observed, began in 1159.[76] It is in this context that we may locate the significance of the siege of Crema. The reduction of Crema should not be seen as a divergence from the general direction of this larger campaign, nor as indicative of inferior methods compared with those employed against Milan. Milan was defeated in a complex campaign which involved a range of siege techniques against a variety of positions. The expulsion of Crema's population from the area between Lodi and Brescia significantly facilitated the long-distance blockade against Milan. Not only was a staunch Milanese ally removed

[74] *Gesta Fed. Lom.* 50.
[75] Ibid. 49–51; Otto of Morena and Acerbus, *His. Fred.* 148–53.
[76] Delbrück, *Kriegskunst*, iii. 346–9.

from this strategic area, but Lodi and Cremona were freed from defending their territories against Cremasce forces. Moreover, Barbarossa's arduous and expensive winter siege of 1159–60 was a demonstration of resolve.

The forces which humbled Lombardy's largest and most powerful city were drawn from both sides of the Alps. Lombard cities provided Barbarossa with bases, manpower, and, especially, manual labour, as well as material and technical resources. The fighting men from German and Bohemian lands which the emperor caused to be brought into Lombardy were important in achieving the aims of cities opposed to Milan and her allies. These ultramontane forces played a prominent role not only in battles and assaults, but also in the doubtless tedious operations of the blockade of 1161–2. In some ways the defeat of Milan by the emperor's Italian and German forces reflects complementary military resources as well as congruent political aims. When the political alliances behind the victory of 1162 altered, the balance of power in Lombardy shifted again.

Alessandria, 27 October 1174–12 April 1175

Milan's submission marked Barbarossa's last large-scale siege operation in Lombardy until the attack on Alessandria in October 1174. The political situation in Lombardy had changed considerably from that of the 1150s. Although Pavia and William of Montferrat continued supporting the imperial cause, patterns of alliance and conflict had altered significantly. With the formation of first the league of Verona and then the Lombard league in 1167, cities traditionally hostile to one another joined in an alliance opposed to the exercise of imperial lordship in Lombardy. Barbarossa's opponents were not only those of earlier wars but also former allies, including Lodi, Cremona, and Como. Moreover, Milan had been reconstituted and had re-emerged as a substantial political and military power in the region. Supported by Alexander III and Manuel Comnenus, the Lombard league profoundly affected the emperor's military capabilities in the region.

Barbarossa's ultramontane forces were probably less numerous and powerful in the expedition of 1173–4 than those of the offensive against Milan. The emperor was assisted by fewer major territorial lords from north of the Alps. Moreover, the disastrous outbreak of malaria among triumphant imperial forces at Rome carried off important lieutenants and supporters, including Rainald of Dassel, Daniel of Prague, Duke

Frederick of Swabia, and perhaps as many as two thousand knights.[77] This catastrophe crystallized opposition to the emperor and triggered a reversal of imperial fortunes in the Italy. Barbarossa's invasion force of 1173–4 included substantial contingents of infamous 'Brabancon' mercenaries, which compensated to some degree for these losses. While particularly adept at large-scale and thoroughgoing devastation, their value in more formal operations may have been less than that of combatants in earlier expeditions.[78]

Several factors motivated Barbarossa's decision to attack Alessandria in 1174. Although imperial chancellor Christian of Mainz had extracted payment from Ancona for abandoning the blockade maintained from April to October 1173, that siege was less than a clear demonstration of imperial might. Christian's forces, in conjunction with substantial Venetian sea power, isolated the city, which was supported in part by Byzantine emperor Manuel Comnenus and his fiscal resources. Although a famine developed, Ancona's defences were never overcome. The chancellor's decision to accept compensation for raising the siege was influenced by the relieving force headed by William of Marchisella and Altrude of Ferrara, which moved to support Ancona in the autumn of 1173. While the departure of Manuel I's representative could be construed as a victory, Ancona's treatment did not serve to intimidate northern Italian communities.[79] Thus a restoration of imperial prestige was a clear incentive for reducing Alessandria. The desire of Pavia and the lord of Montferrat to eliminate a nascent rival was another. Yet one important reason in the decision to attack was that Alessandria appeared a relatively easy position to reduce.

Founded at a strategic location on the Tanaro River in 1168 and named in honour of a firm supporter of Lombard resistance to the emperor, Pope Alexander III, the city had a symbolic as well as a military importance in Barbarossa's war against the Lombard league. Because of its recent foundation, Alessandria's citizens had not erected stone fortifications. Moreover, inhabitants' homes were roofed only

---

[77] Giesebrecht, *Kaiserzeit*, v. 544–64, 726–7.
[78] Romuald of Salerno, *Chron.* 262; H. Grundmann, 'Rotten und Brabanzonen: Söldner-Herre im 12. Jahrundert', *Deutsches Archiv für Erforschung des Mittelalters*, 5 (1941–2), 419–92.
[79] The account of this siege by the late twelfth-/early thirteenth-century humanist Boncompagno illustrates his rhetorical skills more than events at Ancona (Boncompagno, *Liber*, 10–17; BM 59; Romuald of Salerno, *Chron.* 260; Andrea Dandolo, *Chronica*, ed. E. Pastorello (RIS 12; Bologna, 1938), 261.

with straw, and Romuald of Salerno records that the emperor's troops expressed their contempt for the city's defences by naming it 'the city of straw'.[80] Alessandria's defences consisted of a massive earthen rampart topped with a palisade. The dimensions of these earthworks are not known, but from the record of events it is evident that they were a substantial obstacle to personnel assaults. While gates are mentioned in accounts of the siege, it is not clear how they were fortified. The city was surrounded by a deep and wide wet ditch, which was the most formidable of its defences.[81] In addition 150 foot soldiers from Piacenza assisted the city's well-motivated defenders.[82]

Soon after their arrival in late October, the attackers decided to launch a general personnel assault against the city. In facilitating this attack, artillery, small armoured roofs, bridges, and ladders were prepared. However, a downpour prior to the commencement of the assault turned Alessandria's approaches into a quagmire. When the assault forces finally launched their delayed attack, they were rapidly bogged down and easily repulsed. The defenders, moreover, sallied, capturing attacking artillery, which they brought back into the city.[83] Barbarossa declined to abandon operations despite the season and encamped on 27 October, making manifest his decision to undertake another arduous winter siege in Lombardy.[84]

Imperial forces rapidly consumed what could be foraged and, while they were supplied through Pavia, food and fodder was expensive. As living conditions deteriorated during December, the Bohemian contingent and that of William of Montferrat were allowed to leave the siege.[85] Nevertheless, Barbarossa's forces and those of Pavia and Italian territorial lords persevered until more clement weather made offensive operations possible.

As at Crema, imperial poliorcetics initially centred on a mobile siege tower and a large armoured roof supported by artillery. However, the engineers who built these devices were not itinerant experts from Out-

[80] Romuald of Salerno, *Chron.* 263.
[81] Vincent of Prague and Gerlach, *Ann.* 687; Giesebrecht, *Kaiserzeit*, v. 750.
[82] While several writers mention the siege, Godfrey of Viterbo and John Codagnellus provide the most important accounts. The latter, while a firm Guelf supporter, was nevertheless an accurate historian and was writing his *Annales* from 1174, if not earlier (John Codagnellus, *Annales Placentini*, ed. O. Holder-Egger (MGH SRG; Hanover and Leipzig, 1901), pp. v–xii).
[83] Godfrey of Viterbo, *Gesta Frid.*, vv. 901–12, p. 34.
[84] The disposition of encampments is obscure.
[85] Godfrey of Viterbo, *Gesta Frid.*, vv. 914–22, p. 35; Vincent of Prague and Gerlach, *Ann.* 687–8.

remer, nor Lombards who changed sides. Rather the siege tower was built and at least partially manned by Genoese, and it is likely that they were involved in the construction and employment of the armoured roof and some artillery. Godfrey of Viterbo attributes the construction of the tower to Genoese experts, and a Piacenzan writer mentions Genoese crossbowmen serving in the device.[86] It is most unlikely that these men served as part of a formal arrangement between the emperor and the Genoese commune. However, it is possible that Barbarossa recruited his siege engineers, craftsmen, crossbowmen, and even labourers from Genoa and its dependencies, and that professionals came to the emperor and offered their services when news of the war's commencement circulated in northern Italy. In any event, Genoese expertise contributed significantly to imperial siege efforts at Alessandria. Moreover, these imperial siege crews were pre-eminent in the poliorcetics of besieging operations. Pavia did not play a role of corresponding importance in operations at Alessandria, as had Cremona and Lodi at Crema, although Pavia was involved in the blockade and fighting throughout the siege. Whatever their origins, Barbarossa's forces took the lead in neutralizing Alessandria's fortifications.

As the weather improved after the coming of the new year, men labouring under the armoured roof filled a section of the great ditch and led this structure followed by the siege tower in close proximity to the walls. A ram battering underneath the roof broke down one of Alessandria's gates. However, the attackers were unable to exploit their advantage because of accurate defensive missile fire and artillery.[87]

Although our sources do not describe events in detail, it is clear that a protracted battle for control of the walls developed, centred around the siege tower. While imperial forces were unable to achieve a decisive advantage in this fighting, the defenders were equally unable to destroy or substantially damage the siege tower. In order to end the stalemate, Barbarossa attempted to take the city unawares during an Easter truce. After the truce had begun, labourers dug tunnels from the attackers' positions near the walls into the city. Some two hundred German troops entered the tunnels, intending to throw open a city gate to an assault. However, this manœuvre did not remain undetected, as the noise of the men moving through the tunnels into the city alerted defenders. They kept their gates secure and collapsed the tunnels, suffocating the attackers. The defenders immediately counter-attacked

[86] Godfrey of Viterbo, *Gesta Frid.*, v. 924, p. 35; John Codagnellus, *Ann. Plac.* 10.
[87] Godfrey of Viterbo, *Gesta Frid.*, vv. 927–33, p. 35.

and, in the confusion, burnt the siege tower. It may be noted that Barbarossa's attempt to capture the city clandestinely during a truce negotiated to observe the Easter festival is the most widely reported event of the siege of Alessandria.[88] This major set-back and the mobilization of a Lombard relieving army prompted the emperor to abandon the siege.

Although Alessandria initially appeared to be weakly fortified, it proved a formidable defensive position. Thus it is appropriate to comment on some of the special problems that Alessandria presented to imperial methods of attack. In the first place, Alessandria's earthen walls could not be battered, sapped, or damaged by artillery. While they could have been levelled, this would have required truly massive quantities of manual labour and complete control of the walls. Moreover, weather conditions during the winter and early spring would have hampered such an undertaking. Although Alessandria's defences probably lacked the height and certainly the durability of stone walls, they had certain advantages. Troops could be moved easily along them and the whole of the upper surface served as a fighting platform. More importantly, artillery could be located anywhere along them rather than solely atop mural towers, as was usually the case. In connection with this, it should be noted that these walls probably did not need to be reinforced to bear the stress of particularly large rock-throwers. All of these factors complicated the besiegers' attempts to neutralize defensive artillery, and the effectiveness of such artillery in the siege is noted in our sources. Thus Barbarossa's failure may be partially explained by the advantages earthen fortifications offered in thwarting siege techniques geared to stone defences. However, a lack of manpower and the ability to blockade the city, as well as the larger military situation, were important factors in the siege's outcome. Nevertheless, it remains that this relatively simple fortification thwarted the conqueror of Crema and Milan.

There was another year of conflict between the emperor and the Lombard league until the former's defeat at Legnano in 1176. That defeat, his relatively limited degree of support in northern Italy, and political conflicts north of the Alps all contributed to an imperial willingness to seek peace. Divisions within the Lombard league, the wider aims of the papacy, Barbarossa's political skills, and perhaps war wear-

[88] John Codagnellus, *Ann Plac.* 9–10; Godfrey of Viterbo, *Gesta Frid.*, vv. 934–55; Romuald of Salerno, *Chron.* 263–4; Boso, *Vita Alexandri III, Liber Pontificalis*, ed. M. L. Duchesne (2 vols.; Paris, 1892), 427–8.

iness brought Lombard cities and the emperor together in a compromise expressed in the peace of Venice. Regardless of the degree to which Barbarossa succeeded in realizing his objectives in northern Italy, his siege warfare in that region was at an end.

One significant aspect of that siege warfare was the military and logistical partnership between the emperor and Lombard cities in reducing their opponents. If nothing else, the importance of this partnership illustrates the military power necessary in defeating Lombard cities during the twelfth century. The strength of cities, their fortifications, and their alliances compelled attackers to concentrate powerful forces in reducing them. This is demonstrated in Milan's conflict with Como and particularly by the large coalition Milan organized for the siege of 1127 as well as in the wars of Barbarossa. As in other aspects of Lombard history, the strength of cities and the degree of difficulty involved in their capture determined much of the pattern of siege warfare.

A contrast between the campaigns of Barbarossa and Roger II is noteworthy in this regard. Like Roger II, Barbarossa drew considerable military strength from areas independent from the Italian cities he strove to dominate. However, it was not militarily possible to overcome the relatively larger and better-developed cities of Lombardy solely with these forces. For the emperor, the strength of Lombard urban communities required the resources of other Lombard cities in reducing them. For Milan's enemies, imperial military resources and particularly the fighting men brought into the region by Barbarossa and his supporters were essential in offensive action against the region's most powerful city-state and its allies.

To a degree techniques of large-scale siege operations in Lombardy resemble those we have examined in the eastern Mediterranean. In both areas close assaults based around siege towers and their supporting machinery which involved significant casualties were essential in grinding down defences and well-motivated defenders. Moreover, personnel from Italian maritime states contributed to the design, construction, and employment of sophisticated devices in Lombardy. Important as the expertise and experience of these men were, Genoese and Pisan contingents did not play the same role in Lombard operations as they did in Outremer. Lombard cities possessed the personnel and material resources which maritime fleets brought to eastern Mediterranean sieges. Lombard craftsmen and labourers built and manned devices as well as prepared the ground for them when engineers from outside the region designed and directed them. The populations of Lombard cities provided a reservoir of labour as best demonstrated by the filling of the great ditch at Crema. Moreover, the region's terrain and river systems permitted the transport of large timbers and other materials necessary in these forms of siege operations. Thus the similarities in methods of siege warfare were not the result of Italian maritime contingents of similar organization operating in both regions. Rather it had more to do with the similarities in military resources which were brought against well-fortified and -defended urban centres.

Although individual and small groups of engineers were important in specific sieges, they were not part of a transfer of methods and technology developed by Genoa and Pisa and proven in the Crusades before coming to Lombardy. Their contributions were in larger and more sophisticated versions of known machines or in employing devices more effectively, which doubtless resulted from their experience. Moreover, while maritime and Outremer engineers are noted in Lombard siege warfare, so are experts from the region itself.

Engineers and their importance stand out vividly in Lombard siege warfare. That these figures are visible in the sources is significant, and such men are more prominent in Lombard historical works than in those of other areas in the period. One such figure was Milan's great expert Guintelmus, who was an important engineer for the city during Barbarossa's offensive against it. He made artillery pieces for the capture of Stabio in 1156, built a bridge across the Ticino in 1157, and designed the carts and barriers used by Milanese archers in a field battle in 1158. He also directed the blockade of a Pavese castle during the same year.[89] He probably engineered a siege tower brought against Carcano in August 1160 which was burnt in a sally while the main Milanese forces were preoccupied with a relieving force under Barbarossa. He was also probably responsible for the siege tower and artillery which helped Milanese forces capture a fortress at Trezzo in 1159 which contained a significant amount of imperial treasure.[90] Otto of Morena describes him as a 'most talented master' and the man in whom they had placed their greatest hopes during the offensive of 1161. Otto also tells us that it was Guintelmus who handed the keys of Milan to the emperor when that city surrendered in 1162.[91]

[89] *Gesta Fed. Lom.* 22, 23, 26, 40.
[90] Otto of Morena and Acerbus, *His. Fred.* 117–18, 65–6.
[91] Ibid. 163.

Cities and Siege Warfare

It is clear that siege warfare offered opportunities for economic and social advancement in twelfth-century Lombardy. We have already noted the German *strator* who demonstrated his valour before the emperor and who was rewarded as he considered appropriate.[92] Yet the requirements of siege warfare allowed experts like Guintelmus to gain renown and reward through the exercise of skills not traditionally associated with those of the warrior. In this context it is appropriate to note one portion of Otto of Freising's celebrated description of Lombard city-states during the twelfth century: 'Also, in order that they lack not the means of subduing their neighbours, they do not disdain to give the belt of knighthood or honourable offices to young men of lower status, and even some workers of the contemptible mechanical arts, whom other peoples bar like the plague from more respected and honourable pursuits.'[93] Nothing better illustrates the role and importance of these men in the siege warfare of twelfth-century Lombardy.

[92] See above, p. 133.   [93] Otto of Freising, *Gesta. Frid. Imp.* ii. 13, p. 116.

5
SIEGE WARFARE IN THE IBERIAN *RECONQUISTA*

THE siege warfare waged in the twelfth-century phase of the Iberian *Reconquista* involved factors also important in Latin expansion in the eastern Mediterranean: an extension of Christian authority at Muslim expense in which the capture of large urban centres was essential; a military and economic partnership between land-based rulers and maritime powers; the importance of periodic crusading operations in augmenting Christian manpower. Although the context of siege operations in Spain and Portugal differed militarily and politically, there were important similarities in methods and machinery. Thus an examination of siege techniques in Spain and Portugal not only invites comparisons with the crusader states, but may also contribute to a wider understanding of Latin siege warfare in the twelfth century.

As in other facets of the expansion and development of the Christian kingdoms, the *Reconquista* of the twelfth century was considerably influenced by non-Iberians. Knights and nobles as well as mariners and merchants sought renown and their fortunes while aiding their co-religionists militarily. Although such support had been a factor in warfare from the mid-eleventh century, it increased significantly during the hundred years following the First Crusade. Much of this assistance, and particularly maritime forces from areas bordering the North Sea and the English Channel, came in the wake of major European crusading ventures. Some campaigns were deemed worthy of crusading rewards and privileges, whose attractions significantly increased recruits for such expeditions. Individuals and contingents also fought in the service of native rulers for gain as well as for Christian fraternity. Although important in all aspects of warfare, these men were nowhere more decisive than in the major siege operations of the twelfth century.

It has been suggested that twelfth-century Spaniards were generally inexperienced in siege warfare, and consequently found the assistance of veterans in this form of conflict useful.[1] While, given its nature, such inexperience is difficult to demonstrate, there is evidence that parts of Spain lagged behind areas of the Mediterranean in naval architecture. In 1115 the coasts of Galicia were repeatedly raided by Muslim corsairs and virtually depopulated. Diego Gelmirez, bishop of Compostela, realized that his people were unable to strike back because they did not know how to build ships capable of long-distance raiding. Diego sent letters to Genoa, Pisa, and maritime towns in southern France requesting aid. Eventually a Genoese named Eugerio came to Spain's north-western corner and instructed the seamen and shipwrights of Iria. Two galleys were built, and Eugerio commanded them in a raid against Muslim coastal areas. The raids were successful, and much plunder and many captives were brought back. According to the *Historia Compostelana*, the raiding was so effective that Muslim corsairs discontinued their attacks for five years.[2]

This story has important implications for the diffusion of poliorcetics, which we have seen was closely related to the naval engineering of the period. Moreover, it is clear that twelfth-century Iberian operations were centred around the attacks of complex machinery to a substantially greater degree than had been the case earlier in the conflicts of the Christian states. Yet, in assessing the impact of non-Iberian combatants and their skills, we must examine the period's siege warfare in its military and political context.

Sources

Considered as a whole, the Latin narrative works which deal with siege warfare reflect in their authorship the importance of personnel drawn from outside Spain and Portugal. The most informative accounts of siege operations and techniques were written by Italians, Anglo-Normans, and Germans who fought in Spain and Portugal during the twelfth century. This is not to say that Spanish writers were uninterested in the siege warfare of their rulers and warriors. However, peninsular historians tended to record the sieges of their kings and heroes in the course of their political history of rulers, dynasties, and states. It

[1] D. W. Lomax, *The Reconquest of Spain* (London, 1978), 82.
[2] *Historia Compostelana*, ed. H. Florez (Espana Sagrada, 20; Madrid, 1765); i. 103, ii. 21, pp. 197–9, 301.

is from those who came to Iberia to capture cities that the most specific and detailed accounts of campaigns and operations derive.

A joint Pisan–Catalonian offensive against Muslim pirate bases in the Balearic Islands in 1113–15 involved major siege operations on Ibiza and Majorca as well as elaborate preparations in Italy and Spain. These are recorded in considerable detail by a contemporary and probably eyewitness Pisan historian. The anonymous *Liber Maiolichinus* describes the background and preparations as well as the deeds of arms of the campaign in a 3,500-hexameter work. Although the poet's emulation of Virgilian style and concentration on individual Pisan heroes occasionally obscure his account, he provides much information about the sieges of this campaign.[3] Another work emanating from the same circle of Pisan historians, the *Gesta triumphalia per Pisanos facta*, provides a complementary and less-embroidered account with important technical details.[4] Although these sources are biased towards and concentrate on the activities of the Pisan contingent, they provide valuable information about siege operations during this campaign.

A combined Spanish and Genoese attack on Almeria and Tortosa in 1147–8 is also well recorded by an eyewitness Italian historian. Caffaro, Genoese annalist and veteran of expeditions to the eastern Mediterranean, was closely involved in the planning and leadership of the Genoese force which participated in this offensive. He also wrote a history of this campaign entitled *Ystoria captionis Almarie et Turtuose* describing Genoese military operations and achievements.[5] This work, together with Caffaro's description of the background and results of this campaign in the Genoese *Annales*, constitutes a detailed if partisan account of the organization and functioning of this maritime expedition.

Another major siege of the Second Crusade, the attack on Lisbon in 1147, is well recorded by members of contingents which originated outside the Iberian peninsula. The anonymous *De expugnatione Lyxbonensi* provides an observant and very detailed eyewitness account of the expedition from the viewpoint of an Anglo-Norman cleric.[6] A

[3] *Liber Maiolichinus de gestis Pisanorum illustribus*, ed. C. Calisse (FSI 29; 1904).
[4] For a discussion of these works as a group, see below, pp. 194–5; *Gesta triumphalia per Pisanos facta de captione Hierusalem et civitatis Maioricarum et aliarum civitatem et de triumpho habito contra Ianuenses*, ed. M. L. Gentile (RIS 6; Bologna, 1936).
[5] Caffaro, *Ystoria captionis Almarie et Turtuose*, in *AG* i.
[6] *De expugnatione Lyxbonensi*, ed. C. W. David (New York, 1936). The most recent editor argues that the author was a priest connected to the Glanville family (C. W. David, 'The Authorship of the "De expugnatione Lyxbonensi"', *Speculum*, 7 (1932), 50–7.

shorter account in letter form emanating from the Flemish–Rhineland contingent has been preserved in several German chronicles and provides important information as well as a balance of perspective.[7]

Although there are many references to Spanish and Portuguese raids, battles, and sieges scattered throughout Latin chronicles of the twelfth century, the works mentioned here are important in their detail and interest in siege warfare. It should be noted that these are among the best accounts of major sieges to be found in twelfth-century historical writing. Moreover, these Italian, German, and Anglo-Norman writers clearly depict the struggles of their comrades as equal to similar deeds done in the Holy Land.

Even though the above-mentioned accounts constitute the most important body of material for this enquiry, twelfth-century Spanish historians provide useful information as well as a wider political context. Often references to siege operations are occasional in the narrative of major political, dynastic, military, and religious events. The *Gesta comitum Barcinonensium* is an important narrative history of Catalonia from the eleventh century onwards, although the final form dates from the early fourteenth century.[8] While the first redaction was written at the monastery of Ripoll in 1276, it was based on a genealogy of the counts of Barcelona composed between 1162 and 1184. Considerable attention is given to the deeds of Count Raymond-Berenguer IV (1137–62), including passing references to his conquests of Tortosa and Lerida in 1148 and 1149. The work also describes a floating siege tower employed against the count's southern French enemies near Arles in 1154 or 1155. This tower contained two hundred men and was manœuvred along the Rhone by small boats and crews pulling from the river banks. It was employed against the castle of Trencatrala, which surrendered and was levelled soon after the tower's appearance.[9] This appears an impressive example of an adaptation of the technology of siege towers to particular topographical and military circumstances. But it is difficult to assess the sophistication of the count of Barcelona's siege crews, as this siege is described briefly and is the only account of siege machinery contained in this section of the work.

[7] While three versions of this letter survive, they are very similar and derive from the same source. I have cited the letter of Duodechinus, except when another contains information not recorded in the former (*Ex. Lyx.* 48–9; Duodechinus, 'Epistula Cunonis', *Annales Sancti Disibodi* (MGH SS 17), 27–8).

[8] *Gesta comitum Barcinonensium*, ed. L. Barrau-Dihigo and J. Masso Torrents (Barcelona, 1925), pp. xxvi–xxvii.

[9] Ibid. v. 8–9.

However, there are two major twelfth-century historical works which are important to an understanding of the patterns of siege warfare and operations. The *Historia Compostelana*, although a complicated text, is one of fundamental importance for Galician and Leonese history in the late eleventh and early twelfth centuries.[10] Written at the request of Compostela's first archbishop, Diego Gelmirez, the work gives a vibrant account of the politics and warfare of north-western Spain, in which Diego was a major participant.[11] Although highly partisan on behalf of St James of Compostela and Archbishop Diego, the work gives considerable information on naval warfare, fortifications, and siege operations, as well as the conflicts which affected Galicia and Leon during the minority of Alfonso VII (1128–57).

The *Chronica Adefonsi Imperatoris* is a history of central importance for the twelfth-century *Reconquista*.[12] Although anonymous, it is clear that this chronicle was written by a cleric with a strong Leonese bias and may be considered almost as an official history of Alfonso VII of Leon–Castile. The work's most recent editor has suggested that the author was Arnold, a Cluniac bishop of Astorga from 1144 to 1152 and a close associate of Alfonso VII.[13] Whoever the author, the chronicle gives detailed accounts of the campaigns of Alfonso VII and some of his chief lieutenants as well as his bellicose stepfather, Alfonso I of Aragon. The work ends with a verse account of the siege of Almeria in 1147 which, while detailed in its description of the marshalling of forces, mentions little about the siege's assaults and complex negotiations. Nevertheless, this chronicle is fundamental in developing an understanding of the siege warfare of twelfth-century Spain.

For the political and social context of Iberian warfare the reader is referred to the surveys of O'Callaghan and MacKay, and Bishko provides a fundamental narrative of the *Reconquista*.[14] Lomax discusses the *Reconquista* as a theme throughout Spanish history, and

[10] *His. Comp.*; B. F. Reilly, 'The "Historia Compostelana": The Genesis and Composition of a Twelfth Century Spanish "Gesta"', *Speculum*, 44 (1969), 78–85.

[11] For a recent biography of Diego see R. A. Fletcher, *Saint James's Catapult: The Life and Times of Diego Gelmirez of Santiago de Compostela* (Oxford, 1984).

[12] *Chronica Adefonsi Imperatoris*, ed. L. Sanchez Belda (Madrid, 1956).

[13] Ibid., pp. ix–xxi.

[14] J. F. O'Callaghan, *A History of Medieval Spain* (Ithaca, NY, and London, 1975); A. MacKay, *Spain in the Middle Ages: From Frontier to Empire*, 1000–1500 (London, 1977); C. J. Bishko, 'The Spanish and Portuguese Reconquest', in H. Hazard (ed.), *A History of the Crusades*, iii. *The Fourteenth and Fifteenth Centuries* (Madison, Wis., 1975).

Lourie summarizes relationships between social institutions and military experiences.[15] Fletcher discusses aspects of recent historiography which are pertinent to the study of twelfth-century military affairs and the idea of the *Reconquista*.[16]

The Context of Iberian Siege Warfare

Important as French, Italian, Anglo-Norman, German, and other combatants were in twelfth-century Iberian siege warfare, they did not operate independently from the conflicts of the peninsula. In setting out this context, it is appropriate to note some of the major political factors which influenced the patterns of siege warfare, as well as summarize the nature of siege operations in the eleventh century.

The political as well as religious divisions of the Iberian peninsula were important in the impulses of the *Reconquista*. Regarding the Christian states, it should be noted that the kingdoms of later medieval Spain were still taking shape during the twelfth century. Portugal emerged as an independent kingdom under Afonso I (1128–37), who continued the work of his father Henry of Burgundy. Although Aragon and Catalonia were united dynastically in 1137 when Raymond-Berenguer IV married Petronilla of Aragon, these states conducted military affairs independently during the early twelfth century. The separate kingdoms of Leon and Castile were ruled jointly during most of the first half of the twelfth century. However, Alfonso VII divided them between his two male heirs, and the kingdoms were not united until 1230. Moreover, Alfonso VII enjoyed his inheritance only after debilitating civil conflicts, waged intermittently from 1109 to 1127, which engulfed not only Leon and Castile but also Aragon and nascent Portugal. Central in these conflicts was Alfonso VII's mother and regent during his minority, Urraca, whose adversaries included her sometime husband, Alfonso I of Aragon.[17] These conflicts between rulers, claimants, and their partisans not only diverted military re-

---

[15] Lomax, *Reconquest*; E. Lourie, 'A Society Organised for War: Medieval Spain', *Past and Present*, 35 (1966), 54–76.
[16] R. A. Fletcher, 'Reconquest and Crusade in Spain, c.1050–1150', *Transactions of the Royal Historical Society*, 5th ser., 37 (1987), 31–48.
[17] For an account of these conflicts and the career of this remarkable woman, see B. F. Reilly, *The Kingdom of Leon–Castilla under Queen Urraca, 1109–1126* (Princeton, NJ, 1982).

sources, but also gave scope to ambitions of territorial lords and families. The suppression of their revolts also affected Christian military strength.

Although sharing similar aims of territorial and financial expansion, the separate institutions and traditions of the Christian states were such that common military ventures were usually rare in the twelfth century. While there were exceptions, such as the attack on Siguenza in 1123 or Almeria in 1147, major campaigns were usually conducted by individual states. Moreover, on occasion the ambitions of Christian states conflicted and they competed against one another in their common attempt to restore Christian rule throughout the peninsula.

The divisions of Muslim Spain were equally important in determining the pace and direction of the *Reconquista*. The period 1031–85 was one of political instability, bordering on disintegration. With the fall of the Caliphate of Cordoba in 1031, *al-Andalus* fragmented into a number of independent *taifa* kingdoms, each with its own allies and adversaries, both Christian and Muslim. Moreover, some *taifa* states were bitterly divided between some of the ethnic, military, and dynastic groupings which made up *al-Andalus*.[18] The situation offered the proportionally stronger Christian states opportunities for financial as well as territorial gain, and a system of tribute payments—*parias*—which enriched Christian leaders at Muslim expense developed during this period. One modern writer, in distinguishing the eleventh- and twelfth-century phases of the *Reconquista*, has characterized this aspect of eleventh-century Christian–Muslim relations as one of 'protection rackets' as opposed to a twelfth-century pattern of 'crusaders'.[19] These payments, as well as political conflicts, facilitated Christian territorial expansion, which culminated in the capture of Toledo by Alfonso VI of Leon–Castile in 1085. This Christian advance to the Tagus was a major achievement in the *Reconquista* and compelled the inhabitants of *al-Andalus* to seek external support.

The aims, ambitions, and perhaps techniques of non-peninsular combatants affected *al-Andalus* in the twelfth century at least as much as they did the Christian states of the north. The advent of the Almoravids—a fundamentalist sect which drew its military strength from large, disciplined formations of north African Berbers—significantly altered the balance of power in the Iberian Peninsula. Although

---

[18] D. Wasserstein, *The Rise and Fall of the Party-Kings: Politics and Society in Islamic Spain, 1002–1086* (Princeton, NJ, 1985).

[19] MacKay, *Spain*, 16–31.

they came initially to save *al-Andalus*, the Almoravids gradually replaced native dynasties with their own regime, and this resulted in conflicts between native and north African Muslims, which occasionally provided opportunities for Christian powers. However, the Almoravids stemmed the Christian advance and, with their manpower resources and battlefield skills demonstrated at Zallaca (1086) and Ucles (1108), forced the Christian states on to the defensive and threatened Toledo repeatedly.

Although the sources of Almoravid power were distant from areas of Christian siege operations, the mobilization of a powerful field army was a major threat to any besieging force. In 1132–4 Alfonso I of Aragon attempted to capture Muslim positions north of the Ebro between Zaragossa and Lerida. Having taken Mequineza, Alfonso I settled into a year-long siege of Fraga in 1133. The Almoravids collected a powerful force from their north African and Iberian territories and concentrated near the Aragonese in July 1134. The battle which developed annihilated Alfonso I's forces and the king died soon afterwards.[20]

Despite its powerful armies, the Almoravid state disintegrated during the decade of the 1140s and the vacuum was briefly filled by native rulers. As this coincided with the Second Crusade, Christian leaders were presented with opportunities for military and territorial expansion. While the cities of Tortosa, Lerida, and Lisbon were permanently brought under Christian authority, some of the gains of the 1140s were lost to another fundamentalist Berber sect, the Almohads, who also eventually took full control of *al-Andalus*. Their military power was considerable and they constituted a major threat until the great Christian victory at Las Navas de Tolosa in 1212, which inaugurated a new period of Christian expansion, the so-called 'Great Reconquest' of the middle thirteenth century. Between the coming of the Almohads and their defeat in this battle the primary military concern of the Christian states was the protection of their territory and possessions. Consequently there was limited opportunity for expansion, and relatively little large-scale siege warfare.

Important as great battles and sieges were, it should be noted that the military problems which usually confronted the societies of Spain and Portugal were those of the frontier. Providing security for livestock, agricultural areas, and habitations, as well as the raiding of Muslim communities for all forms of wealth, were the staples of much 'frontier

[20] *Chron. Adefonsi*, i. 51–7, i. 42–7; OV xiii. 8–10, pp. 409–19.

warfare'. Lands between the Duero and Tagus rivers constituted a sparsely populated borderland between Christian and Muslim centres of authority in the eleventh century. The 'repopulation' of this frontier by Christian settlers capable of making a livelihood under these conditions was a notable factor in the shift in the strategic balance towards the Christian kingdoms. The challenges involved in organizing this area and its inhabitants for this 'low-intensity' warfare was a primary impetus behind characteristic Spanish military institutions, including the 'villein knight'.[21] The urban military structures which developed in the twelfth and thirteenth centuries were also principally responses to military concerns geared to raiding and counter-raiding.[22]

This fluid military situation was conducive to long-term 'softening-up' operations against intended targets. This included not only small-scale depredation but also much larger *razzias*, such as that led by Alfonso VII of Leon–Castile in 1138. On that occasion the king's forces raided along the Guadalquivir and deep into Andalusia, carrying off or destroying much wealth. During this raid captured Almoravid religious and legal experts were executed, perhaps as a gesture to Alfonso VII's native Muslim allies, who, it is reported, were particularly hostile to these personnel of the Almoravid regime.[23]

While such operations prepared objectives for attack and may have assisted in persuading inhabitants to negotiate rather than fight, important and well-fortified positions were usually subjected to a more specific attack. While we shall discuss these in detail below, the possible consequences of such attacks for frontier settlements and fortifications may be noted here. The concentration of forces required in major attacks weakened areas of an exposed frontier and left them vulnerable to raiding or major offensives. During Alfonso VII's siege of Oreja in 1139 Muslim troops from Seville and Cordoba attempted to draw his men away by attacking some of the key fortifications which defended Toledo. Alfonso VII's queen, however, took charge of a vigorous defence which maintained these positions without requiring her husband's assistance.[24] Although Toledo's defences were adequate on this occasion, the example illustrates the problems that a concentration of forces, particularly in prolonged siege operations, presented to commanders and rulers.

[21] Lourie, 'Society Organised for War', 55–60; MacKay, *Spain*, 36–9.
[22] J. F. Powers, 'The Origins and Development of Municipal Military Service in the Genoese and Castillian Reconquest', *Traditio*, 26 (1970), 91–111.
[23] *Chron. Adefonsi*, ii. 103–4.   [24] Ibid. ii. 147–50, 114–16.

Regarding the tenor of eleventh-century siege warfare, it may be noted that operations were influenced not only by these factors but also by the complex political and economic relations of Muslims and Christians during the period. Eleventh-century siege warfare was characterized by raiding, blockade, and negotiated surrender which often permitted captured populations to retain their religion, usually in return for a poll tax. The most significant Christian military advance of the century, the capture of Toledo, was achieved through a three-year campaign of depredation and isolation which culminated in a close blockade of the city centred on nearby bases and counter-forts. Yet these operations were not the only major factors in the city's capitulation. Toledo's Mozarab inhabitants and a party of Muslims who preferred Alfonso VI's (1065–1109) rule to that of the lord of Seville played a role in negotiating surrender terms, and Zaragoza was preoccupied with an Aragonese offensive during much of Alfonso VI's attack. Toledo was also divided by civil strife, and elements of the population were hostile to the rule of al-Qadir, whom Alfonso VI had helped take control of the city in 1081. Moreover, Alfonso VI enjoyed a close relationship with al-Qadir, who was promised the lordship of Zaragoza or Valencia, should either fall into Alfonso VI's power after Toledo surrendered.[25]

Not all positions surrendered to terms after having been cut off from supplies and military assistance. Alfonso VI's father, Fernando I (1035–65), is reported to have employed crossbowmen, battering-rams, and probably artillery in his offensive against Muslim positions in what would become Portugal. However, that campaign also involved a six-month blockade of Coimbra and widespread devastation.[26]

An episode from the career of eleventh-century Spain's most celebrated historical figure, Rodrigo Diaz de Vivar, illustrates something of the nature of siege warfare in the political and religious context of the period. In 1092 the sometime vassal of Alfonso VI had established himself as protector of Valencia's new ruler al-Qadir against the Almohads and his own subjects. Al-Qadir, whose rule in Valencia owed much to the support of Alfonso VI, soon made himself as unpopular in that city as he had been in Toledo. Some of his subjects revolted in

[25] E. Lévi-Provençal, 'Alphonse VI et la prise de Tolède', in *Islam d'occident études d'histoire médiévale* (Paris, 1948), 110–35; B. F. Reilly, *The Kingdom of Leon-Castilla under King Alfonso VI, 1065–1109* (Princeton, NJ, 1988), 161–72; Wasserstein, *Party-Kings*, 253–5.
[26] *Historia Silense*, ed. J. Perez de Urbel and A. Gonzalez Ruiz-Zorrilla (Madrid, 1959), 85–9, pp. 188–93.

October 1092 and executed him, installing their leader as the city's ruler. The Cid responded with his second siege of the city, building counter-forts in close proximity to Valencia, whence his forces conducted a difficult twenty-month close blockade which was successful in mid-June 1094. After executing the revolt's leader as a regicide and both intimidating and conciliating Valencia's population, Rodrigo defended the city until his death in 1099.[27]

Regardless of whether the Cid is perceived as a champion of Christian Spain or as an opportunistic warlord, his accomplishment at Valencia owed much to the inability or unwillingness of the Almoravids to concentrate against him as well as to the generally unstable political conditions of eastern Spain. A willingness of a portion of Valencia's inhabitants to acquiesce to his lordship may also have played a role. These ingredients, as well as the steadfastness of purpose of the Cid and his well-rewarded followers, account for his remarkable success in establishing his lordship.

Patterns of siege warfare were also influenced by non-Iberian combatants, who played an increasing role in the peninsula's military affairs from the mid-eleventh century.[28] The most notable campaign involving such men was the attack on Barbastro of 1064, in which a force recruited from Catalonia and north of the Pyrenees blockaded the city. An Arabic source distinguishes between Spanish and 'foreign' troops, and it is not clear whether all of the non-Iberians fought as one group, or whether the Aquitainean, Burgundian, and Norman contingents each operated separately with its own leader, although the latter seems more likely. While Barbastro's defenders repulsed Christian attacks effectively, their water supply was impeded when a stone lodged in a civic aqueduct, and they surrendered forty days after the blockage. A series of massacres ensued, perhaps at the command of one of the northern leaders, or as a result of the attacker's longing for plunder and fears that the inhabitants were concealing wealth. Those citizens not killed were enslaved and transported north or to Constantinople. Most of the non-peninsular forces departed soon after the conquest and Barbastro was retaken by Zaragoza in 1065.[29]

Although it is difficult to be certain given the sources, it is very likely

[27] R. Menendez Pidal, *The Cid and his Spain*, trans. H. Sutherland (London, 1934), pp. 288–311; Lévi-Provençal, 'La Prise de Valence par le Cid', in *Islam d'occident*, 185–235; Wasserstein, *Party-Kings*, 262–4; Reilly, *Alfonso VI*, 227–33.
[28] M. Defourneaux, *Les Français en Espagne aux XIe et XIIe siècles* (Paris, 1949).
[29] A. Ferrerio, 'The Siege of Barbastro, 1064–1065: A Reassessment', *Journal of Medieval History*, 9 (1983), 129–44; Defourneaux, *Espagne*, 133–40.

that these non-Iberian combatants did not bring any specialized poliorcetic skills to the siege. Rather, their primary role was augmenting manpower. While some scholars have interpreted this campaign as a precursor of the First Crusade, the evidence for this opinion is inadequate.[30] Perhaps the treatment of Barbastro's inhabitants is the aspect of this campaign most similar to patterns of the First Crusade. But, regardless of its relation to the crusading movement, the attack on Barbastro foreshadows important developments in Iberian military affairs. Spanish rulers continued to utilize non-peninsular troops periodically during the later eleventh century. The period soon after the capture of Jerusalem saw a significant increase in the scale and significance of such forces, and the resources which crusading impulses brought to Spain and Portugal offered opportunities which were not missed.

The Capture of the Balearics, 1113–1115

In August 1113 a powerful and well-equipped Pisan fleet, reported to have numbered three hundred ships, set sail for the last independent *taifa* kingdom of Majorca ruled by Nasir al-Dawla. The expedition was generated by the depredations of Majorca's sea raiders, the exhortations of Pope Paschal II, and Pisan commercial interests in the western Mediterranean.[31] The Pisans' elaborate preparations had included providing considerable quantities of timber and iron fastenings for whatever siege machinery might be required.[32] However, the Pisan ships were buffeted by a storm and landed on the coast of Catalonia in disarray. Pisan representatives met with Count Raymond-Berenguer III, who joined the venture and granted Pisa commercial privileges in his territories. Yet, as the campaigning season had almost passed, it was decided to wait until the following spring before attacking. In March of 1114 part of the fleet returned home to collect reinforcements, while remaining Pisans gathered material and conducted a reconnaissance of Ibiza.

---

[30] Ferrerio, 'Barbastro', 130–4; Fletcher, 'Reconquest and Crusade', 42.
[31] *Gesta triumph*. 90–1; *Lib. Maiol.*, vv. 5–55, pp. 1–7; BM 8.
[32] *Lib. Maiol.*, vv. 120–9, p. 11.

Ibiza, June–July 1114–February 1115

During the period between September 1113 and June 1114 more participants from areas of southern France with close Catalan ties were recruited. Forces from Arles, Roussillon, Narbonne, and Montpellier joined the expedition. Among these was William VI of Montpellier, who had participated in the First Crusade and fought atop the siege tower at Marrat in 1098.[33] Pisan and Catalan ambitions, as well as rekindled crusading spirit, resulted in a concentration of experienced military and naval forces under the overall leadership of Raymond-Berenguer III. While this was an excellent opportunity, Catalonia might be left exposed to the Almoravids then moving into the Ebro valley. However, Raymond-Berenguer III's possessions were not immediately at risk in June 1114, when the entire force concentrated at Salou near Tarragona before sailing against its first target, Ibiza.[34]

After disembarking and skirmishing, the Christians established themselves outside the island's chief city and commenced their siege. The city's defences at this time included three circuit walls protecting a citadel perched on high ground on the city's southern side. Each wall was protected by a wet ditch with palisades, and the approaches consisted of marshes.[35] The attackers' operations centred on relentless assaults. Maximum pressure was maintained by frequent attacks based on rams and artillery and on occasion only upon scaling ladders. More sophisticated and cumbersome devices and the terrain over which they were to attack were prepared during these assaults.

Having probed the city's defences, the attackers made the first of several assaults within a few days of their arrival. Bridges, secured to the banks of the outermost ditch by large hooks, were thrown across the ditch, permitting troops to cross and approach the walls. A battering-ram whose crew was protected by rope-and-osier armour was then brought up to the walls and was perhaps intended to be employed against a gate. However, the device's supporting forces were driven back and the assault abandoned.[36]

Skirmishing and attempted escalades continued, while the attackers prepared a large rock-thrower and armoured roof to permit continued battering and sapping.[37] These devices were brought into play and

---

[33] See above, pp. 42–3; *Lib. Maiol.*, vv. 417–24, p. 23.
[34] *Gesta triumph.* 90–1; *Lib. Maiol.*, vv. 1170–242, pp. 51–2.
[35] Ibid., vv. 1251–69, p. 53   [36] Ibid., vv. 1295–324, pp. 54–5.
[37] These devices are termed *tormentum* and *testudo* respectively (ibid., vv. 1360, 1362, p. 56).

some damage is reported. To counter the effects of the artillery piece, defenders protected the upper works of their walls with cloths and carpets as shock absorbers.[38] However, the walls were sufficiently damaged by artillery fire and the efforts of the crews under the armoured roof to encourage another major assault on 21 July. Although the attackers gained entry into the city, they were unable to exploit their advantage. Assaults and the efforts of artillery and labouring crews were maintained, and small groups fought each other during the night. Finally on 28 July attacking troops made their way into the city and the Muslim commander surrendered unconditionally. Prisoners were taken and portions of the city's fortifications levelled.[39]

Majorca, August 1114

With Ibiza neutralized, the expedition sailed for its primary objective of Majorca and the island's key city of Palma. Men were landed within sight of Palma along the coast, south-east of the city, on 22 August 1114. After they had been organized and exhorted by their leaders and preachers, troops moved the short distance to positions facing the city's eastern walls on 24 August, after skirmish with defenders.[40]

Although Palma was most easily attacked from its eastern approaches, its defences were considerable. Palma consisted of three cities in terms of its fortifications: a citadel, an old city, and a new one, each with towered walls. The new and old cities were also protected by ditches. Moreover, a small fortified area and two towers further protected the citadel from the old city. The city was well supplied with provisions and materials and defended by a large garrison.[41] In describing the strength of defenders and the quantity of artillery and other weapons available to them, the *Liber Maiolichinus* mentions that the garrison consisted of more than sixty thousand defenders, including four thousand archers. These figures are clearly exaggerated and serve only to illustrate the poet's desire to emphasize the formidable challenges which attacking the city presented.[42]

Having encamped near the eastern gates and probed the city's defences, the attackers began preparing for a major assault based on two

[38] Ibid., vv. 1369–71, p. 57.
[39] Ibid., vv. 1386–507, pp. 57–61; *Gesta triumph.* 91.
[40] *Gesta triumph.* 91; *Lib. Maiol.*, vv. 1545–2029, pp. 62–80.
[41] *Gesta triumph.* 91–4; *Lib. Maiol.*, vv. 2036–64, pp. 80–1; W. Heywood, *A History of Pisa* (Cambridge, 1921), 68.
[42] *Lib. Maiol.*, vv. 128–9, 2076–83, pp. 11, 82.

mobile siege towers. Although these devices were not described in detail, it is clear that they overtopped the walls and were protected by hide-and-osier armouring, and they appear to be similar to devices employed in the eastern Mediterranean during the same period.[43] Two large armoured roofs and many smaller ones and portable shields were constructed to prepare terrain for the towers' advance. After ditches and approaches had been filled and levelled, the larger roofs were equipped with rams and moved towards the walls.[44]

A full-scale assault, involving all of these devices as well as supporting artillery and missile fire, was launched sometime during the autumn. Although the siege towers made progress and facilitated an escalade, they were unable to seize control of a section of the walls, partially because a wooden defensive tower erected opposite them checked their advance. Although crews operating under the armoured roofs created a breach, a vigorous defence based around the Muslim counter-tower stymied Christian progress.[45] The coming of winter halted offensive operations, and during the withdrawal of the Christian machines, one of the armoured roofs was burnt during a night-time sally.[46] The attackers, however, maintained their encampments and prepared for another major assault.

The second assault commenced on 2 February 1115, with two more siege towers added to the Christian attack. Once all four towers were in position the attackers ignited the Muslim counter-tower with incendiaries delivered from a shipmast or spar extended from an attacking siege tower.[47] Although it is not clear if this operation was carried out in exactly the same manner as the dropping of jars of incendiaries on to an attacking Christian tower at Tyre in 1111, close similarities seem likely.[48] Our sources also do not mention if Pisan incendiaries in any way resembled Greek fire or a naphtha-based combustible. In any event, the resultant conflagration burnt the counter-tower and spread along the fortifications, facilitating the storming of the city's outer fortifications, which was completed by 6 February.[49]

This success, and an Almoravid attack on Catalonia at the end of

---

[43] *Gesta triumph.* 92; *Lib. Maiol.*, vv. 2134–40, p. 84.
[44] *Gesta triumph.* 92; *Lib. Maiol.*, vv. 2151, 2202, pp. 84, 86.
[45] *Gesta triumph.* 92; *Lib. Maiol.*, vv. 2148–384, pp. 84–92.
[46] Ibid., vv. 2428–64, pp. 94–5.
[47] 'Factum est de ingenio Pisanorum ignis pennatus de castello Christianorum per antennam porrigeretur in castellum Sarracenorum' (*Gesta triumph.* 92).
[48] See above, p. 81.
[49] *Gesta triumph.* 92–3; *Lib. Maiol.*, vv. 3125–64, pp. 117–19.

1114, opened divisions in the Christian ranks. Wishing to return to protect their lands, Raymond-Berenguer III and his men negotiated a settlement with Nasir al-Dawla which the Pisans refused to accept. Although relations between the allies were strained, the impasse was apparently solved when Pisan forces battered and stormed Palma's next line of defence, after filling a section of its protective ditch and moving up their towers with artillery support. Although the count of Barcelona remained at the siege, its rapid resolution was a necessity.[50]

The final stages of the siege consisted of overcoming each successive fortification by large-scale assaults based around the attackers' four siege towers and supporting artillery. The citadel was finally seized on 3 April, during an attack which involved two siege towers and a massive bridge thrown across from one tower to the citadel's fortifications.[51]

While the sources emphasize the Pisan role in events, this probably accurately reflects the importance of Pisan combatants, craftsmen, and engineers in operations. It should be noted that Pisan warships were not required to fight a battle, nor to maintain a blockade. This may illustrate both Pisan strength and Muslim naval weakness of this period. It remains clear that the principal Pisan maritime contribution was in transport and the provision of war materials. Yet Catalonian and southern French manpower was also important, especially in the protracted close-quarters combat that this campaign involved. The Balearic campaign of 1113–15 involved a considerable range of military and technological resources in highly complex and protracted operations. However, neither Pisa nor the count of Barcelona could maintain an adequate garrison in the Balearics to resist an Almoravid fleet based in Denia which easily captured the islands in 1119.

Alfonso I of Aragon and the Capture of Zaragoza, 1118

Three and a half years after the Balearic campaign Alfonso I of Aragon completed a major achievement in the *Reconquista* with the capture of Zaragoza. The city was not only the key to the Ebro valley and a northern bulwark of Islam, but also an important Almoravid possession since 1110. Zaragoza's capture confirmed Alfonso I's military reputation and his position as a leading figure in the war against Islam.

[50] *Gesta triumph.* 92–3; *Lib. Maiol.*, vv. 2386–426, 3124–338, pp. 92–4, 117–25.
[51] *Gesta triumph.* 93–4; *Lib. Maiol.*, vv. 3345–510, pp. 127–31; BM 8.

Important as Zaragoza's capture was, the deeds of its conquerors were not recorded in any detail by Latin writers. However, Lacarra's reconstruction of events from Latin and Arabic sources permits us to examine an important aspect of Alfonso I's victory: the contribution of trans-Pyrenean combatants and veterans of the First Crusade with experience in that expedition's siege operations.[52]

The first stages of Aragonese operations against Zaragoza involved the establishment of fortified positions, whose garrisons interdicted lines of communication and harassed agricultural areas. This was a gradual process taking almost a quarter of a century, as Zaragoza lay in the midst of powerful Muslim strongholds, to the north-east and north-west as well as to the south.

The capture of Huesca in 1096 by Pedro I of Aragon marked an important phase in the offensive against Zaragoza. Although Sancho I Ramirez had died besieging the city in 1094, his son continued pursuing this key strategic objective of the Aragonese kingdom, subjecting Huesca to a close blockade in 1096. In the course of this investment Pedro I's forces, including non-Iberian recruits, defeated a Zaragozan relieving force in November which was supported by two prominent vassals of al-Mustain's protector, Alfonso VI of Leon–Castile. Zaragoza was the sole remaining *taifa* kingdom paying tribute to Alfonso VI, who must have envisaged eventual Leonese–Castilian expansion down the Ebro.[53] The Huesca campaign illustrates how the expansion of one Christian state might become embroiled in the ambitions of another. Yet, with Huesca's capture and Leon–Castile's internal difficulties after Alfonso VI's death in 1109, the Aragonese were able to move against Zaragoza without serious Leonese–Castilian intervention.

However, the Ebro valley in the early twelfth century remained a complicated military frontier where Muslim and Christian forces held cities and castles, but the countryside was left largely as a no man's land. Aragonese positions situated between Zaragoza and lines of support were bases for raiding and reconnaissance, but by themselves could not maintain a secure blockade. However, they could provide necessary assistance to a force attempting a close siege of Zaragoza. By 1117 the establishment of such fortifications along all major lines of communication was completed, and Alfonso I of Aragon began organizing a major attack on Zaragoza itself.

[52] M. J. Lacarra, 'La conquista de Zaragoza por Alfonso I', *Al-Andalus*, 12 (1947), 65–96; Defourneaux, *Espagne*, 156–8.
[53] Reilly, *Alfonso VI*, 282–3.

Alfonso I's military strength was considerably enhanced by southern French knights, who fought with him from 1117, if not earlier. Chief among these was Gaston IV of Béarn, a distinguished veteran of the First Crusade who had directed the construction of siege machinery for the northern contingents at the siege of Jerusalem in 1099.[54] In July 1117 Gaston and his brother, Centule II of Bigorre, raided Zaragoza's agricultural areas and investigated the city's defences, before returning to their lands to prepare for an attack in the next year.

In 1118, during Alfonso I's preparations and recruitment, a major ecclesiastical council met at Toulouse, attended by important southern French and Spanish prelates and Pope Gelasius II, then journeying through the region. The pope proclaimed crusading indulgences for those who participated in the imminent offensive against Zaragoza, and the council called on Christian warriors to assist their Spanish brethren. With this impetus to recruitment Alfonso I collected a powerful force, not only from his own lordships and the Pyrenean counties, but also from areas of southern France which had contributed significantly to the First Crusade. While an accurate assessment of the size of contingents is not possible, some contemporary sources emphasize the importance of the *exercitus francorum* in Alfonso I's expedition.[55]

Siege operations began on 22 May 1118 with the arrival of Béarnais forces outside Zaragoza's northern walls. As more forces arrived, a close blockade was established and preparations for an assault were made by Gaston of Béarn's men, who had brought their own materials. Artillery and at least two siege towers were built by the Béarnais under Gaston's direction and moved against Zaragoza's fortifications. However, these devices did not destroy or seize control of a section of the walls, and the city did not fall to a full-scale assault. A combination of famine, the defeat of an Almoravid relieving force in December 1118, as well as the threat presented by these machines and especially the towers led to the city's capitulation later that month.[56] Because of the scarcity of information, it is difficult to assess the importance of Gaston IV's forces and their poliorcetic skills in the reduction of the city. Whatever their role in wearing down the defenders, it is clear that Gaston IV provided a major contribution to Aragonese success, as he was rewarded with the lordship of Zaragoza soon after the city's surrender.[57]

[54] See above, p. 51. [55] *His. Comp.* ii. 4, p. 262.
[56] Lacarra, 'Conquista de Zaragoza', 83–5.
[57] Ibid. 94–5.

Alfonso I's possession of Zaragoza and Almoravid weakness enabled him to compel the surrender of Muslim cities between Zaragoza and Castile, most notably Tudela in 1120. His reputation and the opportunities his campaigns offered encouraged men from northern France and especially Normandy to seek their fortunes in his service.[58] These men strengthened all forms of Aragonese military endeavour and were not exclusively recruited for siege operations.[59] Moreover, the capture of Zaragoza was the centre-piece of an Aragonese offensive of long duration and not simply a result of the presence of non-peninsular forces. Nevertheless, it is clear that the crusading impulses and the experience of one of the First Crusade's veterans was of some importance in Alfonso I's conquest of Zaragoza.

The Conquests of Alfonso VII

With the death of Alfonso I of Aragon in 1134, Alfonso VII of Leon–Castile became Christian Spain's dominant political figure. Ruler of Leon, Galicia, and Castile, his major vassals included the king of Navarre, the counts of Barcelona and Toulouse, and the lord of Montpellier. He was also protector of a group of Spanish Muslims opposed to the Almoravids centred around Alfonso VII's vassal Sayf al-Dawla, son of the last native Muslim ruler of Zaragoza. As an expression of his position, Alfonso VII was crowned emperor in 1135, which ushered in a period of Leonese predominance in the *Reconquista*.[60] Alfonso VII's long minority and reign encompassed considerable military activity which ranged from the suppression of Castilian rebels to major offensives against Almoravid power in central and southern Spain. The disintegration of the Almoravid state during the period 1139–48 was exploited by Alfonso VII and other Christian rulers to their territorial advantage. Alfonso VII's Spanish Muslim supporters were important in his political and military offensive against the Almoravids which was central to his bid to dominate southern Spain. Alfonso VII's ambitions and the Leonese hegemony they involved were thwarted by the coming

---

[58] OV xii. 2–8, 394–411.
[59] For a revision of the career on one such figure, see L. H. Nelson, 'Rotrou of Perche and the Aragonese Reconquest', *Traditio*, 26 (1970), 113–33.
[60] Lomax discusses aspects of Alfonso VII's military operations (Lomax, *Reconquest*, 86–91; for political developments, see O'Callaghan, *Medieval Spain*, 222–33).

of the Almohads, whose military challenges threatened Alfonso VII and his descendants for the next half-century.

Alfonso VII's siege operations may be discussed in terms of three categories: the reduction of Christian opponents in Leon and Castile after he assumed his majority in 1126; attacks against Almoravid positions in the Tagus valley; and participation with the Genoese and Catalans in the siege of Almeria in 1147. Unlike Roger II, the king of Leon–Castile did not consider siege warfare the principal form of military activity. Sieges were undertaken not only as part of specific campaigns, but also in the context of rebellions, Portuguese and Almoravid invasions, and Leonese–Castillian *razzias*, which accounted for many of the military affairs in the mid-twelfth century. In focusing on siege warfare we should not lose sight of the range of military challenges which confronted this monarch.

Attacks against partisans of Aragonese or purely local interests who refused to recognize Alfonso VII's authority were waged between 1126 and 1130 in Leon, Galicia, and Castile. These operations often involved the close blockade of positions which capitulated after being subjected to personnel assaults supported by missile weapons. In the case of Valle, defenders surrendered rapidly to Alfonso VII after his siege crews had damaged the fortification's walls with rock-throwers and battering.[61]

In north-western Spain Alfonso VII was greatly assisted by Archbishop Diego Gelmirez of Compostela, whose considerable military resources had supported the king since 1110. At royal request Diego conducted a campaign against the holdings of Arias Perez in 1126, which centred on Tabeiros and its fortifications. Diego's forces pressed an assault on a key castle, sapping and toppling the fortification's main tower under protection of an armoured roof. This brought the remaining defenders to surrender, and an end to the rebellion.[62] It may be noted that this archbishop of Compostela, a prelate well experienced in military affairs, reduced several fortifications by blockade and artillery, and simply by storming them, during his conflicts in Galicia.[63] As we have seen, Diego solicited and received Italian assistance in constructing and employing warships in Galicia.[64] There is no evidence that his siege technology came from a similar source,

[61] *Chron. Adefonsi*, i. 20–1, pp. 20–1.   [62] *His. Comp.* ii. 84, pp. 443–4.
[63] Ibid. i. 76, ii. 30, pp. 132–6, 314–15. Fletcher illuminates Diego's temporal responsibilities and achievements as archbishop (Fletcher, *Saint James's Catapult*, 246–8).
[64] See above, pp. 154–5.

although this is possible. However, it is also possible, especially considering the relatively small scale of operations, that Diego may have drawn only on local talents, and, given his own extraordinary abilities, may have taken a leading role himself in directing activities. While it is difficult to measure the sophistication of Diego, or Alfonso VII's siege crews from these examples, their poliorcetics were clearly successful if waged on a small scale.

In July of 1138 Alfonso VII, having completed a large-scale and destructive raid into Andalusia and Granada, turned his attention to the Almoravid city of Coria in the central Tagus valley, with forces which included the urban militia of Salamanca and the men of the count of Leon. Coria was a well-defended Almoravid position, important as a protection against Castilian encroachment and as a base for harassing Toledo. Alfonso VII's troops ambushed Coria's mounted garrison as it pursued a small Spanish force which had raided crops and livestock in close proximity to the city. The ambush was successful and Coria's garrison suffered substantial casualties. Alfonso VII decided to take advantage of the situation and concentrated his forces around the city, while formally requesting the support of his subjects in Leon and Estremadura in prosecuting a siege.

Alfonso VII's men constricted the city from encampments close to its walls while reinforcements arrived and machinery was constructed. Although our principal source does not inform us in detail of the response to the emperor's appeal, important nobles joined the siege and built or at least financed siege machinery, including wooden towers, artillery, and armoured roofs.[65] Although these devices operated within missile range of the walls, it is not clear if they were moved up in an attempt to seize control of the fortifications. There is no description of major assaults or even of moving the wooden towers. While it is conceivable that the besiegers never achieved the domination necessary to attempt to seize the walls, it is also possible that these towers were not mobile but simply firing platforms to aid in harassing the enemy with missile fire. A story of the death of the count of Leon by an arrow which penetrated his tower's armouring supports an interpretation that these devices were never brought within close proximity to the walls, as his mortal wound was received while Alfonso VII was absent from the siege on a hunting expedition. The emperor is reported to have abandoned the siege out of grief, and it seems likely that this decision reflects not only bereavement and a desire to minimize casualties but also the

[65] *Chron. Adefonsi*, ii. 136, p. 107.

siege's lack of success.[66] The city's investment represented an attempt to capitalize on a successful ambush rather than an integral part of well-planned and thoroughly organized offensive. In any event, the siege of Coria illustrates siege operations in the fluid frontier warfare of the Tagus valley.

In 1139 Alfonso VII shifted his focus to the eastern Tagus valley and attacked an important Almoravid fortification and advance base at Oreja. While Alfonso VII gathered knights and foot soldiers from Galicia, Leon, and Old Castile, his commander in Toledo concentrated forces from nearby areas at Oreja. With the arrival of the emperor and his northern forces, a close blockade for the well-fortified and -defended position commenced. As the garrison's water supplies became exhausted, the attackers positioned men protected by an armoured roof at a location nearby to where Muslims were drawing water. Although this device, which was presumably intended to protect men from garrison missile fire, was burnt during a night sally, the Spaniards maintained their blockade.

In order to draw Alfonso VII from the siege, the governors of Cordoba, Valencia, and Seville mobilized their forces and attacked Toledo and the fortresses guarding its approaches. Although the monastic fortress of San Servando on the south bank of the Tagus had one of its towers destroyed, it remained secure, and a vigorous defence of Toledo and its dependent fortifications led by the empress thwarted the attackers, who abandoned their attempt to support Oreja.

During this period Alfonso VII's artificers began demolishing some of the fortification's towers by artillery and battering. The garrison's commander asked for a month's truce in which to request and receive succour from the Almoravids in north Africa. The commander agreed to surrender Oreja if no aid was forthcoming. A month after this truce was agreed the garrison surrendered and marched out.[67]

In 1142 Alfonso VII returned to Coria with a large and well-organized force, ringing the city with camps in blockade. The emperor's siege crews built rock-throwers, armoured roofs, and a siege tower, and attacked the fortifications. On this occasion Coria's defences were dominated by the siege tower and, when labourers undermined a section of the walls, including a key mural tower, the Muslim commander requested terms similar to those granted to Oreja's defenders in 1138. Thirty days later Coria surrendered.[68]

[66] Ibid. ii. 135–8, pp. 106–9.   [67] Ibid. ii. 145–53, pp. 113–19.
[68] Ibid. ii. 159–61, pp. 123–5.

Although an understanding of Alfonso VII's siege warfare is limited by the nature of the sources, several points about it merit comment. While the terms involved in the surrenders of Oreja and Coria were not extraordinary in twelfth-century siege warfare, they reflect more than a desire among warriors and professionals to reach an 'honourable compromise'. This policy, which had obvious advantages for isolated and probably demoralized defenders, also benefited Alfonso VII and his forces. It undoubtedly reduced casualties from what they would have been had assaults been carried to a conclusion. Neither Oreja nor Coria were rich cities with the opportunities for plunder available in the great urban sieges of the eastern Mediterranean. Moreover, Alfonso VII's forces were drawn from his kingdom of Leon–Castile and fought out of loyalty and obligation in campaigns which aimed to augment Christian territorial possessions and security in the strategically important Tagus valley. They were not combatants directly seeking new lordships or the means to finance further participation in their expeditions. The terms Alfonso VII honoured may also have been part of a wider policy of dividing Spanish Muslims from their Almoravid overlords. Whatever the factors behind such negotiated surrenders, a successful commander clearly needed to be able to control his own force and prevent them from taking revenge or despoiling defenders. Moreover, a reputation for maintaining discipline was an asset in negotiating such terms. Unlike some of the forces we have examined, particularly in the eastern Mediterranean, the composition of his forces facilitated Alfonso VII's ability to keep his promises.

Whatever the origins of his siege engineers and crews, they played a role in the maintenance and extension of Alfonso VII's kingdom. They constructed and employed artillery and armoured roofs effectively and provided military alternatives to blockade or personnel assaults. The reduction of Coria in 1142 may reflect not only better organization and logistics, but also perhaps experience gained in design and utilization of siege towers. However, while royal engineers may have improved their operations, it is clear that Alfonso VII's resources were capable of waging siege warfare successfully, albeit on a modest scale. Attacks on larger or more exposed positions required allies.

In the Wake of the Second Crusade

Almeria, 1147

With the collapse of the Almoravid state during the decade of the 1140s, a brief period of *taifa* independence and Christian opportunity ensued. That this occurred during the period of the Second Crusade was fortuitous for peninsular rulers, who channelled crusading impulses against key cities.[69] Of the important positions taken during this period only Almeria was lost again to the Almohads.

The attacks against Almeria and Tortosa resulted from the synchronous interests of Leon–Castile, Catalonia, and Genoa and the willingness of Pope Eugenius III to recognize these campaigns as having a crusading status. The offensive against the important port city and pirate base of Almeria was an extension of Alfonso VII's recent successes in upper Andalusia, and his possession of key passes across the Sierra Morena. Tortosa was the principal Muslim bastion remaining in the Ebro valley and had been an important military objective of Raymond-Berenguer IV throughout the 1140s. Although these cities were discrete strategic objectives of separate Spanish states, their conquest was linked by at least one common factor: Genoese expansion in the western Mediterranean.

Between 1146 and 1148 Genoa mobilized extensive naval and military resources to establish tax exempt, judicially and ecclesiastically independent trading enclaves in Spanish port cities similar to those founded in Syria and Palestine during the early twelfth century. In 1146 a Genoese fleet of twenty-eight ships carrying a hundred knights and their horses, as well as siege machinery, attacked Minorca. Although Port Mahon was not taken, the island was raided and an indemnity forced from its inhabitants.[70] In September of that year representatives from this force negotiated a complex series of alliances with Alfonso VII and Raymond-Berenguer IV for attacks first on Almeria and then on Tortosa. In return for commercial privileges and one-third of the cities and their spoils, the Genoese agreed to assist Alfonso VII in taking Almeria and Raymond-Berenguer IV in taking Tortosa.[71] In these agreements the Genoese promised to provide not only ships but also siege machinery.[72] While modern writers have

[69] For the Second Crusade and its scope, see G. Constable, 'The Second Crusade as Seen by Contemporaries', *Traditio*, 9 (1953), 213–79; Mayer, *His. Crus.* 96–109.
[70] Caffaro, in *AI* 33.   [71] *CDGR* i, nos. 166–9, pp. 204–17.
[72] Ibid. i, nos. 166, 168, pp. 205, 211.

noted the republic's maritime assistance in these campaigns, the importance of Genoese poliorcetics should be emphasized.[73] Genoese responsibility in siege operations was not confined to providing the materials and technology for siege machines. The Genoese also brought with them the labourers and combatants necessary to bringing the machines to the walls.

Although the attack on Almeria was planned for May 1147, it was not until June that five hundred knights on board fifty-three ships from Catalonia arrived to support a Genoese squadron patrolling the southern Spanish coast while the main fleet waited on Minorca.[74] The main fleet, which Caffaro recorded to have consisted of sixty-three galleys and 163 other ships when it left Genoa, then joined the advance parties and beached its ships near Almeria.[75] Although Almeria's naval forces had battled the Genoese patrolling squadron, the strength of the main fleet and the Catalan ships ensured undisputed control of the sea.[76]

Camps were established outside the city and the construction of siege machinery commenced. Although the defenders made sallies against the beached Genoese vessels while construction was under way, the attackers maintained the security of their boats. On 1 August Alfonso VII arrived with a force recruited from his Spanish and southern French followers and allies. Estimating the size and importance of the major contingents at the siege of Almeria is difficult, given the information available.[77] The verse account of the siege which concludes the *Chronica Adefonsi Imperatoris* emphasizes the important personages which served in this expedition and Alfonso VII's role as its leader.[78] Caffaro is highly partisan towards Genoa and tells us very little about other contingents and their efforts. However, it is clear that this attack marshalled forces significantly beyond the usual capabilities of Leon–Castile, Catalonia–Aragon, or Genoa.

With the arrival of Alfonso VII's forces and the completion of machinery, the besiegers began moving against the city, which was

---

[73] Lomax, *Reconquest*, 91.
[74] *Gesta comitum Barc.* 5, pp. 6–7, Caffaro, *Almerie et Turtuose*, 81.
[75] Caffaro, *Almerie et Turtuose*, 80, 83.   [76] Ibid. 80–2.
[77] Caffaro numbers Alfonso VII's forces at 400 knights and 1,000 footmen. However, this is the same figure given for Alfonso VII's initial concentration at Baeza, and, if accurate, may represent only his own contingent and not the whole force. Caffaro does not specify the number of Genoese involved nor give a breakdown of the types of vessels which set sail (ibid. 81–3).
[78] This work gives little information about other contingents or detail about operations (*Chron. Adefonsi*, vv. 1–372, pp. 181–6).

most easily attacked along its eastern and southern approaches. The assault was based around two Genoese siege towers supported by artillery and armoured roofs. Although the defenders concentrated artillery and incendiary attacks against the Genoese towers, the attackers were able to dominate a section of Almeria's defences. Two mural towers were taken and a breach of approximately eighteen feet was made in the city's walls. Fearing that they could not hold against a personnel assault, the defenders made contact with two of Alfonso VII's important followers, the king of Navarre and count of Urgel, offering a substantial sum if a negotiated surrender could be arranged. Upon learning this, the Genoese feared that their interests were being abandoned, and they formed a force of a thousand men, while beseeching Alfonso VII and Raymond-Berenguer IV not to negotiate but to assault the city. According to Caffaro, Alfonso VII agreed and, as a formal assault was being organized, the select Genoese force moved silently into the breach and made its way into the city, inflicting significant casualties on the city's garrison and population. Although Almeria's main defences were ruptured, some inhabitants retreated into inner defences located on elevated ground in the north-west corner of the city. After four days of continued fighting, on 17 October the besiegers accepted the survivors' surrender and preserved their lives in return for a substantial payment. The citadel was garrisoned by Alfonso VII's forces and the city's remaining wealth seized by its conquerors.[79]

As the expedition broke up and contingents departed for their homelands, the Genoese left a garrison in Almeria to maintain the commune's interests and sailed north to winter in Barcelona. After dividing their spoils among themselves, Genoese leaders sent some of their recent gains back to Genoa to meet debts incurred by the commune in financing this expedition.[80]

Tortosa, 1148

Preparations for the attack on Tortosa began in the early part of the campaigning season when Genoese labourers gathered timber for siege machines from forests near Barcelona. These activities were completed by the end of June 1148 and a reinforced Genoese fleet sailed up the Ebro to Tortosa, joining Raymond-Berenguer's contingent of two thousand men on 1 July 1148.[81] As at Almeria, it is difficult to establish

---

[79] Caffaro, *Almerie et Turtuose*, 83–4.   [80] Ibid. 84–5.
[81] Ibid. 85; *Gesta comitum Barc.* 5, p. 7.

the relative strength of contingents, although it is clear that Raymond-Berenguer IV enjoyed considerable support from allies and crusaders. Besides the Genoese, southern French veterans of Almeria under William VI of Montpellier also fought at Tortosa. Crusaders who had participated in the capture of Lisbon during the previous year continued their conflict with Iberian Islam at Tortosa. These men were joined by Templars and representatives of the Holy Sepulchre pursuing their interests in the region, which derived from the will of Alfonso I of Aragon.[82]

The attackers established encampments around the city, which lay between the east bank of the Ebro and a range of hills which formed Tortosa's northern and eastern boundaries. The Genoese took up positions next to the river on the city's southern side. Raymond-Berenguer IV's forces joined these camps, with the positions of William of Montpellier and his followers situated on elevated ground at the northeast. The Templars and Lisbon veterans held the ground between these camps and the river.[83] Although the main walls were accessible from the west and south-east, a well-fortified inner city located atop elevated ground in the north-eastern area could be approached only with considerable difficulty. Poliorcetic operations, which were primarily conducted by the Genoese, were directed initially at seizing lower areas of Tortosa from its south-eastern approaches. Two siege towers were brought to the walls and dominated them, permitting labouring crews to make an extensive breach. One tower was led into the city and was employed in taking buildings and fortifications. It seems that Tortosa's defenders prepared and maintained a thorough defence and that attackers faced organized resistance throughout. The secure and elevated firing platform provided by the siege tower, as well as its mobility, were important in this operation. Although the second siege tower assisted in clearing defences from the initial point of attack to the inner walls, the defenders successfully retreated into their citadel.[84]

Thwarted in attempts to force these defences, the Christians shifted their attack to higher ground, probably on the citadel's eastern side. Defensive fortifications here were protected by a massive ditch, described by Caffaro as approximately 140 feet wide and 105 feet deep.[85]

[82] R. Hiestand, 'Reconquista, Kreuzzug und heiliges Grab: Die Eroberung von Tortosa 1148 im Lichte eines neuen Zeugnisses', *Gesammelte Aufsätze zur Kulturgeschichte Spaniens*, 31 (1984), 149–54; G. Constable, 'A Note on the Route of the Anglo-Flemish Crusaders of 1147', *Speculum*, 28 (1953), 525–8.
[83] Caffaro, *Almarie et Turtuose*, 85–6.   [84] Ibid. 86.   [85] Ibid. 87.

Whatever its precise measurements, filling a section of this ditch was a major undertaking, requiring the efforts of knights and foot soldiers, rich and poor. It is not clear whether all contingents participated in this activity or whether it was solely the Genoese, although the latter seems more likely. While personnel deposited earth, stones, and wood into this ditch, Genoese craftsmen built a third siege tower near the intended area of assault.[86] That the attackers built a third tower rather than attempt to move one of their completed ones up from the city illustrates not only the extent of their materiel preparations, but also the difficulty involved in manœuvring these devices over anything but level ground.

Caffaro's account emphasizes the efforts of the Genoese and doubtless obscures the actions of other contingents. It is clear that Raymond-Berenguer IV mobilized extensive credit in prosecuting this siege, and providing for this third tower may have been one of the objectives of the count-prince's financial activities during this period of the siege.[87] However, the problems of providing for his followers as the autumn wore on outside the city may account for these expenditures. Raymond-Berenguer's financial resources, as well as political leadership in organizing these disparate combatants into a single endeavour, were in any event important in the outcome of the siege.

Once erected, the third tower was manned by a crew of three hundred and moved over a filled and levelled section of the great ditch towards Tortosa's remaining defences. The attackers encountered accurate and destructive artillery, which substantially damaged the siege tower. The device was withdrawn and its shattered corner section repaired before being layered with intertwined ropes as protection against artillery. Although moved back into an attacking position, the besiegers did not attempt to overwhelm defenders but rather kept their tower in a threatening position, while waiting for an opportunity to close for a personnel assault. Sometime in mid-November Tortosa's defenders sought terms offering to surrender within forty days if help requested from Muslim Spain and north Africa was not forthcoming. Even though the political situation and Alfonso VII's establishment in Almeria doubtless rendered such assistance unlikely, the Genoese still required a hundred hostages to ensure good faith. On 30 December 1148 Tortosa surrendered, with the surviving inhabitants paying an

[86] Ibid. 87.
[87] *Fiscal Accounts of Catalonia under the Early Count-Kings (1151–1213)*, ed. T. N. Bisson (2 vols.; Berkeley and Los Angeles, 1984), ii, nos. 142–3, pp. 259–62.

indemnity keeping some of their real property and commercial rights within the city.[88]

During 1149 Raymond-Berenguer IV cleared the lower Ebro of Muslim positions with the capture of now isolated Lerida and its dependent fortifications.[89] These conquests completed an important stage in the extension of Catalan–Aragonese authority, and, with his marriage to Petronilla of Aragon in 1150, Raymond-Berenguer IV maintained a pre-eminent position in peninsular and southern French affairs. For Genoa the results of these campaigns were less enduring. The advent of the Almohads and a financial crisis partially brought about by the costs of these expeditions undermined the successes of Genoese arms and naval power.[90] Nevertheless, Genoa made a notable contribution to the Aragonese reconquest.

Lisbon, 1147

The mobilization of European military resources generated by the preaching of the Second Crusade contributed significantly in one of the great sieges of the twelfth-century *Reconquista*. After a seventeen-week siege, the city of Lisbon surrendered to crusaders from Flanders, the Rhineland, England, and Normandy, as well as the forces of Afonso I of Portugal. Lisbon's capture was a milestone in the early history of Portugal, and rivalled the taking of Toledo and Zaragoza in the extension of Christian authority in the Iberian peninsula. While Lisbon had been an occasional objective of Christian commanders since the early twelfth century, its conquest was achieved only when the resources of the nascent kingdom were combined with a powerful force of seaborne crusaders.

In May of 1147 crusaders from the Rhineland, Flanders, Normandy, and the British Isles concentrated at Dartmouth before setting sail for the Holy Land. These diverse contingents bound themselves by individual oath into a common association which established its own judicial procedures and officials charged with arbitrating disputes and distributing booty.[91] The expedition numbering approximately 170 ships left Dartmouth on 23 May, reaching Oporto in mid-June, where the bishop of that see invited the crusaders to join Afonso I in an attack

---

[88] Caffaro, *Almarie et Turtuose*, 85–8; *Gesta comitum Barc.* 5, pp. 5–7.
[89] *Gesta comitum Barc.* 5, pp. 7–8.
[90] See below, p. 196.   [91] *Ex. Lyx.* 56.

on Lisbon.[92] Although there was some resistance within the Anglo-Norman contingent, primarily from veterans of an earlier abortive attack on Lisbon, the crusaders ultimately decided to participate in this offensive. While a number of factors influenced the crusaders' decision to assist their fellow Christians, the generosity of Afonso I's offer is notable. As well as a remission of royal tolls and rights of settlement, the northerners were entitled to the whole of the city's spoils, including captives' ransom, rather than the one-third portion customary in Mediterranean sieges of a similar kind. Moreover, Afonso I provided at least the Flemings with food and drink during these deliberations and also while they were waiting for the advent of Arnold of Aerrschot.[93] These terms, and the Portuguese hostages required by the crusaders, may reflect the degree of mistrust of the Portuguese king within the expedition. They also illustrate how important the military and naval power of these forces were to Afonso I.[94]

While Lisbon had been exposed since Afonso I's successful surprise attack on Santarem in March 1147, a full-scale siege commenced only with the arrival of the crusaders. With the alliance concluded on 28 June, the crusaders disembarked and began taking up positions around the city. As much out of enthusiasm as from planning, Anglo-Norman forces stormed Lisbon's suburb on the western side of the city on 30 June, and defenders abandoned positions outside the eastern walls soon afterwards.

With the Anglo-Normans established on the western side of the city, the Flemings and Germans encamped to the east of the city, with the Flemings situated near Lisbon's south-eastern corner.[95] Afonso I's forces remained in the hills to the north of the city on guard against relieving forces and completed the blockade of Lisbon. Most of these troops took no direct part in operations, and one source reports that some were disbanded when captured letters indicated that no relief would be forthcoming. The only Portuguese forces involved in operations for the duration of the siege were those of the bishop of Oporto.[96] Afonso I, however, remained at the siege, maintaining a small household force on guard above the city, facilitating supplies and financing operations throughout the siege.[97]

Lisbon's natural and man-made defences were formidable and in-

[92] Ibid. 56, 69–84; Duodechinus, 'Ep.' 27.
[93] Arnulf, 'Epistola ad Milonem', *Recueil des historiens de Gaules et de la France* (24 vols.; Paris, 1869–94), xvii. 326.
[94] *Ex. Lyx.* 110–12.   [95] Ibid. 124–30.   [96] Ibid. 140.   [97] Ibid.

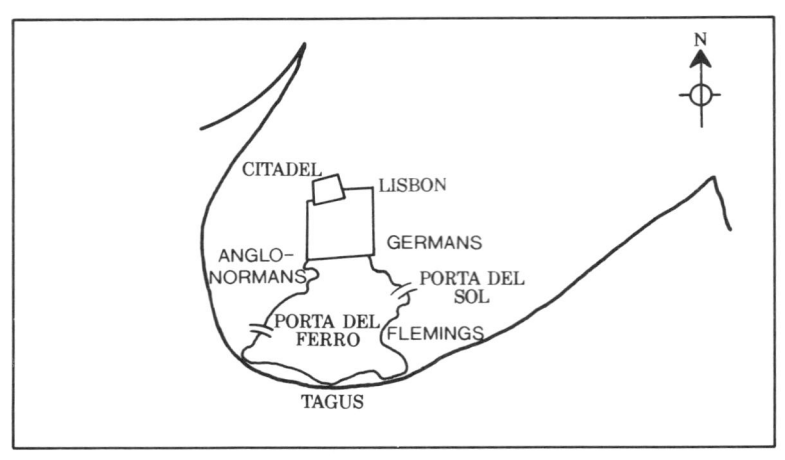

MAP 2. The siege of Lisbon, 1147

cluded a citadel and inner city located atop a hillock near the Tagus. The walls of the outer city extended southwads down to the Tagus. Although the south-eastern and south-western walls presented the most accessible targets, an attacker's approach was complicated by the tides of the Tagus estuary.[98] The city's defenders were numerous, as were refugees from dependent fortifications, although the figure of 154,000 given by the Anglo-Norman source is doubtless greatly exaggerated. Whatever their true number, provisioning and sanitation became more difficult as the siege wore on.[99] The attackers experienced few problems in provisioning themselves, as they had captured foodstuffs in Lisbon's suburbs soon after arriving and were able to forage widely.[100]

Although skirmishing and probing attacks continued during July, the first major assault was launched in early August and based around machines built separately by the major contingents on each side of the city. The Germans and Flemings concentrated on destroying the city's walls with a battering-ram and five artillery weapons. They also employed an undermining device, which was probably an armoured roof equipped with an apparatus for extracting foundation-stones. The besiegers on this side of the city also built a mobile siege tower to attack the outer city's south-eastern corner. As part of this assault, attempts

[98] See Map 2.   [99] *Ex. Lyx.* 94, 142.   [100] Ibid. 124–6.

were made to ascend the southern walls facing the Tagus by German and Flemish forces from half a dozen ships equipped with four bridges.[101] Although there are few details, this appears one of the earliest such assaults and prefigures events at Acre in 1190 and Constantinople in 1204.[102] A ninety-five foot siege tower was the centre-piece of the Anglo-Norman attack against the south-eastern sector of the outer city near the riverbank.

Despite the range of machinery brought against the city, this onslaught failed. Of the devices brought against the eastern walls, only the ram escaped burning. The Anglo-Norman tower was moved near to the walls before it became stuck in the sandy soil of the south-western approaches. After damaging the tower with artillery, Lisbon's defenders sallied out and burnt this device also.[103]

Despite this set-back, operations continued, as each contingent attempted to destroy sections of the city's defences. The Anglo-Normans tried to sap the walls near the Porta del Ferro, but were discovered and thwarted by the defenders. Two artillery pieces were erected by the Anglo-Normans against the walls between the Porta del Ferro and the south-western corner tower. Crews of one hundred men worked these weapons in shifts so that five thousand rocks—doubtless of relatively small size—were cast in ten hours. Despite these efforts, Lisbon's defences in this area remained intact.[104]

The Germans and Flemish concentrated on undermining a section of the inner city's walls on Lisbon's eastern side. An extensive gallery mine was dug first into the hillside and then along the circuit of the walls. After a month's effort the sap was fired on 16 October and a breach of some fifty feet was made. The defenders built a makeshift retaining wall from the rubble and held off repeated attacks up the steep incline of the inner city's approaches. Anglo-Norman forces offered their services in assaulting the breach, but were refused by the Germans and Flemings. Presumably they believed that the breach might be carried by the Anglo-Normans, allowing them first access to the city's spoils. Although participants had sworn to an equal division of booty, the Flemings and Germans must have wished to be certain of this as well as to enjoy the honour of carrying the breach. However, the defenders held firm.[105]

[101] Duodechinus mentions seven ships, and Arnulf, the Flemish source, six (Duodechinus, 'Ep.' 27; Arnulf, 'Ep.' 326).
[102] See below, pp. 223–4, 233–4.
[103] *Ex. Lyx.* 134–6; Duodechinus, 'Ep.' 27.
[104] The artillery crews averaged one rock cast every eight seconds (*Ex. Lyx.* 142).
[105] *Ex. Lyx.* 142–6; Duodechinus, 'Ep.' 28.

By 19 October the Anglo-Normans had completed a second siege tower whose construction had commenced around 8 September. The tower was eighty-three feet high and covered with osiers and animal hides. After a benediction, probably by the author of *The Conquest of Lisbon*, the tower and supporting armoured roofs were moved to within eight feet of the walls on 20 October. After a night at this position the tower was moved to within four feet and its bridge extended. As the tower's crew appeared to have established control of the walls near their bridge, the garrison, worn out by a long and arduous defence, sought terms. After considerable wrangling among the attackers, which almost undermined negotiations, terms were granted which allowed defenders to depart with their lives after they had been thoroughly plundered, and the city was entered on 24 October.[106] As we have examined siege-tower assaults, and particularly battles for the control of defensive walls, in connection with Levantine warfare, there is no need for a detailed description here, especially since such an account could not in any way improve upon that of *The Conquest of Lisbon*.[107] However, certain points concerning this tower and its assault should be noted.

Although built at the Anglo-Norman encampments and drawing substantially upon their men and material, this tower and its assault drew upon the resources of other contingents. According to German sources the tower was financed at least partially by Afonso I, and knights in his service—either Anglo-Norman stipendiaries or Portuguese—who manned and guarded the tower during its assault.[108] Galicians are also mentioned among the tower's escort, and these could have been royal forces or drawn from the contingent of the bishop of Oporto.[109] German and Flemish sources emphasize the communal efforts of the entire expedition in the construction of this machine. Moreover, some of those escorting the tower became exhausted by their efforts and were replaced by German and Flemish troops.[110] Nevertheless, Anglo-Normans provided much of the support for the tower's advance and manpower for its movement.

This tower was designed and directed by a Pisan engineer described as a man of considerable talent.[111] While the Anglo-Norman account does not mention his origins, it does make clear his importance in commanding the advance of the tower until he was severely wounded

[106] *Ex. Lyx.* 146–78; Duodechinus, 'Ep.' 28.   [107] *Ex. Lyx.* 146–8, 160–4.
[108] Duodechinus, 'Ep.' 28.
[109] These men are termed Galletiani (*Ex. Lyx.* 164).
[110] *Ex. Lyx.* 162–4; Duodechinus, 'Ep.' 28; Arnulf, 'Ep.' 326.
[111] Duodechinus, 'Ep.' 28.

on 21 October.[112] How such an expert came to serve at Lisbon is not clear, although the most plausible explanation is that he joined it as an individual or with a small group after news of the siege began to circulate.

While comparing the two siege towers brought against Lisbon's south-western defences would be informative, this is difficult because the first is not described in any detail. Although somewhat shorter, the second tower was manœuvred and defended more effectively than the first. The second siege tower was initially brought against the city's south-western corner, where the first device had attacked. Although the second tower did not become embedded in Lisbon's approaches, its initial attack was thwarted when defenders massed in their mural tower against the attacking machine. The Christians then moved their device to a different point of attack along the south-western wall which dominated the rear of the mural tower. Although the passage describing this manœuvre is difficult to interpret, given the topography, it seems most likely that, in changing direction, the attackers first turned their tower away from the city towards the Tagus and then moved it back against the south-western wall.[113] Whatever its precise course, the tower's advance was carefully prepared by labouring crews to prevent it from becoming immobilized, as had the first tower. A small armoured roof manœuvred by a crew of seven is described in the defence of the tower at Lisbon's walls, and this device doubtless facilitated preparations for the tower's advance.[114] The second tower was also well protected against incendiaries, with hides suspended in such a way as to permit liquid poured from the tower's summit to flow over the surface of one side of the device.[115]

Undoubtedly the attackers' greatest challenge in manœuvring and defending their siege tower was the tidal system of the Tagus estuary, which not only limited periods of time in which the device could be moved, but also isolated it from the main Anglo-Norman positions on several occasions during the assault. It was during these periods that the siege tower and what may be considered its garrison were subjected

[112] *Ex. Lyx.* 162.
[113] 'Nam nostri machinam contra fluvium ad dextram declinantes, turrim quasi cubitis viginti preterierunt iuxta murum fere ad Portam Ferream que turrim respicit.' As David points out, this movement is difficult to comprehend, and suggests that an erring scribe substituted *dextram* for *sinistram*. However, what is to be made of the siege tower passing the mural tower? While it is possible that the Anglo-Norman device was manœuvred in such a way that it initially approached the mural tower from the south, this seems unlikely given the location of the Anglo-Norman camps (ibid. 160 n. 3).
[114] Ibid. 160.   [115] Ibid. 162.

to intensive incendiary, missile, and personnel attacks. The author of the Anglo-Norman source gives eloquent testimony as to how the escort of a hundred knights and a hundred foot soldiers, together with archers, crossbowmen, and labourers, successfully defended their tower for two and a half days.[116] In understanding the intensity of the fighting around the siege tower some eight feet from Lisbon's walls, we can better appreciate how the importance of the attacker's ability and willingness to relieve the escort contributed to the grinding down of Lisbon's defenders.

The success of the attackers in manœuvring and maintaining their second tower in a position from which a decisive assault could be launched may have been due to superior design or construction. In particular, the second tower may have had a more stable structure and better mobility than the first. However, the success of this device may also have been the result of knowledge of Lisbon's approaches and defences gained during the course of the siege. Despite the passion of the author of *The Conquest of Lisbon* to record the story of the second tower and the bravery of its crew, Lisbon's capture was not simply a result of the effective employment of this device. The cumulative effect of blockade and repeated assaults on the garrison's resilience was essential in the eventual capitulation.

The coalition behind the siege and the solidarity of the Anglo-Norman contingent were strained by the terms sought by Lisbon's defenders. The central issue appears to be the fear of some that inhabitants would be able—perhaps with the connivance of Christian leaders—to conceal riches.[117] However, a compromise was reached, as northern crusaders despoiled inhabitants and pillaged the city before it was handed over to Afonso I and former citizens were allowed to depart. The conquest of Lisbon was the most successful undertaking of the Second Crusade and, despite the fate of the expeditions in the Holy Land, the religious and economic forces behind that movement contributed to the extension of Christian authority in the Iberian peninsula.

---

[116] It is likely that the author was the same priest who blessed the tower and preached to its crew at the commencement of the assault and who probably served as the tower's chaplain (ibid. 147–64).

[117] Ibid. 164–78.

The Impact of the Almohads and the Third Crusade

The establishment of the Almohads in the 1150s altered the balance of power in the peninsula, limiting opportunities for Christian expansion and large-scale siege operations. Almeria was lost in 1157, despite Alfonso VII's vigorous defence, and the emperor died that same year unsuccessfully trying to hold the area around Baeza and its key passes across the Sierra Morena. Although partially protected by the *taifa* state of Ibn-Mardanish in Valencia and Murcia which they supported, the Christian kingdoms were primarily concerned with matters of defence and consolidation during much of the third and fourth quarters of the twelfth century. Conquered lands were settled and organized, municipal military obligations were more closely defined, and military orders assumed a greater role in defending vulnerable frontiers.[118]

However, there were siege operations of some importance during the period. Alfonso VIII of Castile captured the strategic city of Cuenca in 1177 after a close blockade of three months. Although the king's forces attacked the fortifications during the investment, these operations are not reported in any detail, and the principal factor in the city's fall was the inability of the Almohads to raise a relieving expedition.[119] Afonso I of Portugal exploited the confused military situation by moving into the Alentejo while the Almohads were preoccupied with Castile and the *taifas*. The remarkable Gerald 'the Fearless' and his followers played a prominent role in these conquests by employing techniques of surprise and subterfuge against unwary garrisons. Gerald's capture of Badajoz in 1169 prompted a Castilian response on behalf of the Muslim garrison still holding out in the citadel.[120]

The Third Crusade brought military and naval resources to the Portuguese kingdom similar to those of the Second.[121] In July of 1189 crusaders from areas around the North Sea and the English Channel combined with the forces of Sancho I in a close siege of Silves.[122] In

---
[118] Powers, 'Municipal Military Service', 104–8; A. J. Forey, 'The Military Orders and the Spanish Reconquest in the Twelfth and Thirteenth Centuries', *Traditio*, 40 (1984), 197–234.
[119] Roderigo Ximenez de Rada, *De rebus Hispaniae*, ed. A. Wechel (Rerum Hispanicarum Scriptores; Frankfurt, 1579), i, vii. 26, p. 242.
[120] For an account of this remarkable military leader, see Bishko, 'Reconquest', 414–15.
[121] For the scope and scale of the Third Crusade, see below, pp. 212–13.
[122] The expedition is described by a north German clerical participant and recounted in Anglo-Norman accounts of the Third Crusade, although not in as much detail (*Narratio itineris navalis ad Terram Sanctam: Quellen zur Geschichte des Kreuzzuges Kaiser Friedrichs I*, ed. A. Chroust (MGH SRG NS 5; Berlin, 1928), 183–90; Ralph of Diceto, *Ymagines Historiarum: The Historical Works of Ralph of Diceto*, ed. W. Stubbs (2 vols.; RS 68, 1876), ii. 65–6.

June part of this seaborne force had overwhelmed a dependent fortification of Silves at Alvora, slaughtering almost six thousand inhabitants and refugees.[123] This doubtless increased the numbers who took refuge in Silves and exacerbated problems of supply. Although subjecting the city to a close blockade, the northern crusaders were not content to wait, and undertook to neutralize the city's defences. While siege towers were under construction, the northerners battered and undermined the walls of the second ring of fortifications, having stormed the suburbs and outer city. Although these fortifications were breached and seized, the inner city and citadel resisted until water supplies were exhausted, and a negotiated surrender was arranged on 3 September.[124] Although northerners seem to have resented the Portuguese disinclination to prosecute the siege with the assaults of men or machines, this combination of northern and Portuguese naval and military resources once again brought an important fortress under Christian control. However, Silves was captured by the Almohads two years later and not recaptured until 1249.

Non-peninsular forces were manifestly important in twelfth-century siege warfare and it is appropriate to comment upon their role in concluding this chapter. In the first place these forces provided manpower for Spanish and Portuguese rulers, whose forces and those of their followers faced a number of military commitments. Even relatively small contingents of knights permitted a concentration of forces in one area without excessively weakening defences in another. Major seaborne expeditions brought large numbers of highly motivated combatants, who significantly altered the numerical balance between Christian attackers and their Muslim urban antagonists.

Although Italian and northern European sea power contributed to successful operations, the exercise of that sea power rarely involved naval combat, and blockade by sea was usually ancillary to the main operations. This reflects the naval strength of Christian expeditions and the relative weakness of the *taifas* and Almoravids. However, shipping was essential in the transportation of men and materials to their objectives.

Peninsular rulers and their followers clearly benefited from the technical expertise of men who had learnt their skills elsewhere. In focusing on this point we should not lose sight of the engineers,

---

[123] *Narratio itineris navalis*, 182–3.
[124] Ibid. 185–8; Ralph of Diceto, *Ymagines*, 65–6.

craftsmen, and labourers who served Spanish and Portuguese rulers. They enabled commanders to employ armoured roofs, siege towers, and artillery in the reduction of fortified positions and on occasion with considerable sophistication, as demonstrated by Raymond-Berenguer IV's river-borne tower of 1154. Yet the scale of these operations was usually far less than was involved in the sieges of major urban centres, and it was in attacks on large, well-fortified cities that Italian and northern European poliorcetics played a decisive role. Expeditions which sailed to Iberian sieges brought with them the materials and expertise with which to build sophisticated machinery against difficult objectives. Members of these expeditions performed the considerable manual labour necessary in employing devices and undermining fortifications. These aspects of Iberian siege warfare are similar to those we have examined in the eastern Mediterranean, as the capture of well-defended cities in both areas was made possible by a partnership between land-based rulers and maritime forces with significant poliorcetic resources.

The military contribution of these seaborne forces was not only in building and manœuvring siege machines. Men from these expeditions also fought in them and supported their advances against determined defenders. The importance of crossbowmen, archers, and infantry, as well as assault troops drawn from these contingents, is manifest in the sources. The willingness of these men to endure costly and protracted fighting in reducing well-defended fortifications should not be underestimated in any assessment of their contribution to successful siege operations. This may reflect a lack of sufficient manpower in twelfth-century Spain and Portugal and perhaps a disinclination of mounted combatants habituated to open warfare to engage in siege operations. In any event, the importance of non-peninsular fighting men in siege warfare is clear.

Sieges in which these participants played a prominent role were prosecuted relatively rapidly and with an unwillingness to compromise in negotiated surrenders unless the opportunities for plunder were unrestrained, and in some cases populations were massacred out of hand. This contrasts with some of the campaigns of the eleventh century, and has prompted the suggestion that crusaders perceived their Muslim foes and their warfare against them in significantly different terms from peninsular forces. In this interpretation crusader attitudes and actions were influenced by a narrow religious outlook, whereas Iberians viewed these conflicts in broadly secular terms.[125] It is

[125] Bishko, 'Reconquest', 406, 420, 421.

quite possible and even probable that northern European and Italian fighting men had a much less sophisticated perception of their Iberian Muslim foes than their peninsular co-religionists. However, to explain the determination of non-Iberian forces to reduce positions rapidly and with as much plunder as possible primarily in ideological terms misses important factors at the heart of twelfth-century sieges.

Members of major seaborne expeditions fought in foreign countries at considerable expense, and often with much loss of life, and they expected a return for their efforts and sacrifices. In the case of those who fought on the west coast of the peninsula in 1147 and 1189, their involvement in the *Reconquista* was only the first stage of an armed pilgrimage. While Italian maritime forces did not call into Spain while *en route* to the Holy Land, these communities made sizeable expenditures in outfitting fleets and providing them with equipment. One of the clearest descriptions of the ways in which these factors influenced events comes from an account of the wrangling among Lisbon's besiegers in response to the terms sought by the city's defenders. The author of *The Conquest of Lisbon* was convinced that Flemings and Germans insisted on a complete and thorough plundering of Lisbon's inhabitants as well as their dwellings partially out of an 'innate cupidity'. Yet he noted that, in pressing their demands in a general council, they cited 'the expenses of their long journey, the death of their comrades, and the unpredictable voyage yet to come'.[126] Whatever their religious motivations, these men of Flanders and the Rhineland and those like them who fought in Spain and Portugal knew the costs of siege warfare, and strove to ensure that the efforts and expenditures of one siege did not hamper their involvement in another campaign.

The military and technical assistance which men from outside Spain and Portugal provided aided in enabling Iberian rulers to exploit the opportunities presented by *taifa* and Almoravid weaknesses. However, unlike the situation in Outremer, durable commercial and legal enclaves outside royal jurisdiction did not result from the efforts of these expeditions. Although the Almohads arrested the expansion of Christian states, their rulers were able to consolidate the conquests of the first half of the twelfth century. When the stalemate was broken, Christian states moved forward with notable superior governmental and military resources, compared with those of the early twelfth century. The siege warfare of the thirteenth-century *Reconquista* was conducted largely by peninsular authorities and their forces.

[126] *Ex. Lyx.* 170.

6

SEABORNE SIEGE WARFARE: THE ITALIAN MARITIME STATES AND LATIN EXPANSION

THROUGHOUT this study we have observed the importance of Italian maritime forces and individual engineers in Latin siege warfare. Having examined siege warfare from the perspective of Latin states and rulers, it is appropriate to shift our focus to the arsenals and counting houses of the Italian maritime republics. In so doing we may not only assess the importance of Italian maritime powers in Latin siege warfare, but also develop an understanding of Latin expansion in the twelfth-century Mediterranean.

Sources

Historical sources which describe the siege operations of Italian maritime states have been important throughout this enquiry and merit discussion here. Of the three maritime states which participated in large-scale overseas military ventures during the twelfth century, Venice is the one whose achievements are least well reported by its own historians. The *Historia Ducum Veneticorum*, written sometime during the middle thirteenth century, narrates events of twelfth-century Venetian political history, although in little detail.[1] The fourteenth-century Venetian historian Andrea Dandolo drew upon the above chronicle and other material, some of which has not survived, in his history of Venetian deeds.[2] While these works provide useful information, they do not compare with histories of Pisa and Genoa written during the twelfth century.[3]

---

[1] *Historia Ducum Veneticorum* (MGH SS 14).   [2] Andrea Dandolo, *Chron.*
[3] The activities of the major Venetian expedition in the siege of Tyre in 1122–4 are well reported by historians of the Latin states (see above, pp. 68, 82–3).

Pisan historical writing of the early twelfth century was concerned particularly with the overseas expeditions of that city and consequently relates much about the siege operations involved. The *Annales Pisani* record events from 1004 to 1192 in what is almost an official history of the maritime state.[4] While the authorship and dating of the work covering the eleventh century is not clear, from approximately 1135 onwards it was written by Bernard Maragone, a Pisan notary connected with one of the city's important families. He provides a lucid contemporary account of events in which Pisa was involved, including its siege operations.

An important account for an understanding of the patterns of seaborne siege warfare is the *Carmen in victoriam Pisanorum*.[5] This work recounts an attack on the Zirid city and pirate base of Mahdia in north Africa in 1087 by an Italian seaborne expedition led primarily by Pisa. While the date of composition is not clear, the work's most recent editor attributes it to a period before the First Crusade. Cowdrey, following Erdmann, suggests that it was written between 1087 and 1096 and that its author may have been connected with the circle of Anselm II of Lucca. Fisher argues that the poem in its present form was written in approximately 1120 by a canon of the cathedral chapter at Pisa. Fisher also argues that this work is part of a school of Pisan history intended to glorify Pisan military accomplishments and the Pisan Archdiocese. In any case, both agree that the *Carmen* is based closely on the experiences of an eyewitness.[6]

Two works written during the decade of the 1120s and mentioned above in connection with the Balearic expedition of 1113–15 should be located in this context. The *Liber Maiolichinus* and *Gesta triumphalia per Pisanos facta* were written in Pisa between 1118 and 1127 and celebrate the deeds of Pisans in the conflicts of the early twelfth century.[7]

All of these works serve to illustrate Pisan achievements in a struggle against Islam for control of the Mediterranean. The author of the *Liber Maiolichinus* clearly places the Balearic expedition in such a context, and prefaces his account of the 1113–15 expedition by discussing Pisan

---

[4] Events discussed after 1174 are related only in an Italian translation.
[5] *Carmen in victoriam Pisanorum*, ed. H. E. J. Cowdrey, in 'The Mahdia Campaign of 1087', *English Historical Review*, 94 (1977), 1–29.
[6] Cowdrey, 'Mahdia Campaign', 2–3; C. B. Fisher, 'The Pisan Clergy and the Awakening of Historical Interest in a Medieval Commune', *Studies in Medieval and Renaissance History*, 3 (1966), 144–5, 186–7.
[7] Fisher dates the *Gesta triumphalia* between 1118 and 1119, and the *Liber Maiolichinus* at approximately the same period (Fisher, 'Awakening', 153–5).

campaigns against Muslim pirates and possessions throughout the eleventh century.[8] These works share an interest in the deeds and sometimes deaths of particular individuals and a concern for detail which suggests a close relationship to events and audiences which appreciated them. Moreover, in recounting the Pisan deeds, these works relate much about the conduct of siege operations.

The most important historical writer of twelfth-century Genoa was the veteran crusader, diplomat, and political leader, Caffaro. Born in approximately 1080, and a member of the important Visconti family, Caffaro enjoyed a long career in the service of his city and in expeditions to the eastern and western Mediterranean, and recounted the contributions of Genoa to Christian victories over Islam in several works. The *Annales Ianuenses*, which commence with the expedition of 1099–1101 which assisted in the capture of Arsuf and Caesarea, is the basic source for twelfth-century Genoese history. After Caffaro stopped writing in 1162, the work was continued as an official Genoese history throughout the thirteenth century. Caffaro also wrote a contemporary account of the capture of Almeria and Tortosa discussed above, as well as a short work entitled *De liberatione civitatum orientis*. This latter history records Genoese activities in the Latin conquest of the eastern Mediterranean and was written towards the end of his life.[9] Caffaro's lively narrative of events in which he was often a participant provides an especially important if highly partisan account of Genoese military activities in the twelfth century.

Patterns of Seaborne Siege Warfare

During the twelfth century Genoa, Pisa, and Venice put together far-flung trading empires of notable commercial importance. These establishments were not organized around control of territory and its agricultural products, nor for the direct exploitation of conquered populations. Rather they were based on control of or at least proximity to markets and access points for long-distance overseas trade. To further these ends, Italian merchants found it expedient to have bases

---

[8] The author mentions the Pisan activities in capturing Sardinia and raiding Bona, Palermo, and Mahdia (*Lib. Maiol.*, vv. 20–39, pp. 6–7).

[9] For a discussion of Caffaro and his work, see *AI*, pp. lxx–xcix; R. Face, 'Secular History in Twelfth Century Italy: Caffaro of Genoa', *Journal of Medieval History*, 6 (1980), 169–80.

exempt from the taxes, legal jurisdiction, and religious authority of local rulers in important trading areas. Large Levantine commercial cities, whose conquest was made possible by the First Crusade and Latin establishment in the eastern Mediterranean, presented maritime trading states with opportunities for establishing such enclaves, which they exploited during the early twelfth century.[10] While the maritime and commercial aspects of this development have been discussed, it is appropriate to focus on military concerns which were central in the creation of these enclaves. Long-term commercial enterprises depended upon the capture of major Mediterranean cities.

Urban conquest was also important to Italian maritime states for more immediate financial reasons. Major communal military ventures involved significant capital investment, for which a return was expected. A number of accounts refer to finding funds from captured cities to compensate for the costs of expeditions. However, the most significant example of a relationship between plunder and meeting the expenses of large ventures is a series of expeditions whose costs were not met from wealth acquired during them. From 1146 to 1148 three large-scale Genoese expeditions sacked Majorca and captured Almeria and Tortosa in conjunction with Spanish rulers, thereby winning significant enclave and trading privileges in Aragon and Castile.[11] Although Majorca and Almeria provided some material wealth in the forms of booty and cash payments made by defenders, the commune was unable to repay creditors all the money advanced in financing these expeditions. While the commune attempted to raise sufficient funds by seeking and exercising rights of lordship of certain territories under its control, it eventually went bankrupt. While this collapse was the climax of a long spiral of inflation and debt, it precipitated the removal from communal authority of the expansionist party which had ruled Genoa and directed its foreign policy since the First Crusade.[12]

Despite this example, it is clear that rich rewards from conquered cities promoted future operations. Caffaro proudly records the considerable spoils brought back by the Embriachi from Jerusalem in 1100, and the forty-eight *solidi* and two pounds of pepper each member

[10] Heyd's work remains the basic discussion of maritime expansion; Byrne tries to relate technical and commercial details of Genoese shipping to that city's expansion (Heyd, *Commerce du Levant*; E. H. Byrne, *Genoese Shipping in the Twelfth and Thirteenth Centuries* (Cambridge, Mass., 1930).
[11] See above, pp. 177–82.
[12] H. C. Kreuger, 'Postwar Collapse and Rehabilitation in Genoa, 1149–1162', in *Studi in Onore di Gino Luzzatto* (2 vols.; Milan, 1950), i. 122–4.

of the Caesarea expedition received in 1102.[13] The support and enthusiasm for future campaigns that the wealth brought back by returning veterans generated cannot be overestimated in explaining the motivation of those who participated in overseas expeditions. Moreover, some of this wealth was probably important in financing commercial ventures in the nascent capitalistic Italian city-states.[14]

Thus, in the short as well as the longer term, it was in the economic interest of maritime trading states to effect the capture of lucrative Mediterranean commercial centres. And so it was on taking cities as quickly and economically as possible that Italian maritime military activity focused.

While looting rich cities provided immediate reward, long-term trading opportunities could be guaranteed only by durable political independence and security. Yet the insufficient manpower resources of the maritime cities made extensive colonial states an impossibility.[15] The support of a land-based Latin power which would guarantee both enclave independence and security was paramount. This was equally important for military operations. Most large-scale sieges involved a convenient partnership between a land-based Latin power wishing to extend its authority over another community, and Italian merchants and seamen eager to share in the sack of a city and establish a trading centre. This partnership provided land-based rulers with the naval, material, and technological assistance they lacked or found difficult to obtain. Not only did Italians gain military assistance, especially heavily armoured knights, in their operations, but also allies who could maintain the security of their enclaves. It is into this pattern that Italian maritime military activity throughout the Mediterranean may best be fitted.

While the establishment of such trading centres is usually associated with Latin attacks against Muslim possessions in the Mediterranean, Italian city-states were willing to ally their fleets' technical skills with any power attacking lucrative positions. In 1093 Pisan and Genoese ships arranged to support Alfonso VI's attack on the Cid's Valencia.

---

[13] Caffaro, in *AI* i. 13; *Lib. civ. orient.* III.

[14] Lopez suggests that wealth gained in plundering activities provided a basis for Genoese investments in trade (R. Lopez, 'The Trade of Medieval Europe: The South', *Cambridge Economic History of Europe* (4 vols., ii; Cambridge, 1952), 306.

[15] The Genoese colonies in Syria were small-scale establishments, closely dependent on Genoa for finance and support (E. Byrne, 'The Genoese Colonies in Syria', in L. Paetow (ed.), *The Crusades and Other Historical Essays Presented to D. C. Munro* (New York, 1928), 139–46.

When that attack failed to materialize, the Italian fleet then assisted the count of Barcelona in prosecuting an unsuccessful siege of the Cid's dependency of Tortosa.[16]

In 1143 Pisa and Genoa helped William VIII of Montpellier recapture his city from communal rebels in the course of a long siege. In this action maritime republics were not only responding to a papal initiative to support William VIII, but also reducing a nascent commercial rival.[17] It may be noted that large-scale military and naval activity was often religiously sanctioned in some form. That is, Pisan and Genoese communes usually attacked either unbelievers or excommunicates. Yet this was not always the case. In 1167, as part of their *rapprochement* with Alfonso of Aragon, the Genoese agreed to assist him in taking Aubord from the count of Toulouse. In return, the Genoese were to have enjoyed once more all their customary privileges in Catalonia. Moreover, Genoa was then engaged in a war with Pisa, and Alfonso agreed to refuse anchorage to Pisan ships from Nice to Tarragona, and prohibited them from trading in Barcelona except to transport pilgrims. Genoa, moreover, honoured her agreement by sending eight galleys, whose crews built siege machines to assist Alfonso.[18] In 1174-5 the Genoese commune assisted Raymond V of Toulouse with men and ships in retaking the rebellious city of Marseilles in return for trading privileges.[19]

However, not all such alliances were successful. From 1135 to 1137 Pisa participated in the coalition organized against Roger II of Sicily by Innocent II, which included the Emperor Lothar III as well as Roger II's southern Italian adversaries.[20] While the evidence is not conclusive, it is likely that Pisa was offered commercial privileges in Sicily and southern Italy by Lothar III in return for support.[21] During August 1135 Pisa dispatched a force of forty-six galleys which were intended to raid territories loyal to the Sicilian king then blockading Naples. The Pisans sailed to Salerno but, finding it well protected, landed at

[16] Caffaro, in *AI* i. 13; R. Dozy, *Recherches sur l'histoire et la littérature du Espagne pendant le Moyen Age* (2 vols.; 3rd edn., Leyden, 1881), ii, Appendix xx, pp. xxiii–xxiv; Pidal, *The Cid and his Spain*, 287.
[17] Caffaro, in *AI* i. 31–2; C. Devic and J. Vaissete, *Histoire générale de Languedoc* (10 vols.; Toulouse, 1874–1914), iii. 727–8.
[18] *CDRG* ii, no. 25, pp. 66–7; Obertus, in *AI* i. 206–7.
[19] *CDRG* ii, nos. 92, 100, pp. 197–200, 225–7.
[20] For this conflict, as well as Lothar III's campaign in southern Italy of 1137, see above, pp. 116–18.
[21] D. R. Clementi, 'Some Unnoticed Aspects of the Emperor Henry VI's Conquest of the Norman Kingdom of Sicily', *Bulletin of the John Rylands Library*, 36 (1954), 337.

Amalfi. That city was rapidly stormed, and heavy casualties were inflicted on its inhabitants. The Pisans captured seven galleys and two other ships, and took much plunder. Pisan forces then moved against Amalfi's dependent fortifications and met with little resistance until they attacked the well-fortified position of Ravello. While the Pisans and their siege engines were engaged before the fortress, Roger II concentrated military and naval forces drawn largely from his base at Salerno. The Pisans were scattered in the ensuing attack, and those not killed or captured fled to their ships at Amalfi and returned home.[22]

Despite this set-back Pisa continued to support the coalition against Roger II. In 1137 Pisa participated in the great offensive against the Sicilian king led by Lothar III and Innocent II.[23] While Lothar III was still on the Adriatic coast, allied forces, led by Rainulf of Avellino and Robert of Capua and supported by a powerful Pisan fleet, commenced a close blockade of Salerno in July. In preparation for this attack, Amalfi and her dependent fortifications were captured by Pisa and her allies. Salerno's citizens, under Roger II's chancellor, Robert of Selby, and supported by artillery, successfully resisted initial allied attacks. The Pisans decided to employ siege machinery, building artillery, armoured roofs, and probably a siege tower during August.[24] These devices were brought against the city, which appeared unable to resist. However, with Robert of Selby's approval and after he had retired into the citadel with his own men and forty knights from the city, Salerno surrendered. The citizens paid Lothar III a large sum and were received into imperial protection immediately, which obstructed a Pisan sack of Salerno.[25] Although the sequence of subsequent events is far from clear, it appears that an attack on the citadel did not develop and that, during the delay, Salerno's citizens burnt Pisan machines. Lothar III apparently refused to pay compensation or allow the Pisans to exact their own recompense. Believing that their efforts and machinery had compelled Salerno's surrender, the Pisans grew in-

[22] BM 9–10; Alexander of Telese, *De rebus gestis Rogerii*, iii. 23–4, pp. 140–1; Romuald of Salerno, *Chron.* 221.

[23] For this campaign, see above, pp. 116–17.

[24] Maragone mentions *manganis castellis* and *gattis*, but his account of operations is brief; Romuald of Salerno refers to a Pisan *ligneum castellum* burnt at the end of the siege; the accounts of the *Annalista Saxo* and Falco of Benevento mention devices which broke down walls and whose height was considerable, but neither writer is specific (BM 11; Romuald of Salerno, *Chron.* 222–3; *Ann. Saxo*, 774–5; Falco of Benevento, *Chron.* 232–3).

[25] According to Falco of Benevento, the surrender negotiations were conducted in secret (Falco of Benevento, *Chron.* 233; Romuald of Salerno, *Chron.* 222–3).

censed, negotiated a peace with Robert of Selby, and sailed home.[26] This deprived the emperor not only of a fleet but also of proficient and well-supplied siege crews. Although an offensive against Roger II was already disintegrating, the Pisan departure ensured its collapse.

The Pisan–imperial alliance against Roger II was similar to those which facilitated the capture of Muslim coastal cities in the eastern and western Mediterranean. Operations against Salerno were also similar, as the Pisan fleet isolated the city from Roger II's ships and provided materials, expertise, and men with which to attack it. The breakdown of the alliance over the recompense the Pisans expected for their efforts illustrates not only the importance of the costs of sophisticated machinery, but also something of Pisan motivation.

When these Pisan and Genoese military operations are examined in the light of the major expeditions discussed in earlier chapters, a clear pattern emerges which emphasizes the importance of Italian maritime forces' capabilities in siege warfare. Ships were clearly a necessity for transporting men and materials in achieving the concentration of force essential in overwhelming coastal fortifications. In many cases sea battles were not necessary in neutralizing enemy sea power. Some positions were not supported by friendly shipping, and the presence of a large number of beached galleys which could easily be put to sea served as a deterrent. This does not undermine the importance of sea power, but rather amplifies an understanding of the contributions of Italian maritime forces to Latin expansion in the twelfth-century Mediterranean.

While not every position taken required the construction of sophisticated siege machinery, many important ones did. It is clear from previous examples that Italian maritime siege technique centred on large-scale assaults, which were frequently based around siege towers. While other types of machinery supported them, mobile towers were the hubs of assaults and the defensive responses they generated. Although most dramatically employed in the great sieges of the Levant and Iberia, these devices were also used in conflicts among Italian city-states. The Italian maritime penchant for siege towers is illustrated from their unusual employment in a field battle.

During the period 1165–73 Genoa and Pisa were engaged in a war which involved their allies and dependencies in Tuscany as well as in Sardinia and Corsica. Much of the conflict consisted of raiding and attacks on shipping, and neither principal was able to organize a major

[26] BM II; Romuald of Salerno, *Chron.* 223; Falco of Benevento, *Chron.* 233.

concentration against its rival.[27] In order to facilitate communications with their Luchese allies and to establish a forward base for operations against Pisa, the Genoese established a fortress at Motrone near Viareggio in 1169–70. In November 1170 a powerful Pisan force marched against Motrone, while Lucca raised a relieving army. The Pisans encamped near Motrone and commenced building machinery to facilitate an assault, including six siege towers of unknown height. The Luchesi arrived and encamped near a wooden fortification originally built as an outpost of Motrone. On 27 November the Pisans moved against the Luchese camps, and the Luchesi went forth to meet them, resulting in pitched battle. The six Pisan siege towers were employed in the attackers' front rank, doubtless to provide commanding archery and crossbow fire.[28] After crushing the Luchesi, who withdrew, Pisan forces captured Motrone, after an attack of four days employing these towers as well as artillery.[29] Although this deployment of siege towers in a field battle may be unique, it underscores the importance of these devices in Italian siege warfare.

Shipbuilding and the Technology of Siege Warfare

The primary reason for Italian maritime pre-eminence in constructing and utilizing siege machinery was the close similarity between naval engineering and the poliorcetics employed by these besiegers. The problems of designing and building large wooden structures were common to both naval and military architecture, and abilities in one area could be transferred to another. Moreover, skills of carpentry, joinery, and other crafts necessary in shipbuilding were those required in preparing and making complex wooden devices. Even the handling of massive corner beams which made up the frames of siege towers had a parallel in naval engineering in the masts from which siege towers were occasionally made. Although elementary, this connection between shipbuilding and siegecraft should be emphasized in understanding the impact of Italian maritime states on Latin siege warfare.

The organization of shipbuilding in these communities was particu-

[27] For an account of this conflict, see Heywood, *His. Pis.* 170–94.
[28] 'In prima acie fuerunt pedites omnes et sagittarii et milites octingeni, et sex castella lignea fortissima ...' (BM 51).
[29] BM 50–1; Obertus, in *AI* 237–40.

larly developed, and this may also have been true of military engineering. Twelfth-century practices should be discussed in this regard, as there may have been parallels in poliorcetics. Genoese evidence shows that different tasks in shipbuilding were carried out by different experts, all of whom were organized by one master builder and subcontractor, who often also engineered masts. The *magister axiae* designed the ship and contracted with craftsmen, carpenters, nail-makers, and cordwainers as well as masters in hull-making, decking, and castellation.[30]

While siege-machinery construction was not as highly organized, there were specific experts responsible for it in twelfth-century Genoese operations whose specializations may have reflected those of shipbuilding. Besides the *artifex* who engineered and supervised the building of machinery, there was also another kind of expert associated with siege warfare in later twelfth-century Genoa, termed *magister lignaminis*. Such men were dispatched from Genoa with the garrison of Motrone in 1169 in order to complete Motrone's wooden tower.[31] An arrangement for mutual defence between Genoa and Alessandria in 1181 illustrates that these experts, in conjunction with Genoese archers, crossbowmen, and engineers, were valued by other communities. The agreement stipulated that, should Alessandria be attacked, such personnel were to be sent to Alessandria's aid. It may be noted that *magistri lignaminis* are distinguished from chief engineers. It should be added that Alessandria promised knights should that city be attacked.[32]

While it is clear that *magistri lignaminis* were experts in wooden construction and associated with siege warfare, their specific talents remain obscure. Caffaro used the term *lignamen* in his histories to describe timber specifically prepared for siege machinery and including the wood he claimed William Embriaco brought to Jerusalem in 1099.[33] Thus, it appears likely that the masters of these materials were specialists in the building methods which Genoese poliorcetics usually involved. It is also possible that they were adroit in working with large pieces of timber, and structures which required them, and so closely associated with the production of siege machinery. At the least,

[30] Byrne, *Shipping*, 24–6.
[31] 'Quam turrim Ianuenses armia, victualibus balistariis. xxii. lignamine omnibus que ibidem necesariis aptauerant, et magistros lignaminis illic mandauerant qui turrim prenominatam cooperuerunt' (Obertus, in *AI* i. 222).
[32] 'cc. arciferos et tres magistros lignaminis et unum ingeniosum artificem et balistarios x' (*CDRG* ii, no. 131, p. 266).
[33] Caffaro, in *AI* i. 33; *Almarie et Turtuose*, 85; *Lib. civ. orient.* 110.

our knowledge of these men demonstrates the level of organization of large-scale wooden construction in Genoa. This necessarily gave the community a considerable advantage in poliorcetics. There is little information about such men serving Pisa and Venice during the twelfth century. It is likely, however, that both commanded similar levels of construction organization.

Not only did maritime republics have the expertise necessary for large-scale siege warfare concentrated in their communities, but also the materials. The timber, nails, fastenings, and ropes necessary in constructing siege towers and armoured roofs were also required in shipbuilding. This is well illustrated in the *Liber Maiolichinus*' description of Pisan preparations for the Balearic campaign of 1113–15. Wood collected from Corsica was brought back to Pisan workshops and there prepared for use in building the expedition's ships. The poet mentions that this timber had many uses and especially in the full range of siege machinery. He also mentions that all of Pisa's ironmongery was used in building ships and making the fastenings and nails to be used in siege machinery.[34]

The easy acquisition of materials was particularly important for the straight, long timber necessary for masts and practical for siege towers. While siege towers were sometimes built from composite frames, it was more economical and less laborious to base them around such timbers. These beams were, on occasion, difficult to obtain in some areas and usually expensive. As Byrne has shown, they were usually the single most expensive part of a vessel, and captains occasionally met the cost of their ship and first voyage by selling a mast at the right time and place.[35] Thus, in erecting siege towers rapidly along the shores of the Mediterranean, Italian seaborne expeditions had a considerable advantage in materials.

Not only were masts used for mobile towers, but occasionally whole ships were broken up for machinery and residences during long sieges. Although this practice may initially appear somewhat wasteful, it may have been economic as well as expedient. Moreover, this practice may have been more economic than it initially appears because of developments in shipbuilding techniques.

Although there were variations within each technique, ships were built in one of two different ways in the eleventh-century Mediterranean. In the more traditional method, workers began with the exterior of the hull, building a shell and then assembling an internal

[34] *Lib. Maiol.*, vv. 97–129, pp. 10–11.   [35] Byrne, *Shipping*, 26.

frame before moving on to the superstructure and masts. While this technique was slow and costly, its products were sturdy and particularly durable. In the second method, builders began with an internal frame and nailed whatever was available on to it, in effect constructing the hull from inside out. While ships assembled in this fashion were appreciably less durable, they were also more rapidly and economically built.

Archaeological evidence indicates that the second or 'frame-first' method was utilized in northern Italy around the time of the First Crusade, prompting White to conjecture that this was a newly introduced technique, important in generating the rapid Italian maritime expansion associated with the crusading age.[36] However insightful, White's observation must be qualified. It is apparent that 'frame-first' techniques were known in the Mediterranean before 1096. Furthermore, it has been argued that the stimulus to utilize 'frame-first' techniques was not so much the realization of the potential of this method, as a need to produce a large number of ships rapidly and inexpensively.[37] Whatever the precise timing of the advent of the 'frame-first' method in northern Italy, this technique and the particular Italian practice of using ships for siege machinery were closely connected. While it is not possible to ascertain how many or what type of ships were built according to which method, vessels used in machinery construction were very likely built according to the second method. Whether or not certain ships were deliberately constructed in anticipation that they would be broken up for materials remains conjectural. However, regarding major expeditions specifically intended for overseas military operations, this appears likely. In any case, an understanding of these different styles of shipbuilding and their distinct products helps explain the Italian practice of breaking up ships for siege machinery.[38]

The launching of powerful seagoing expeditions during the twelfth century by Italian maritime states was partially a result of recent developments in naval technology. This was an important factor in successful Italian maritime siege operations, as the economical transportation of men, materials, and sometimes horses was fundamental in the first phase of these expeditions. Subsequent poliorcetic operations were facilitated by the same technology and reservoir of personnel which put together expeditions. The ability to transport and employ

[36] White, 'Thrust', 105–6.
[37] R. Unger, *The Ship in the Medieval Economy* (London, 1980), 120–1.
[38] See above, pp. 72–3.

complex machinery economically in siege operations was the result not of a single technological innovation or new techniques, but of developments and refinements in a number of areas. Yet together these factors made the Italian maritime states exceptionally powerful and significant in the warfare of the twelfth-century Mediterranean.

The organization and financing of siege operations was only one aspect of the overall arrangements involved in seagoing expeditions. It is appropriate to consider how the organization and financing of expeditions may have affected the cohesion and motivation of maritime besiegers. Because of the survival of considerable Genoese commercial records, such considerations will concentrate on Genoese practices.

An understanding of Genoese financing of military ventures begins with possible parallels in commercial shipbuilding and trading. The ownership and value of a Genoese ship was subdivided into shares—*locae*—usually equal to the number of mariners' places in the ship.[39] Controlling partners often sold places to raise capital for a ship's construction and perhaps provide for part of the crew. In some cases crewmen were either small-scale investors or their representatives. It is possible that some of the capital for military expeditions was partially raised in a similar fashion by selling shares in the venture. Caffaro's statement that every participant in the Caesarea fleet in 1100 received an equal share in the plunder supports this, although the division of spoils may simply have been the result of a very successful expedition.[40]

Thirteenth-century Genoese citizens invested in the raiding ventures of their city's corsairs, and it is likely that twelfth-century enterprises were financed in a similar way.[41] While small-scale endeavours were organized privately, larger ones involved the Genoese commune and its elected leaders in directing, organizing, and doubtless financing expeditions. Moreover, it appears that the foundation of the commune was closely connected with arming and equipping naval expeditions and protecting the rights of their investors.[42] The importance of the commune in managing and financing operations is most clearly evident in the Balearic–Spanish expeditions of 1146–8. Communal officials conducted negotiations on behalf of the expedition and directed military operations. Moreover, the inability of the

---

[39] Byrne, *Shipping*, 12–21. [40] Caffaro, in *AI* i. 13.
[41] Byrne, *Shipping*, 62–3.
[42] E. Bach, *La Cité de Gênes au XIIe siècle* (Copenhagen, 1955), 33.

commune to meet its costs in launching these expeditions resulted in financial collapse.[43]

This means of organization gave Genoese forces unity of purpose and cohesiveness as well as incentive. Major military decisions were taken in common, usually in the course of what Caffaro terms a *parlamentum*. The decision to assault Caesarea in 1101 before the completion of siege engines was made communally by Genoese participants after exhortations from Archbishop Daimbert of Pisa and the Genoese leader, William Embriaco.[44] After an over-enthusiastic group of Genoese combatants were mauled during an unauthorized foray from camps outside Tortosa in 1148, a *parlementum* was held in which all swore not to engage in any military contact without specific approval of the consuls.[45] Thus it seems that Genoese expeditions were closely organized around communal military and political institutions and manned by personnel with a financial stake in the venture. This provided a means for maintaining discipline as well as solidarity. These factors should be taken into account in understanding the willingness of Genoese besiegers at all levels to undertake the laborious and often dangerous activities of siege warfare on the shores of the Mediterranean.

Whatever their precise form of organization, it is clear that Genoese military ventures channelled a significant degree of the community's wealth in financing them. Yet parallels from commercial practices suggest that military expeditions may have been to some degree self-financing. Not infrequently a shipowner would get his vessel built partially on credit advanced for shares in the vessel, which would be paid for from the venture's first profits.[46] It is plausible that some similar means were involved in launching military ventures. While it is unclear whether individuals offered credit directly or worked through the commune, it is possible that the costs of shipbuilding and siege machinery were met by credit advanced on the expected plunder to be won. Thus a successful expedition could to a degree be self-financing, and a hapless one result in financial failure. As we have seen, the commune collapsed in 1149 because it could not meet the costs of the 1146–8 military ventures in Spain.

Although speculative, this discussion underscores the degree to

[43] See above, pp. 177–82, 196.
[44] Caffaro, in *AI* i. 10–12.
[45] Caffaro, *Almarie et Turtuose*, 86.
[46] Byrne, *Shipping*, 12–21.

which the Genoese community was involved in overseas military expeditions. While there is less information about practices in Venice and Pisa, the institutions and resources of these city-states provided the basis for organizing seaborne expeditions. The well-known Italian maritime ardour for plunder, which included despoiling populations which had already surrendered to terms guaranteeing them what they could carry, may in part be explained by methods of financing expeditions which depended on substantial credit. Not only were participants propelled by the costs of their own involvement, but the expedition as a corporate group may have had a considerable investment to return. In any event, the means of finance and credit available to these nascent capitalistic city-states were important in their ability to dispatch fleets and wage siege warfare.

The reasons for the Italian maritime pre-eminence in siege warfare on the coasts of the Mediterranean and in the conduct of siege-tower assaults are evident. The same technological and economic factors which sustained and promoted maritime and commercial activities in Genoa, Pisa, and Venice provided the basis of their siege warfare. Moreover, their motives were similar to those which generated the crusading movement. It may be noted that some commentators continue to perceive a fundamental difference in attitudes towards crusading between Italian maritime combatants and men from other areas of Europe.[47] In particular, knights have been contrasted with Italian merchants more attuned to commercial advantage than the business of Jesus Christ. For some, the attack on Constantinople at the heart of the Fourth Crusade is foreshadowed by Italian maritime ventures in the twelfth century. It may be argued that such distinctions rest ultimately on definitions of crusading more than the outlook of participants. Moreover, it is clear that Italian maritime personnel considered their efforts on behalf of Christendom in keeping with crusading traditions, which included renown and the possibility of material reward in pursuit of spiritual objectives. The expeditions which sailed across the Mediterranean carried not only materials, but highly trained experts, craftsmen, labourers, and combatants, eager to win a reward in this world as well as the next.

[47] Mayer, *Crus.* 65–6. Although there is much about this subject in connection with the Fourth Crusade, Queller and Day provide an efficient summary (D. E. Queller and G. W. Day, 'Some Arguments in Defense of the Venetians in the Fourth Crusade', *American Historical Review*, 81 (1976), 717–38).

The Development of Italian Maritime Siege Technique

We have outlined ways in which the organization and aims of seaborne expeditions of Italian maritime states facilitated siege operations as well as the techniques and machinery employed in these operations. Because of the importance of Italian maritime forces in Mediterranean siege warfare of the twelfth century, we should address questions concerning the development of these expeditions and their techniques. As demonstrated above, Italian maritime forces contributed to successful operations in the First Crusade and launched large-scale expeditions during the early twelfth century. It is appropriate to examine a seaborne operation which involved some of the means and techniques noted above which took place before the First Crusade.

Mahdia, 1087

The Italian maritime attack on Mahdia in 1087 anticipates what was to occur along the Syrian and Palestinian coasts during the first decade after the First Crusade. The organization, justifications, propaganda, and attitudes of participants of the Mahdia campaign were similar to those of the early Crusades.[48] Moreover, the Mahdia campaign involved siege operations which anticipate those of the early twelfth century, which are described to some degree in a source based closely on an account of participants, the *Carmen in victoriam Pisanorum*.[49]

In attempting to restore the power of the Zirid state after the devastations of *Banu al-Hilal* tribesmen of north Africa, Tamin, lord of Mahdia, encouraged sea raiders to attack Christian lands and shipping in the central Mediterranean. The resultant misery and destruction were the ostensible justifications for a Christian counter-attack. Mahdia's position as a wealthy trading centre made it a target for plundering, and offered a suitable base for merchants hoping to establish new markets. The city appeared vulnerable, as Tamin and his principal forces were often occupied in other parts of Ifrikiya, combating the *Banu al-Hilal*. Moreover, the Normans were completing the conquest

---

[48] Cowdrey discusses similarities in outlooks and attitudes towards Muslims and the idea of holy war between the *Carmen in victoriam Pisanorum* and the *Gesta Francorum* (Cowdrey, 'Mahdia Campaign', 21–2).

[49] See above, p. 194; Cowdrey discusses references to the campaign in Latin and Arabic histories; however, it is the *Carmen* which describes siege operations and the details of attack (Cowdrey, 'Mahdia Campaign', 3–6).

of Zirid Sicily, offering a potential alliance with a land-based power and, at the least, another diversion from the defence of Mahdia.

These three broad considerations resulted in a papally approved expedition, drawing men and ships from Pisa, Genoa, Rome, and Amalfi, setting sail in the summer of 1087. Although Italians from many cities participated, Genoa and Pisa were the senior partners. The *Carmen* tells us that a thousand ships were built in three months for the attack.[50] While undoubtedly exaggerated, commitment to the attack mobilized appreciable commercial, naval, and military resources.

The expedition struck first at the formidable Zirid fortress on Pantelleria, which appeared a formidable objective.[51] Daunting as this fortress may have been, the Italians had come prepared. Engineers erected siege towers, around which a successful assault was conducted, capturing the position and inflicting substantial casualties upon defenders.[52] The construction of these towers should be discussed here. The Italians must have prepared in advance by transporting the necessary nails, ropes, and armouring for their towers. As Pantelleria was likely to have lacked sufficient seasoned timber for such towers, the Italians probably brought suitable materials with them. While we do not know how tall these towers were or their degree of sophistication, their employment against Pantelleria foreshadows future operations.

From Pantelleria, the expedition moved on to Mahdia, some ninety miles from Tunis, located along the Tripolitanian coast between Sousse and Sfax. Mahdia at that time consisted of two sections, an old city situated on a narrow peninsula jutting out into the Mediterranean, and the walled suburb of Zawilla on the mainland. The attackers chose to strike first at Zawilla. As the only viable approaches were from the sea, attacking troops moved from transports into boats and headed for the beaches near Zawilla. As they landed, they encountered resistance, but drove it off and established a beach-head. The Italians then attacked this defensive force, and drove it back to the gates of Zawilla, which were shut in the faces of friend and foe alike. However, Zawilla's fortifications did not provide a durable defence. Soon after their landing Italian forces broke down the gates and stormed the suburb, sacking it and massacring its defenders.[53]

With Zawilla's fall, Mahdia's garrison launched a sortie which killed the Pisan commander, Viscount Ugo. While the chronology of

[50] Cowdrey discusses the background to the campaign (Cowdrey, 'Mahdia Campaign', 8–15).
[51] *Carmen*, 25.   [52] Ibid.   [53] Ibid. 26.

the conflict consists only of a sequence of events, according to the *Carmen* Ugo's death inspired the Pisans and Genoese to launch an assault on Mahdia itself as rapidly as possible.[54]

Mahdia in the late eleventh century was a substantial fortification in the central Mediterranean. Situated on a peninsula five hundred yards wide and a mile long, surrounded by a thick towered wall on all sides, and dominated by a citadel-palace atop the city's only hill, Mahdia's defences presented a major challenge to besiegers.[55] However, the defenders themselves may not have been in a sufficient state of preparedness to resist a major seaborne onslaught. Not only had the depredations of the *Banu al-Hilal* weakened their state over the years, but they had recently suffered a major attack. Tamin was probably absent from the city on an expedition against the *Banu al-Hilal* when the Italians landed, and only returned after the initial attack.[56] In any event, Mahdia's defenders were unable to stave off the Italian onslaught.

The attackers stormed the city in a mass assault after shattering the gates, presumably from landward approaches.[57] While there are no details concerning this attack, ladders, axes, and small battering-rams as well as enthusiasm and determination carried the city. The attackers poured into Mahdia, looting and killing indiscriminately, including men, women, children, religious officials, and even royal horses and mules. This represents another prefiguration of the crusading epoch. An elegiac *kasad*, written probably by an eyewitness, although surviving only in a fragmented form, expresses a sense of amazement at the ferocity of the Christian assault.[58] The sack and looting came to a halt before the royal citadel and palace, in which Tamin and the garrison's remnants prepared a last stand. Yet the garrison and Mahdia's final defences were not to be seriously tested. According to the *Carmen*, the palace was considered too well fortified to be assaulted.[59] No serious attempt at close blockade or assault supported by complicated machinery was undertaken. This was likely a result of fears for the attackers' security. *Banu al-Hilal* raiders shadowed the Italians and moved into Zawilla, attacking the ship guards while the Italians were

[54] Ibid. 27.
[55] I am indebted to Dr Michael Brett of the School of Oriental and African Studies, London, and his paper on 'Fortification Methods of Warfare in Ifrikiya, 7th-11th Centuries', for this information.
[56] Cowdrey, 'Mahdia Campaign', 8.   [57] *Carmen*, 27.
[58] Heywood, *His. Pis.* 27.
[59] *Carmen*, 27.

busy inside Mahdia. Although the tribesmen were repulsed and may have been more interested in attacking Tamin than the Italians, their presence must have convinced the Italians not to undertake a long siege. The lord of Mahdia offered payment if the Italians would abandon their enterprise and leave north Africa. Moreover, Roger I of Sicily could not be persuaded to assist in operations nor assume the overlordship of Mahdia. Thus the Italians filled their ships with plunder, Tamin's indemnity, and rescued captives, and returned home.[60]

Assessing the Mahdia campaign in terms of later crusading ventures illustrates that the basic ingredients of maritime expansion were active and had come together before the First Crusade. Mahdia and Pantelleria were the first of a long series of profit-making ventures which were based on sea power and a concentration of well-motivated and -supplied assault troops. The attack on Pantelleria involved the first use of the kind of siege machinery which would characterize Italian maritime poliorcetics during the twelfth century. Like the early crusaders, Italian troops at Mahdia were particularly zealous, enthusiastic, and bloodthirsty. In these ways at least, the Mahdia campaign may be compared to operations in Palestine during the decade after the First Crusade.

Yet it would seem that there was a fundamental difference in strategic and commercial intentions. Italian participation in crusading sieges had as one of its goals the establishment of commercial privileges in valuable market areas, as well as plunder. Initially it appears that the attack on Mahdia cannot be placed into this pattern—at least from the Pisan source. Yet it is probable that Italians seriously considered the possibility of opening a trading enclave at Mahdia in 1087. Geoffrey Malaterra mentions that, after the city was taken and presumably before Tamin came to terms, the Italians offered Mahdia's lordship to Roger I, who refused.[61] Although the lord of Sicily declined, it is clear that the Italians hoped to establish the kind of political, military, and economic relationship between themselves and Roger I over Mahdia as they were to achieve with Bohemond and Baldwin I in the Levant. The long-range objectives and economic and religious motivations of the attack against Mahdia were the same as those of Italian crusading ventures. The organization and importance of formidable expertise in siege warfare were also similar. These expeditions were part of a larger pattern of Italian maritime military activity in the twelfth-century

---

[60] For Roger I, see above, pp. 98–100.
[61] Geoffrey Malaterra, *De rebus gestis Rogerii*, iv. 9, pp. 86–7.

Mediterranean. In this regard, the Mahdia campaign was the first manifestation of forces which played a major role in the Latin expansion of the age.

The Italian Maritime States and the Military Impulses of the Late Twelfth Century

Acre, 26 August 1189–12 July 1191

Having treated a number of Italian seaborne siege operations in earlier chapters, it is appropriate at this point to discuss one of the major sieges of the twelfth century which included significant Italian maritime participation. The attack on Acre during the Third Crusade was a military undertaking of the first order, involving a two-year isolation of the city and its besiegers. Although the core of the attack on Acre was one of the great blockades of the Middle Ages, the city was subjected to personnel- and machine-supported assaults throughout the siege. Thus a review of operations permits us to summarize aspects of Latin siege warfare and poliorcetics in the twelfth century. Moreover, it is hoped that, by approaching events at Acre in the context of Italian seaborne siege warfare, an aspect of the siege that is sometimes overlooked will be highlighted, as modern accounts tend to be dominated by the activities of European monarchs and their contingents. Such an approach also allows us not only to review patterns of twelfth-century operations in the Mediterranean, but also to introduce some of the salient features of the next period of pre-gunpowder siege warfare.

Because of the duration of the siege and the international scope of the Third Crusade, information about events at Acre is reported in a considerable number of western chronicles and histories. However, the number of independent accounts relating events at the siege in any detail is much smaller. Foremost among these is the *Itinerarium peregrinorum*, which has provided the basis for medieval accounts of the siege as well as modern ones.[62] The *Itinerarium* is an important source for events until November 1190, and reports much about the sufferings of Acre's besiegers. While useful in relating machinery-supported assaults in the context of the siege, its utility for crusader poliorcetics is

[62] Mayer's edition provides a thorough discussion of the work and its relationship to the historiography of the Third Crusade (*Itinerarium peregrinorum*, ed. H. E. Mayer (MGH Schriften, 18; 1962), 52–77).

limited by its reliance upon Vegetius and Isidore of Seville for descriptions of siege machines and their operation.[63]

L'Estoire de la Guerre Sainte, written by an Anglo-Norman poet known as Ambroise, provides a detailed eyewitness account of the Crusade of Richard I and consequently of events at Acre after 8 June 1191.[64] Ambroise's account is central for the final phase of the siege, which involved a great amount of military activity against the city. His discussion of the siege before June 1191 is clearly based on another written account, as Ambroise himself acknowledged.[65] In any case, Ambroise remains an important source for the latter stages of the siege.

Although few participant accounts span the full duration of the siege, one important work which does is the *De expugnatione civitatis Acconensis* of Haymar. While the author may have been bishop of Caesarea and later Patriarch of Jerusalem, he was clearly associated with the contingent of the bishop of Verona, which was involved in the siege of Acre from its inception.[66] Arabic accounts give much information about the siege warfare waged at Acre. While primarily concerned with the efforts of Saladin's and Acre's defenders, Muslim historians give detailed reports of crusader assaults and machinery. Among these, Saladin's biographer and companion Beha-ed-din provides the most specific accounts of the conflict over the control of Acre's fortifications.[67]

As this discussion concentrates upon the siege warfare waged at Acre, the reader is referred to modern accounts which illuminate other aspects of the siege and the Third Crusade. Röhricht provides well-documented narrative accounts of the siege important for chronology and particularly for the arrival of the diverse contingents which joined the Crusade outside Acre.[68] Prawer illuminates events at the siege in the light of detailed topographical knowledge in his discussion of the Third Crusade.[69] Mayer and Runciman give general accounts of the

[63] *Itin.* 63. The Latin Continuation of William of Tyre is also of limited use for understanding siege operations, as it provides few military details concerning the siege of Acre (*Die Lateinische Fortsetzung Wilhelms von Tyrus*, ed. M. Salloch (Leipzig, 1934)).

[64] Ambroise, *L'Estoire de la Guerre Sainte*, ed. G. Paris (Collection de Documents Inédits sur l'Histoire de la France; Paris, 1897).

[65] Ibid., vv. 2401–6, col. 65.

[66] Haymar, *De expugnatione civitatis Acconensis*, in Roger of Howden, *Chronica*, ed. W. Stubbs, iii (RS 51; 1870), Appendix to the preface, pp. cv–cxxxvi.

[67] Beha-ed-din, *The Life of Saladin, or What Befel Sultan Yusuf*, ed. and trans. C. W. Wilson and C. R. Conder (Palestine Pilgrims' Text Society, 13; London, 1897).

[68] R. Röhricht, 'Die Belagerung von 'Akkā (1189–91)', *Forschungen zur deutschen Geschichte*, 16 (1876); *Geschichte*, 500–71.

[69] Prawer, *His. roy.* ii. 42–68.

Third Crusade and the place in it of the siege of Acre.[70] Lyons and Jackson provide an account of operations from Saladin's camps which also discusses some of the other problems which the sultan faced in 1189–91.[71]

For the Italian maritime republics, the loss of most of the Latin Kingdom of Jerusalem was an economic as well as a religious disaster. The extensive trading privileges built up through years of fighting, bargaining, and political dealing in Outremer were eliminated. Consequently, the European attempt to restore Latin rule and military security in Outremer which generated the Third Crusade also included attempts by the maritime powers to regain their old commercial privileges.[72]

While Genoa, Pisa, and Venice sent expeditions in 1189, only the Genoese and Pisans are clearly manifest in accounts of military activities.[73] A Pisan fleet of 50–60 ships led by the papal legate Archbishop Ubaldo departed from Pisa in 1188, wintering in Messina before reaching Tyre in April 1189. Ubaldo and the Pisan seamen and combatants played an important role in the siege of Acre from its inception.[74] The author of the *Annales Ianuenses* describes his city's expedition, making clear that this force was a powerful one led by the consul Guido Spinola and involving many of the city's important leaders and was well equipped for siege warfare.[75] In that these expeditions hoped to win commercial privileges in return for naval and military assistance, they resembled Italian maritime expeditions which assisted in the first Latin capture of the coast of Outremer. Moreover, there was a close similarity in poliorcetics.

An attempt to re-establish Latin authority in the eastern Mediterranean and the siege of Acre began on 27 or 28 August 1189, when forces under the leadership of Guy of Lusignan encamped outside Acre. Guy's headquarters and initial dispositions were made on and around the hill of Toron, approximately three-quarters of a mile east of Acre. The Belus River provided a water supply, and the position was defensible, which was a not unimportant consideration because of

[70] Mayer, *Crus.* 134–44; Runciman, *His. Crus.* iii. 18–53.
[71] M. C. Lyons and D. E. P. Jackson, *Saladin: The Politics of the Holy War* (Cambridge, 1982), 295–330.
[72] Heyd, *Commerce du Levant*, i. 313–14.
[73] Dandolo claims that a Venetian squadron joined the Pisans on their voyage in 1188 (Andrea Dandolo, *Chron.* 270–2).
[74] Röhricht, 'Belagerung', 485.
[75] Ottobuono, in *AI* ii. 32–3; Röhricht, 'Belagerung', 486–7.

crusader vulnerability to Acre's garrison and Saladin's army.[76] Guy's forces were supported by Ubaldo's Pisan squadron, which held a coastal strip immediately north of the city, vital in maintaining sea communications. Although exposed throughout the early stages of the siege, the Pisans maintained this key position and attempted a sea blockade.[77]

Three days after their arrival the crusaders, carrying scaling ladders and protected only by their shields, assaulted the city. Despite the determination and intensity of this attack, it proved unsuccessful.[78]

The crusaders' objective had been one of the kingdom's chief cities before it fell to Saladin in 1187. Situated on a peninsula jutting southwest towards Haifa, Acre's southern and western approaches were guarded by the sea and sea walls. The harbour was protected by a fortification known as the Tower of Flies, connected to the city by a broken mole. Northern and eastern approaches were defended by a strong wall and deep ditch which ran east–west and north–south, meeting at the city's north-east corner and key fortification known as the Maledicta Tower. Acre's twelfth-century fortifications probably consisted of only one large wall, which did not include the suburb of Montmusard at the city's north-western corner, later encompassed by thirteenth-century walls. Although Saladin had Acre's fortifications placed in good repair after its capture, he did not substantially alter the crusader city.[79]

The crusaders' initial establishment at Toron had some similarities to the siege operations based on counter-forts which we have already examined. However, Guy's encampment was not so much a forward base as the first step in a close blockade, and in this regard a comparison with the siege of Antioch in 1097 is more apposite. As in that siege, the timing of the blockade's commencement was in part due to considerations which were not strictly military. The king, who had been vanquished at Hattin, had little left to lose, and the attack organized under his leadership was also an attempt to revive his political fortunes. Acre's besiegers, unlike those of Antioch, received numerous re-

[76] Estimates of Guy's numerical strength vary: Prawer suggests 600 knights and 7,000 foot soldiers, Röhricht estimated about 9,000 at most (Prawer, *His. roy.* ii 43; Röhricht, *Geschichte*, 502 n. 3).
[77] *Itin.* 307.   [78] Ibid.
[79] See Map 3; N. Makhouly and C. N. Johns, *Guide to Acre* (2nd edn., Jerusalem, 1946), 26–8; D. Jacoby, 'Montmusard, Suburb of Crusader Acre: The First Stage of its Development', in B. Z. Kedar, H. E. Mayer, and R. C. Smail (eds.), *Outremer Studies in the History of the Crusading Kingdom of Jerusalem Presented to Joshua Prawer* (Jerusalem, 1982), 205–13; Prawer, *His. roy.* ii. 43–4.

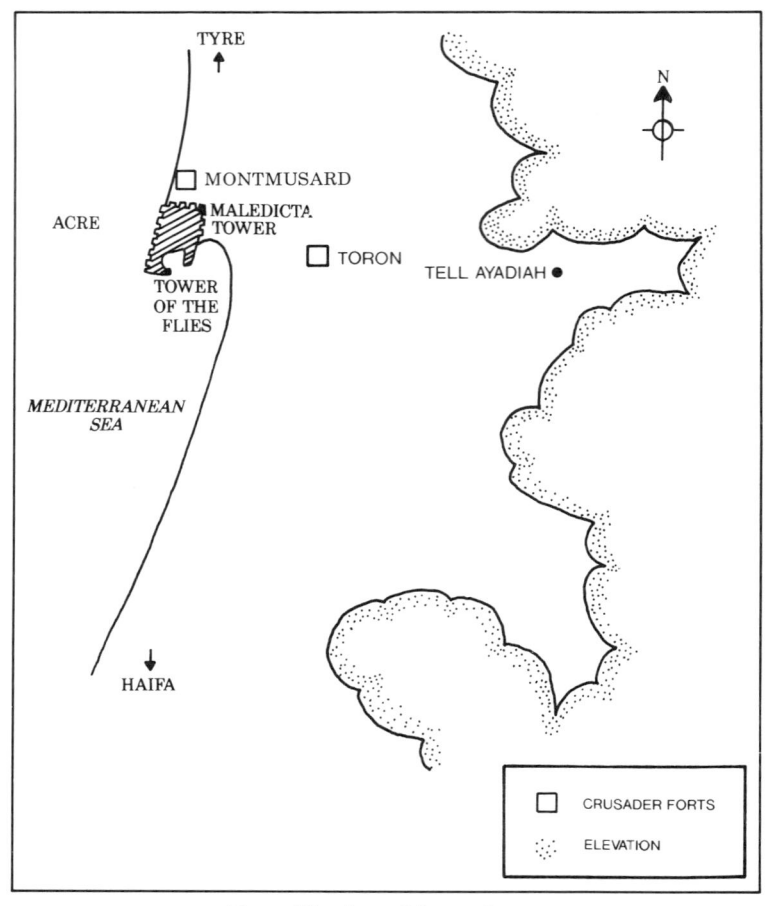

MAP 3. The siege of Acre, 1189–1191

inforcements, which began arriving in September 1189.[80] As we have seen, Antioch's besiegers were vexed by supply difficulties, a vigorous garrison, and attacks by Muslim relieving forces. Those who blockaded Acre also faced logistical problems and active, well-armed

---

[80] That these men joined a major siege begun by Guy of Lusignan influenced his attempts to reassert his own authority after the disaster at Hattin and the challenge of Conrad of Montferrat.

opponents inside the city. Moreover, Saladin maintained a field force in close proximity to crusader positions for the entirety of the siege.

The blockade of Acre and its besiegers. September 1189 saw the arrival of substantial Christian forces as well as Saladin and his troops, who established positions along a line of low hills to the east of Turon.[81] Major conflicts developed over control of the city's northern approaches near Montmusard. Although established there from the outset of the siege, the crusaders were unable to hold their positions against Tâki-ed-din's attacks of mid-September. While supplies and reinforcements were brought into the city, the crusaders launched an attack on 4 October which succeeded in its initial objective of driving Muslim forces from this area. The crusaders were less fortunate in attacking the centre of Saladin's position at Tell Ayadiys. Despite initial successes, the crusaders were eventually driven back, suffering substantial casualties.[82] However, this defeat did not have a decisive effect on the siege, as the crusaders were able to maintain their positions, including those on the city's northern side.

Because of the vulnerability of their positions to Acre's garrison as well as Saladin's field force, the besiegers began erecting a double line of trenches across Acre's peninsula which isolated their encampments from the city and hinterland. Considerable effort and co-ordination were required to protect the labourers from Muslim attacks and skirmishing while they were digging the fortifications. Nevertheless, strong trench lines were completed by mid-November.[83]

While these earthen fortifications provided security, they also isolated Acre from Saladin's army. From this stage of the siege onwards Acre was subjected to a close landward blockade based on crusader encampments.[84] However, communications between the garrison and Saladin were never completely interdicted. Individual messengers and munitions experts and carriers found their way in and out of the city for most of the siege. Crusaders fishing near Acre

---

[81] For a discussion of the topography of Acre's environs and its influence on events, see Prawer, *His. roy.* ii. 44–7.
[82] Prawer, *His. roy.* ii. 48–50; Smail, *Warfare*, 187–8; Röhricht, 'Belagerung', 493–4.
[83] *Itin.* 316–17; Haymar, *Ex. Acc.*, vv. 93–100, p. cix.
[84] For an account of the disposition of groups around the city in 1190 see Ralph of Diceto, *Ymagines*, 71–2; Prawer, *His. roy.* ii. 58–9.

are reported to have captured individual Muslims attempting to swim naphtha into the city.[85]

More importantly, Acre's besiegers were unable to cut off communications between the city and Muslim coastal bases. Ships which made their way through the crusader blockade brought provisions, war materials, and reinforcements fundamental in sustaining the garrison's ability and will to resist. The hardships of close blockade adversely affected Acre's inhabitants, and particularly in December 1190 when they were near starvation and offered surrender terms. However, the arrival of twenty-five ships on 27 December restored the situation and reinforced Acre's garrison.[86] This pattern, in which the effects of crusader blockade, harassment, and assault were neutralized by Muslim blockade runners, was repeated throughout the siege until June 1191.[87]

Seaborne supplies and personnel were equally important in enabling the crusaders to maintain their positions between Saladin's army and Acre for almost two years. Although prevented from foraging or pasturing their horses in Acre's hinterland, the crusaders were provided with provisions and war materials by Christian shipping from the Mediterranean and north-western Europe.[88] Tyre served as the principal base and staging post for these operations, although Roger of Howden reports that ships also brought men and supplies from Apulia.[89]

While a substantial force was sustained by sea during the sailing season, the dwindling of Christian shipping in the Mediterranean during winter caused serious supply problems. Food shortages resulted in high prices, considerable suffering, and famine.[90]

Although supply difficulties affected crusaders during the siege's first winter, famine was most acute during the winter of 1190–1. Efforts were made to alleviate the plight of the poor, including episcopally

[85] *Itin.* 342–3. What is most noteworthy in this story is the importance of naphtha for Acre's defence.
[86] Haymar, *Ex. Acc.*, vv. 225–32, p. cxiv; *Itin.* 319–20; Röhricht, 'Belagerung', 499.
[87] Acre received important sea relief in June, September, and October of 1190 as well as throughout the winter of 1190–1 (Röhricht, 'Belagerung', 499, 501, 505–6, 511).
[88] Prawer notes that Richard I even brought rocks for his artillery from Sicily, this may reflect the origins of Richard's artillery more than a shortage of rocks outside Acre (see below, pp. 228–9; Prawer, *His. roy.* ii. 46 n. 1).
[89] Roger of Howden, *Chronica*, ed. W. Stubbs (4 vols.; RS 51; London, 1868–71), iii. 21.
[90] The privations endured are recounted in some detail (Haymar, *Ex. Acc.*, vv. 150–92, cxi–cxii).

directed collections in the camps. However, substantial relief came only with the arrival of seaborne foodstuffs.[91]

The European response to the call to the Third Crusade provided Acre's besiegers with powerful reinforcements throughout the course of the siege. Although most of these troops arrived by sea, the remnants of Frederick Barbarossa's crusade under Frederick of Swabia reached the Holy Land through the overland route.[92] This somewhat sporadic flow of personnel was one of the main features of the conflict outside Acre and a factor over which Saladin and, for that matter, the besiegers had little control.

The presence of Saladin's field army in the hills around Acre was another major influence on the siege. The blockade imposed on the crusaders aggravated logistical and sanitation problems, and doubtless increased hardship and casualties. The size of Saladin's army varied with the season, and he faced significant difficulties in keeping a powerful force together in the winter of 1190–1.[93] Whatever the military capabilities of this force, it and Saladin's presence indicated a commitment to Acre's defenders and doubtless contributed to the preservation of morale inside the city.

Although unable to attack crusader siege machinery and their support troops directly, Saladin's forces assisted in thwarting major assaults against Acre. When signalled by messengers or drums and incendiaries, Saladin's forces attacked the crusaders' eastern positions to distract them and draw off forces. While it is difficult to assess the importance of these operations, they cannot be discounted among the factors in Acre's ability to resist the ever-growing forces concentrated against it.

Despite their position between a field army and a vigorous garrison, the crusaders never lost control of their main encampments. Saladin was unable or perhaps unwilling to assault crusader fortifications and foot soldiers on a sufficiently large scale to alter the military balance decisively. The crusaders were never successful in shattering Saladin's army, and their major expeditions from the camps suffered

---

[91] Although similar to events at Antioch during the winter of 1097–8, there are no references to mass propitiations or corporate religious ceremonies (see above, p. 35; Haymar, *Ex. Acc.*, vv. 673–720, pp. cxxix–cxxx).

[92] Prawer summarizes the threat Barbarossa's expedition presented to Saladin before the emperor's death in June 1190 (Prawer, *His. roy.* ii. 36–8, 56).

[93] For Saladin's political and military difficulties, see H. A. R. Gibb, 'The Achievement of Saladin', in *Studies in the Civilization of Islam* (London, 1962).

appreciable casualties.[94] Thus beleaguerment and its concomitant attrition were central in the tactics of Saladin, and Acre's besiegers. However, the logistical situation was such that neither side was able to grind down its opponent sufficiently to alter the stalemate. The crusaders attempted to resolve this impasse by overwhelming Acre's defences, and it is to these matters that we now turn.

Crusader assaults and siege machinery. Crusaders attacked Acre with a number of different types of assaults, from simple escalades to those centred on complex machinery. Reference has already been made to the attempt to storm Acre's walls at the siege's outset.[95] A similar large-scale personnel assault, although probably supported by artillery, was made by German crusaders in November 1190. This attack also proved unsuccessful.[96]

Assaults involving machinery required co-ordination and support, not only in guarding devices from garrison sorties, but also in arranging for the security of the camps from Saladin's army. Major devices were built from the resources of important commanders and their contingents. While it is not clear whether each contingent constructed its own devices or hired experts present in the camps, the central importance of sufficient finance in bringing machinery against Acre's fortifications is paramount.

The first such major assault involved the attack of three mobile wall-dominating siege towers in April–May 1190. Although the construction of these devices had commenced soon after the siege's beginning, the towers and their approaches across Acre's ditch were not prepared until this time.[97]

It is difficult to establish from Christian sources which groups of crusaders were responsible for these devices. The *Itinerarium peregrinorum* reports that towers were built by Louis of Thuringia, the Genoese, and communally by the rest of the besiegers.[98] According to Ambroise, Guy of Lusignan and Conrad of Montferrat, together with the Genoese, were responsible for the towers.[99] A charter of Guy of Lusignan and Sibylla dated 4 May 1190 and probably connected with

[94] For the battle of 4 October, see above, p. 217; crusader foot soldiers raided Muslim camps in July 1190 but were eventually defeated; a provision gathering expedition of November 1190 returned intact but with few provisions (Röhricht, *Geschichte*, 523–4, 536; Prawer, *His. roy.* 61–3).
[95] See above, p. 215.   [96] Haymar, *Ex. Acc.*, vv. 537–47, pp. cxxiv–cxxv.
[97] Ibid., vv. 101–4, p. cix.   [98] *Itin.* 325.
[99] Ambroise, *Est.*, vv. 3401–9, col. 91.

this attack rewarded the Genoese for their efforts at Acre with a confirmation of earlier commercial privileges.[100] Moreover, these descriptions may refer to which groups manned and paid for devices, and not necessarily who actually provided materials or constructed them.[101] However, it is likely that Louis and the Genoese were closely associated with at least one tower, and the latter may have been involved in the construction of others. More importantly, the scale of the assaults and the necessity of defending the trenches against Saladin involved most of the besieging force.

Whoever was responsible for these towers, they were not described in accurate detail by crusader sources. Although the author of the *Itinerarium* used phrases from Vegetius' description of a siege tower in composing his account of these devices, little specific information emerges except that these towers overtopped the walls, were mobile, and armoured with rope against artillery.[102] Beha-ed-din gives the most illuminating account of these towers, emphasizing the threat they posed to Acre's defenders. Each tower dominated the walls, and was covered with vinegar-soaked hides as protection against incendiaries. Each is reported to have had a crew of five hundred men, which must have included all those involved in propelling the devices, as well as archers, crossbowmen, and assault troops. The towers were roofed, and each had a small rock-thrower mounted on its summit.[103]

Although it is not clear which section of Acre's walls was attacked, the three towers and supporting artillery were brought across the city's ditch near the end of April 1190. The threat of imminent assault and its inability to destroy these towers prompted the garrison to offer surrender terms, which were rejected by the crusaders. However, the crusaders were unable to exploit their advantage before Muslim pyrotechnical skill destroyed the structures. According to some Christian sources, the towers were burnt by Acre's defenders, while Saladin's army distracted the crusaders with a diversionary attack.[104] Haymar, however, reports that, after dominating the walls, these towers were burnt by Acre's garrison, in an action which did not coincide

[100] *CDRG* ii, no. 196, pp. 374–6.
[101] Röhricht noted that the wood for these towers came from Italy (Röhricht, 'Belagerung', 497).
[102] *Itin.* 325–6, 326 n. 5; Ambroise, *Est.*, vv. 3401–9, col. 91.
[103] Beha-ed-din, *Saladin*, 178; Ibn al-Athir reports that the towers were composed of five stages and some sixty cubits—approximately a hundred feet in height (Ibn al-Athir, *Extrait de Kamel-Altevarykh* (RHC Orient. 2), 18).
[104] Ambroise, *Est.*, vv. 3410–32, cols. 91–2; *Itin.* 326.

precisely with an external attack.[105] Beha-ed-din relates a story of a certain Damascene coppersmith who claimed to be able to destroy these towers, despite the failure of professional naphtha-men. The man made his way into Acre from Saladin's camp and soaked combustibles in a naphtha solution which may have included elements of gunpowder. These materials were then cast on to the towers by artillery before incendiaries were projected on to the devices. The towers were ignited and destroyed during Saladin's demonstrations of 5 May.[106]

Before discussing subsequent attacks and machinery, some reflection on this large-scale assault is appropriate. Despite the outcome of events, these towers dominated a section of Acre's defences and came close to overwhelming them, and also prompted negotiations concerning a surrender. Although a lack of detailed crusader accounts makes it difficult to verify Muslim descriptions precisely, it is clear that these towers were large and formidable.[107] In any case, the scale of the assault, particularly in terms of the number of siege towers brought against the same fortification at the same time, compares favourably with earlier sieges which we have examined. This assault on Acre represents the single greatest concentration of such devices and their supporting machinery in the period under discussion.

The siege of Acre from August 1189 to May 1190 may be compared in siege technique and style of assault with a number of operations already examined. The closest parallel is with the siege of Tyre in 1112, in which Baldwin I's forces attacked with two siege towers after wintering outside the city protected by earthen fortifications against a Damascene field force. However, when Baldwin's second tower was destroyed, he abandoned his siege.[108] Acre's besiegers, in contrast, maintained their blockade and made further attempts to neutralize the city's fortifications.

Although Acre's besiegers and Saladin's forces battled during the summer of 1190, the next major land assault based around large siege machinery was organized in September of 1190, after the arrival on 27 July of Henry of Champagne, who was placed in charge of operations.[109] Instead of dominating and seizing Acre's walls, the crusaders

[105] Haymar, *Ex. Acc.*, vv. 201–16, p. cxiii.
[106] Beha-ed-din, *Saladin*, 178–9; Röhricht, 'Belagerung', 497–8. Hill and al-Hassan suggest that an early form of gunpowder could have been utilized by Saladin's forces (Hill and al-Hassan, *Islamic Technology*, 109–12).
[107] Beha-ed-din's description may be compared with accounts of Barbarossa's tower at Crema regarding the size of the crew (see above, pp. 138–9).
[108] See above, pp. 79–82.   [109] Röhricht, 'Belagerung', 501–2.

strove to batter them down. The centre-piece of this assault was an iron-tipped battering-ram built from a mast from the resources of the archbishop of Besançon. Although implicit in crusader accounts, Beha-ed-din makes it clear that the ram and its crews were protected by an armoured roof.[110] The *Itinerarium* refers to a second ram built by Henry of Champagne but this is not supported by Haymar or Beha-ed-din, and such a device does not appear subsequently in the *Itinerarium*'s account.[111] However, there are references to supporting roofs and devices, which may have included rams built by Henry and perhaps others. Whatever the number or type of supporting machinery, the attackers were unable to protect this principal battering-ram once it had been moved across Acre's ditch and up to an attacking position underneath its walls. The city's defenders heaped combustibles including naphtha on the device and then ignited their inflammables on the ram. Although the shelter's crew abandoned their device, crusaders attempted to douse the blaze, despite casualties inflicted by garrison missile fire. However, the ram burnt and the assault was abandoned on 4 October.[112]

Not all crusader assaults were directed against Acre's land walls. Shortly before the archbishop of Besançon's attack, an attempt was made to seize the Tower of the Flies, which controlled Acre's harbour and consequently sea communications. On 24 September Pisan sailors and combatants approached the Tower of the Flies in a siege tower built on to two galleys lashed together. The floating tower was armoured with vinegar-soaked hides and presumably built around the masts of the two galleys. The tower was also equipped with a ladder to permit an escalade on to the Muslim fortifications. Although the Pisans manœuvred their ships to be able to dispute control of the Muslim tower, they were unable to dominate it. The Muslim tower's garrison, supported by galleys from the city, waged a protracted conflict with missile weapons and hand-cast stones and beams. Despite its armouring, the Pisan structure was eventually burnt with naphtha projected from the Tower of the Flies. A Pisan attempt at the same time to burn Muslim vessels inside the harbour by means of a fire-ship was also thwarted.[113] Conrad of Montferrat tried to

[110] Beha-ed-din, *Saladin*, 215–18.
[111] This account of the ram is drawn from Isidore of Seville's *Etymologies* (*Itin*. 346–7, 347 n. 1).
[112] Haymar, *Ex. Acc.*, vv. 428–60, pp. cxx–cxxi; Beha-ed-din, *Saladin*, 215–18; Ambroise, *Est.*, vv. 3825–88, cols. 102–4.
[113] Haymar, *Ex. Acc.*, vv. 405–20, p. cxx; Ambroise, *Est.*, vv. 3771–818, cols. 101–2; *Itin*. 345–6; Beha-ed-din, *Saladin*, 210–12; Röhricht, 'Belagerung', 503.

attack this tower by sea in January 1191, but this effort proved ineffectual.[114]

Although the events described above were the main attacks against Acre of 1190, there was considerable skirmishing and desultory fighting throughout the siege. Haymar reports a garrison sally made against a mound built up between the crusaders' camps and the city in January 1191. Presumably this mound had been constructed in order to facilitate missile weapon attacks on the city. Neither the combat which developed from this sortie nor crusader activities based on the mound proved decisive to the course of the siege.[115]

The garrison's ability to resist and the difficulties encountered by the crusaders' logistical problems are well illustrated by events of early January 1191. A substantial breach occurred in Acre's walls on the night of 5 January after a severe winter storm. The effects of the rainstorms on Acre's approaches and the poor condition of the crusaders' underfed men and horses delayed a rapid move against the breach. The crusaders remained unable to exploit their fortuitous advantage, despite the repeated assaults which developed, and Acre's defenders were able to restore the situation after a few days by building a retaining wall.[116]

Despite a blockade of fifteen months, and a number of determined assaults, including several supported by sophisticated and expensive machinery, the deadlock outside Acre remained unbroken. Neither side was able to impose a decisive military solution, and neither had succumbed to a war of attrition. Although crusader supplies improved with the spring sailing season, the military impasse remained. The fresh contingents of Philip II and Richard I, with material, manpower, and financial resources, altered the balance and enabled besieging forces to wage a series of attacks which finally compelled the garrison's surrender.

The final phase of the siege of Acre began with the arrival of Philip II and his troops, who established their camps opposite the Maledicta Tower on 20 April 1191.[117] Although Philip's forces made some efforts against Acre, large-scale operations commenced only after Richard I's arrival on 8 June. Philip's men filled part of Acre's ditch and made an artillery attack, probably on 30 May. However, this attack was unsuc-

[114] Haymar, *Ex. Acc.*, vv. 605–16, pp. cxxvi–cxxvii.
[115] Ibid., vv. 585–92, p. cxxvi.
[116] Ibid., vv. 557–84, pp. cxxv–cxxvi; Beha-ed-din, *Saladin*, 236–7; Röhricht, 'Belagerung', 510.
[117] Röhricht, *Geschichte*, 547.

cessful and the garrison burnt the French devices. Meanwhile, Saladin moved his main camps down to Tell Ayadiah in order to support the garrison more closely.[118]

Although Richard and many of his followers arrived on 8 June, some troops and his siege engines were delayed at Tyre. This and Richard's illness retarded English siege operations.[119]

It should be noted that Beha-ed-din describes the attack and destruction of a crusader siege tower on 14 June, which is not confirmed by any crusader account. Richard is reported to have built a wooden tower, but it is not clear if it was mobile or a defensive structure similar to the one erected at Messina in 1190. Crusader accounts of military operations in 1191 do not reflect the presence of a siege tower. That this event is not mentioned by crusader sources which usually report Christian failures argues that Beha-ed-din's description should be discounted.[120]

Soon after their arrival at Acre, Richard and Philip rivalled each other in bidding for the service of men who had preceded them. Roger of Howden reports that the Pisans and Genoese attempted to associate themselves with Richard. Although he accepted the Pisans, whose commercial privileges he recognized, he declined to ally with the Genoese, who were already associated with Philip.[121] The Pisans and Richard's forces co-operated closely throughout the remainder of the siege.

Despite Richard's lack of preparedness, Philip's forces assaulted Acre in mid-June with artillery, having already filled portions of Acre's ditch in preparation for assault. In response, Saladin's forces attacked a section of the crusaders' eastern trenches which were stoutly defended by Geoffrey of Lusignan. On 17 June Philip's artillery and the portable shields and armoured roofs which protected artillery crews and labourers were burnt. Roger of Howden reports that Philip's devices were inadequately protected, because Richard had hired *servientes*

---

[118] Haymar, *Ex. Acc.*, vv. 733–40; Beha-ed-din, *Saladin*, 240; Rigord, *Gesta Philippi Augusti: Œuvres de Rigord et Guillaume le Breton*, ed. H. F. Delaborde (Société de l'Histoire de France; Paris, 1882), i. 108–9; Röhricht, *Geschichte*, 547–9.

[119] Ambroise, *Est.*, vv. 4615–19, col. 123.

[120] Beha-ed-din, *Saladin*, 251; Ambroise, *Est.*, vv. 4781–6, col. 128; Röhricht, 'Belagerung', 514, *Geschichte*, 555–6. For the structure at Messina, see Richard of Devizes, *Chronicle*, ed. J. T. Appleby (Oxford, 1963), 25.

[121] Philip recognized Genoese commercial privileges in his arrangement for the transportation of his followers from Genoa to Acre in 1190 (*CDRG* ii, no. 198, pp. 378–80; Roger of Howden, *Chron*. iii. 113; Ambroise, *Est.*, vv. 4570–99, col. 120).

responsible for guarding machinery whom Philip had dismissed from his service.[122]

Regardless of these set-backs, the besiegers were in a dominant position with the arrival of the remainder of Richard's forces. Moreover, the crusaders enjoyed a stroke of good fortune during the period of Richard's arrival which had an important bearing on the course of the siege. The celebrated interception and sinking of a Muslim supply ship outside Acre by Richard's ships deprived the garrison of much-needed supplies and above all reinforcements.[123] Thus the crusaders achieved their greatest concentration of men and materials at a time when Acre's defenders were deprived of relief.

The besiegers exploited their advantage with a policy of close investment and frequent assault. Although primarily directed at battering down or sapping Acre's defences, these attacks and the personnel assaults which followed them enabled numerically superior crusaders to wage an effective battle of attrition against the garrison. While the crusaders were not completely successful in destroying Acre's defences, they wore down the garrison and compelled surrender.

The crusader offensive centred on demolishing the Maledicta Tower by undermining and artillery. These efforts were supported by archers and crossbowmen who harassed defenders active on the ramparts. Richard and Philip each built armoured roofs principally intended to protect archers and crossbowmen, including the kings themselves, in range of the walls. Although Philip's roof was destroyed by incendiaries, arrows and bolts continued to be shot at Acre's defenders for the duration of the siege.[124]

The crusaders deployed large and effective stone-throwers against the Maledicta Tower and its adjacent walls. A continual bombardment was maintained by artillery weapons built and operated by Richard—who had also inherited the Count of Flanders's engines after his death—Philip, Hugh of Burgundy, the Pisans, Hospitallers, and Templars.[125] Philip's chief lever stone-thrower was called Male Veisine and was shot at by a garrison device known as Male Cosine, which knocked out its opponent on a number of occasions. Philip's

[122] Ambroise terms shields and armoured roofs *cerclies* throughout his work (Ambroise, *Est.*, vv. 4620–92, cols. 123–5; Roger of Howden, *Chron.* iii. 113; Rigord, *Gesta Phil.* 109; Beha-ed-din, *Saladin*, 251–2).
[123] Röhricht, *Geschichte*, 551–2.
[124] Ambroise, *Est.*, vv. 4815–40, 4930–40, cols. 129, 132. Ambroise terms both of these roofs *cercloie*.
[125] Ambroise, *Est.*, vv. 4737–92, cols. 127–8.

crewmen, however, were able to re-erect their piece and continue bombardment.[126] Presumably the Muslim rock-thrower broke down the stand of Philip's weapon without damaging the lever beam itself and the besiegers simply repaired or replaced the stand and erected their device in a different location.

One artillery device known as the *Periere Deu* was operated communally by the besiegers and financed through pious donations collected by a priest who preached during the weapon's operation. This device not only shattered a section of Acre's walls, but also provided a means for crewmen and non-military pilgrims to earn money necessary in purchasing provisions.[127]

The effectiveness of this artillery in the context of Latin siege warfare in the twelfth century is noteworthy. Beha-ed-din reports that crusader artillery battered sections of Acre's walls down to only the height of a man.[128] Ambroise reports considerable artillery damage at this time, and Roger of Howden tells of a breach made as a result of this bombardment which Philip's forces were unable to seize because of determined resistance.[129] Although it is not possible to quantify the damage done precisely because of the lack of evidence as well as the cumulative effects of crusader operations over almost two years, it is clear that crusader artillery seriously damaged Acre's defences during this stage of the siege.[130] The degree of damage reported, as well as the attention paid to artillery in the narrative sources, has led to the suggestion that these pieces may have included counterweight devices.[131] It is possible that combination counterweight-and-traction weapons which continued to be termed *petrariae* were deployed. Appealing as these hypotheses are, they cannot be conclusively demonstrated from available evidence, and crusader artillery at Acre may have been simply large and effective traction pieces. However, the importance of artillery in destroying fortifications at the siege of Acre is clear.

How this artillery came to be employed at Acre is equally problematic, and it is appropriate to discuss the possible origins of some of these artillery pieces. There is evidence to suggest that Richard may have obtained some of his artillery from Sicily during his stay there.

[126] Ibid., vv. 4745–52, col. 127.   [127] Ibid., vv. 4759–67, col. 127.
[128] Beha-ed-din, *Saladin*, 256.
[129] Roger of Howden, *Chron*. iii. 116–17.
[130] This may be contrasted with the attack on Jerusalem of 1099, in which the defenders protected their upper works with various shock-absorbing materials (see above, p. 56).
[131] Huuri, 'Geschichte', 94.

Roger of Howden reports that Richard procured artillery and other devices in the winter of 1190 after his treaty with Tancred.[132] Ambroise mentions that some particularly effective rocks cast by Richard's artillery had been selected and transported from Sicily.[133] While it is possible that Philip may have obtained artillery from Sicily, this is not reflected in narrative accounts. However, at the least it is likely that Richard drew on Sicilian resources for some of his artillery. Whatever the origins, it is clear that Acre's besiegers deployed large and effective artillery pieces in their operations.

Although important, heavy artillery was not the only component of the crusader assault. Much of the main ditch masking the city's walls was filled during June 1191. Although sections had been filled and levelled for earlier assaults, the task was completed during this period, probably as a result of the intensive effort of many elements of the besieging force.[134] Filling the ditch and making Acre's walls more accessible rendered them more vulnerable to escalade. While the steady damage inflicted by crusader artillery increased the vulnerability of certain sectors, the accessibility of fortifications to storming parties was a major threat to Acre's defence.[135] While the walls were not carried, the requirements of constant vigilance contributed to weakening the garrison's ability to resist.

As some of the besiegers laboured to make the circuit of Acre's walls more vulnerable, others concentrated on destroying the Maledicta Tower and its adjacent fortifications. The artillery deployment described above was made in conjunction with mining and sapping operations against this sector of Acre's defences. In June crews employed by the king of England sapped and brought down a section of walls which assault troops had been unable to seize.[136] Richard's men also undermined part of the Maledicta Tower and its adjoining walls on the night of 5 July, sections of which collapsed the following day.[137]

Philip's men brought down an important section of the walls adjacent to the Maledicta Tower on 2 July, which French forces unsuc-

[132] 'Interim Ricardus rex Angliae fecit parari perarias, et alias machinas...' (Roger of Howden, *Chron.* iii. 72).
[133] The Normans transported artillery ammunition by sea for their attack on Alexandria in 1174 (see above, pp. 120–1; Ambroise, *Est.*, vv. 4793–800, col. 128).
[134] Roger of Howden, *Chron.* iii. 116.
[135] Roger of Howden reports that the garrison offered what proved unacceptable surrender terms as a result of the ditch's filling (ibid.).
[136] Ibid.
[137] Ibid. 118; Ambroise, *Est.*, vv. 4909–14, 4943–93, cols. 131–2; Haymar, *Ex. Acc.*, vv. 809–15, p. cxxxiii.

cessfully attacked the following day. Miners of the king of France also participated in the operations, which left the Maledicta Tower indefensible.[138] Garrison countermeasures included missile weapons, incendiary attacks on armoured roofs, and at least one countermine. Ambroise records a story of Christian captives, forced to dig a countermine, who encountered their co-religionists in a shaft. Acre's defenders subsequently collapsed the mine and countermine.[139] While some crews were protected by armoured roofs, the crusaders' control of Acre's approaches and ditch, as well as their numerical superiority, were the major factors which facilitated these operations.

Despite successes in toppling Acre's defences, the crusaders were unable to overwhelm them. Garrison resistance and the difficulties of negotiating rubble-filled breaches stemmed crusader assaults against them. A major French assault against their breach next to the Maledicta Tower was repulsed on 3 July and again on the seventh.[140] Richard's forces, which had failed to exploit the gap made in June, tried unsuccessfully to seize the ruins of the Maledicta Tower on 6 July. Richard's troops and the Pisans made successive assaults against the same sector on 11 July, while the rest of the besiegers were dining at their customary time. Although the attackers doubtless hoped that the garrison would be unprepared, they were repulsed.[141]

Saladin responded to these onslaughts against Acre by attacking the crusaders' eastern entrenchments. As his army had been significantly reinforced on 25 June, he was able to deliver attacks in support of Acre's garrison. However, crusader foot soldiers, archers, and crossbowmen held their positions, thwarting Saladin's attempts to divert the attackers. In order to exploit the large breach of 3 July effectively, the crusaders divided the responsibilities of attacking Acre and holding the trenches between the kings of France and England. Although Philip's troops failed to penetrate Acre's defences, forces under Richard's direction maintained their positions successfully against Saladin's determined attacks.[142] Despite repeated attacks on crusader positions, launched in response to signals from Acre and

---

[138] Roger of Howden, *Chron.* iii. 117; Ambroise, *Est.*, vv. 4867–908, cols. 130–1; Haymar, *Ex. Acc.*, vv. 789–815, p. cxxxiii; Rigord, *Gesta Phil.* 115.
[139] Ambroise, *Est.*, vv. 4909–26, cols. 131–2.
[140] Roger of Howden, *Chron.* iii. 117, 119; Ambroise, *Est.*, vv. 4862–908, cols. 130–1; Haymar, *Ex. Acc.*, vv. 789–808, p. cxxxiii.
[141] Ambroise, *Est.*, vv. 4984–5040, cols. 133–5; Roger of Howden, *Chron.* iii. 120.
[142] Beha-ed-din, *Saladin*, 158–61.

Saladin's personal commitment to these assaults, his field army was unable to alter the balance turning against Acre's defenders.

While the Anglo-Pisan assault of 11 July failed to seize Acre's defences, the garrison was finally rendered incapable of further effective resistance. The cumulative effects of personnel assaults, the destruction of Acre's north-eastern fortifications, and Saladin's evident inability to bring relief compelled defenders to seek terms.[143] Although complicated negotiations had been conducted intermittently since early July, Saladin had little choice but to acquiesce to a negotiated crusader entry into Acre on 12 July 1191.[144]

In understanding the technique employed in this phase of the siege, the importance of the battle of attrition waged during attacks on Acre's fortifications is manifest. The degree to which men and resources were committed to this action, as well as the intensity of the fighting outside Acre, may be illustrated from one of the incidents of the sustained attack on the Maledicta Tower in July 1191. In order to accelerate the tower's collapse, Richard offered at first two, then three, and finally four *besants* for each stone extracted and brought back from the fortifications. As Richard's increased reward suggests, this activity was dangerous, and Ambroise describes the difficulties involved and losses sustained by the *serjanz* who participated.[145] It is not known how much this endeavour contributed to weakening the Maledicta Tower, or how many survived to claim their reward. However, the activities of these men enabled their comrades to inflict casualties on the garrison. For, in order to shoot at crusaders whose courage and good fortune had carried them to the foundation-stones of Acre's walls, defenders had to lean out over their ramparts, exposing themselves to archery, artillery, and crossbow fire, which took its toll. This story illustrates not only the intensity of the crusaders' assault but also their willingness to suffer casualties in order to inflict them on their opponents. It should be noted that Richard I's strategic insight and skill in military leadership has been rehabilitated by scholars, who have rightly seen his generalship as

---

[143] In his account of their decision to seek terms, Beha-ed-din emphasizes a collapse of morale among defenders due to exhaustion and to the apparent hopelessness of their situation (Beha-ed-din, *Saladin*, 261–5).

[144] For a discussion of the problems regarding claims to Acre and its spoils, including the episode of Leopold of Austria's banner, see J. Gillingham, *Richard the Lionheart* (London, 1978), 176–8; Röhricht, *Geschichte*, 563–5; Prawer, *His. roy.* ii. 66–7.

[145] Ambroise reports that even well-armoured Muslims were killed by the heavy quarrels projected by *arbaleste a tur* set up against this sector (Ambroise, *Est.*, vv. 4948–89, cols. 132–3).

more than an extravagant display of the chivalric ethos.[146] The effective battle of attrition at the core of Richard I's operations at Acre is another illustration of his acute military judgement. That his men waged this battle is a testament not only to their own courage, but also to Richard I's abilities as a leader.

These concerns should not detract from the importance of artillery and mining attacks in debilitating Acre's defences. These operations were components of an offensive which overcame defenders by subjecting them to every pressure which attackers were able to exert. In this manner, the besiegers exploited the advantages in men and resources which the arrival of Philip's and Richard's contingents brought to end the military deadlock at Acre.

The final and decisive period of the siege involved sustained attacks by men and machines, which are reported in some detail by eyewitness accounts. However, we should not let the achievements of Philip's and Richard's contingents overshadow a more general understanding of the siege warfare waged at Acre in relation to Latin operations which we have already examined.

As in earlier Latin siege operations in the eastern Mediterranean, the attack on Acre was based on close investment and the frequent assaults such an establishment made possible. Although thwarted by Saladin's army, Acre's sea communications, and Muslim pyrotechnics, the besiegers were not contented with a passive blockade, but assaulted the city when resources, manpower, and weather conditions permitted.

The organization of operations, and particularly those based on machinery, was similar to that of the sieges of the First Crusade. Machines were built for and manned by the troops of major leaders and groups which possessed the finance necessary for construction. Problems of suitable materials and technical expertise involved in the production of siege engines in the First Crusade were of little concern at Acre. Timber, fastenings, and other materials were provided by fleets and ships which brought men and supplies to the camps. Christian sources say little about specific engineers or craftsmen, but it is clear that such expertise was available from Italian maritime contingents and perhaps others. Moreover, the siege doubtless attracted itinerant

[146] J. O. Prestwich, 'Richard Cœur de Lion: Rex Bellicosus, *Accademia Nazionale dei Lincei: Problemi attuali di scienza e di cultura*, 253 (1981), 3–15; J. B. Gillingham, 'Richard I and the Science of War in the Middle Ages', in J. B. Gillingham and J. C. Holt (eds.), *War and Government in the Middle Ages: Essays in Honour of J. O. Prestwich* (Cambridge, 1984), 78–91.

experts whose skills were available to those who could afford them.[147] The principal factor necessary in bringing men and materials together for the construction and operation of siege machinery was finance.

Although the financial and personnel resources of major leaders were central in bringing siege engines against Acre's defences, a number of important tasks were undertaken by the whole army. Major assaults required co-ordination among crusaders in order to maintain the security of their camps, while portions of the besiegers were involved in assault and protecting machinery situated near Acre's walls. Richard and Philip divided these responsibilities in June and July 1191, and other large-scale operations, such as the siege-tower attack of 1190, must have involved similar arrangements.

Moreover, operations outside Acre required considerable manual labour, which was carried out communally by the besiegers. The crusader entrenchment, filling portions of the civic ditch necessary for the siege-tower attack of 1190, and the completion of this task in June 1191 all involved such effort. As in the First Crusade, much of this labour was performed without payment by non-military personnel present in the camps; and, as in the First Crusade, these besiegers suffered most during famine and periods of high food prices. The contributions, sacrifices, and suffering of humble pilgrims are recorded in crusader and Arabic accounts of the siege of Acre. The story of the dying woman who requested that her body be cast into the city ditch which she had laboured so hard to fill is the most well-known such story.[148] These crusaders also operated artillery, extracted stones from fortifications, and battled with their Muslim opponents. Beha-ed-din recounts the exploits of a woman equipped with a powerful bow who defended crusader positions valiantly during the Muslim attacks of 1191.[149] While the Third Crusade has been described as the 'Crusade of Kings', the role of humbler folk in the siege of Acre should not be overlooked.

Crusader poliorcetics at Acre involved techniques and machinery characteristic of twelfth-century operations and also some which foreshadow developments important for later periods. The attack of three siege towers in April–May 1190 illustrates continuity with methods

[147] Ambroise refers to a large number of engineers employed by Saladin to fortify Acre and prepare artillery and other machines; this passage serves to emphasize the strength of Acre's defences, and is primarily interesting as a stylistic device (Ambroise, *Est.*, vv. 3205–22, col. 86).
[148] Ambroise, *Est.*, vv. 3625–60, cols. 97–8; *Itin.* 339–40.
[149] Beha-ed-din, *Saladin*, 258–9.

employed around the Mediterranean throughout the twelfth century. Moreover, the number of these devices brought against Acre's defences makes the scale of this assault noteworthy. These towers were among the largest and best equipped of the period and managed to dominate a section of Acre's walls despite counter-attacks. Although ultimately destroyed, these towers almost broke the deadlock at Acre.

In some regards this assault marks the end of the pre-eminence of these devices in Latin siege technique. Although towers were on occasion built at later sieges, they were not central as they had been in twelfth-century operations. Although a number of diverse factors influenced the decline of their importance, including the targets besieged and the availability of other means of attack, it should be noted that these devices, and assaults based around them, were well suited to forces deployed against coastal cities and operating in conjunction with Italian maritime forces which provided their chief logistical support.

Important sections of Acre's walls were brought down by sapping during the final phase of the siege. Yet, as in operations at Lisbon, Ascalon, and earlier in the siege of Acre, attackers were unable to exploit breaches in the face of determined and numerous defenders capable of building emergency retaining walls. Although undermining was an important tool of siegecraft, certain types of targets were able to neutralize the effects of successful mining operations. In this regard, events at Acre illustrate not only the effectiveness of sapping in breaching defences, but also some of the difficulties well-defended cities presented to those who employed this technique.

The Pisan assault on the Tower of Flies of September 1190 was one of the earliest examples of a method of attack important in notable sieges of the thirteenth century. Although eventually destroyed, the Pisan floating siege tower enabled troops to attack strong fortifications directly from ships. A similar device was employed successfully against the Chain Tower in the siege of Damietta in 1218.[150]

An ability to attack directly from the sea was important in the capture of Constantinople in April 1204, as this allowed assaults to be delivered simultaneously against the city's land and sea walls. Armoured catwalks permitted assault troops to be placed on the city's sea-walls

---

[150] Oliver of Paderborn describes this floating siege tower and its attack (Oliver of Paderborn, *Historia Damiatina: Die Schriften des Kölner Domscholasters, späteren Bischofs von Paderborn und Kardinalbischofs von S. Sabina Oliverius*, ed. H. Hoogeweg (Tübingen, 1894), i. 181–6.

from the masts of the larger attacking ships. Although this means of attack belongs more to the realm of nautical technology than poliorcetics, it was nevertheless a factor in the crusaders' successful attack.[151] While unsuccessful in its attack, the Pisan floating siege tower at Acre was one of the earliest adaptations of nautical and siege technology to a specific military problem.

The effects of artillery in destroying fortifications provide one of the more notable aspects of the final period of the siege of Acre. Artillery weapons played a role in earlier periods of the siege. Crusader *perariae* located behind earthen fortifications improved the security of their positions after entrenchment in October–November 1189.[152] Presumably crusaders took advantage of the high trajectory of lever artillery to cast rocks over their embankments from weapons placed behind them. This location protected devices and crews from direct missile fire from their opponents.

However, in attacking Acre's fortifications, artillery devices were deployed in closer proximity to the city in positions more vulnerable to garrison missile fire and counter-attack. Before the advent of Philip and Richard, a number of artillery pieces and weapons were brought against the city, some of which were burnt by sally parties. Henry of Champagne is reported to have deployed an artillery weapon costing 1,500 gold pieces, which was burnt on 17 September 1190.[153]

The most sustained artillery attack was waged in June–July 1191. Doubtless the numerical strength of the besiegers, which enabled them to protect their devices, was a factor in the success of this attack, and this permitted the concentration of a number of artillery pieces against a section of the walls which effectively destroyed their defences. While the evidence does not permit certainty, it is a plausible hypothesis that some of these weapons were counterweight or counterweight-and-traction weapons. Whatever their size or destructive power, the bombardment with these devices contributed significantly to the debilitation of Acre's defences. This differs from the role of artillery in earlier eastern Mediterranean siege operations, in which rock-throwers primarily supported escalades or siege-tower assaults by harassing defenders and neutralizing their artillery. Moreover, the effectiveness of artillery at Acre anticipates developments in thirteenth-century polior-

[151] Among crusader sources, Robert of Clari gives the most detailed account of this phase of the attack. For a discussion of the assault, see D. E. Queller, *The Fourth Crusade* (New York and Leicester, 1978), 210–14; Robert of Clari, *La Conquête de Constantinople*, ed. P. Lauer (Paris, 1924), 71–3.
[152] Roger of Howden, *Chron.* iii. 22.   [153] Beha-ed-din, *Saladin*, 203.

cetics, in which bombardment played an important role. Heavy artillery was effective not only against Levantine cities, but also against less accessible positions and ones less vulnerable to other methods of attack. In this regard, the siege of Acre can be seen as ushering in the great age of pre-gunpowder artillery in the medieval west.

The role of Italian maritime powers in the siege of Acre and the nature of seaborne siege warfare illustrate earlier patterns and noteworthy developments. Italian maritime forces played a prominent role throughout the siege, from the initial establishment to the final assault before the city's surrender. Genoese, Pisan, and Venetian ships transported men and materials to the scene of operations. Contingents included not only craftsmen and engineers, but also combatants involved in many operations against the city.

Viewed in its broadest perspective, the siege began with the association of a Pisan fleet and one claimant to Jerusalem's crown in the capture of an important city of the eastern Mediterranean seaboard. The capture of Acre became an international event, involving military and naval forces from many areas of the Latin west. The participation of these diverse contingents, and particularly those from northern Europe which possessed their own sea power, altered the context of this siege from those of earlier operations along the Syro-Palestinian coast.

Acre's capitulation completed the first stage of the Third Crusade. Those who remained in the Holy Land, largely under the leadership of Richard I, re-established Latin control of the littoral in a campaign which has been a measure of military achievement and generalship in the High Middle Ages. The centre-piece of this endeavour was the decisive Latin victory at Arsuf, which led to the establishment of the Second Kingdom of Jerusalem. It may be noted that this late twelfth-century conquest of the coast, achieved rapidly by a relatively cohesive force supported by substantial sea power, differed markedly from the initial Latin acquisition of the area.

The political situation in Outremer in the late twelfth century, the involvement of the kings of France and England in the affairs of the crusader states, as well as the achievement of Saladin made for a different situation from that of the initial period of the conquest of the littoral. Italian maritime strength provided substantial assistance in both campaigns. Yet, however important Italian aid in the early stages of the siege of Acre, it does not compare with the significance of these states in the early twelfth century. Although Philip II transported his men with considerable Genoese assistance, Apulia and Sicily emerged as significant staging posts during the Third Crusade. Moreover, these areas became increasingly important in thirteenth-century crusading ventures. Furthermore, Cyprus, which was taken in the course of Richard's journey to Acre, developed into a central base for subsequent Latin operations in Palestine and Egypt.

These factors, as well as the participation of other seafaring peoples in crusading siege warfare, altered the nature of Christian naval power in the Mediterranean. Consequently, the relationships between sea power and siege warfare of twelfth-century Latin expansion were altered. This coincided approximately with developments in poliorcetics and naval technology which promoted methods of reducing coastal fortifications which differed from those of primary importance in the twelfth century. In this regard, the siege of Acre may be seen to close one chapter of seaborne siege warfare and introduce another.

7

TOWARDS CONCLUSIONS

IN examining Latin siege warfare in the twelfth century we have maintained a close focus on operations, emphasizing how the employment of techniques and machinery was related to the resources and capabilities of besiegers, as well as to particular military conditions. In drawing together some of the strands we have followed, it is appropriate to address questions concerning the spread and development of siegecraft and technology. Such a discussion may highlight the factors through which this took place. Moreover, it will allow reflection upon the relationship between the military experience of the Latin west and this growth and diffusion. Let us first turn to the personnel important in the organizational and technological dimensions of twelfth-century siege warfare.

Experts in Siege Warfare

Rulers

Among those whose expertise in siege warfare was significant in the spread of siege techniques were military and political leaders. Military men who also exercised lordship took particular interest in the details of siege operations for their own or an ally's benefit. Gaston IV of Béarn is one of the most prominent figures of the period. Gaston directed northern building operations at the siege of Jerusalem, and, whatever his knowledge of poliorcetics beforehand, his experiences there made him an expert in the organization of siege-tower assaults as well as the building of these and their supporting devices. Gaston did not remain in Outremer, but returned to Béarn and became involved with the campaigns in the Ebro valley of Alfonso I of Aragon, and particularly with the conquest of Zaragoza, whose lordship he received from the

Aragonese king.[1] Gaston's activities at the siege of Zaragoza are an example of how experience gained in crusading siege warfare was employed in another area of Latin conflict. Gaston brought to the siege of Zaragoza not only his personal experience, but also necessary labourers and craftsmen under his own command. Gaston's activities also provide an illustration of one way in which siege techniques were brought from one area to another.

There were other lords and commanders in twelfth-century Europe with skill at siegecraft which was not a direct result of crusading experience. Geoffrey Plantagenet, count of Anjou, was an experienced besieger and a man noted for skill in carpentry and engineering.[2] Whatever his abilities as a woodworker and builder, his education in the literature of the ancients also assisted him in his siege operations. During the siege of Montreuil-en-Bellay of 1147 Geoffrey consulted a manuscript of Book IV of the *Epitoma rei militaris* of Vegetius to find a solution to problems presented by defensive countermeasures taken against his battering-rams. Although Geoffrey's rams damaged the castle's inner walls by day, the defenders repaired them with massive oak beams by night. Having consulted his text, with the aid of a passing monk of Marmoutier, Geoffrey made an incendiary bomb based on specially prepared highly inflammable oils and had it projected on to the defenders' beams by a lever artillery piece. It may be noted that there is no description of such an incendiary bomb or anything resembling it in Vegetius. However, this does not detract from Geoffrey's reliance on what he believed were Roman traditions, since there were undoubtedly additions to the manuscript. While Murray uses this example to illustrate twelfth-century warriors' respect for Vegetius as well as the literacy of the nobility, Geoffrey's activities may be interpreted in another way.[3]

Geoffrey was closely involved in directing all aspects of the siege of Montreuil, including technical ones. Counter-forts, siege towers, and battering-rams had all been designed by Geoffrey and employed in a complex operation supervised by him.[4] It was Geoffrey who applied whatever pyrotechnical information the manuscript contained to his particular problem, and managed to get a fire-bomb projected accur-

[1] See above, pp. 170-2.
[2] '*Jubente artifice et artis lignarie perito comite precipiente*' (John of Marmoutier, *Historia gaufredi ducis Normanorum et comitis Andegavorum: Chroniques des comtes d'Anjou et des seigneurs d'Amboise*, ed. L. Halphen and R. Poupardin (Paris, 1913), 217).
[3] A. Murray, *Reason and Society in the Middle Ages* (Oxford, 1978), 127-8.
[4] John of Marmoutier, *His. gauf. Andegav.* 215-19.

ately on to the defenders' beams with his artillery. The count of Anjou clearly applied the range of his talents and learning to his siege warfare. Moreover, the recorded experience of the past facilitated successful operations in the twelfth century.

It may be appropriate to consider what this anecdote suggests about the relationship of twelfth-century engineers and commanders to earlier traditions. On the one hand, Geoffrey of Anjou's utilization of Vegetius seems to express the considerable debt twelfth-century besiegers owed to Roman military traditions. It is clear that Geoffrey held what he believed to be the writings of Vegetius in the highest regard. However, Vegetius' discussion of siege warfare is not particularly specific nor geared to the problems of twelfth-century attackers. Moreover, neither the incendiary device employed nor the form of artillery which cast it were known to the Romans. Without realizing it, the count drew on medieval as well as Roman experience. More importantly, it was the ability of Geoffrey and others like him to adapt whatever they had learnt to the specific problems before them. This was an essential factor in the development of Latin siege technique during the twelfth century. Moreover, Geoffrey's interest in the details of siege warfare and his knowledge of it contributed significantly to his conquests.

King Henry I of England may be taken as a third figure illustrating ways in which military and political leaders influenced twelfth-century siegecraft. This Anglo-Norman monarch was neither a returned crusader nor a student of Vegetius. However, as a king and commander he had much experience in military affairs and in motivating personnel. Orderic Vitalis describes the importance of such abilities in Henry I's successful siege of the castle within Pont Audemer in 1123: 'He personally directed all operations skilfully, rushing about everywhere like a young knight, and, by energetically assisting with what needed doing, he inspired everyone. He instructed the carpenters building a siege tower; he mocked those whose work was deficient, and encouraged those who worked well to do better.'[5] Like his Angevin son-in-law, Henry I took a personal interest in ensuring that construction as well as operations were conducted properly, even if he did not design siege engines. The involvement of political and military leaders in the details of siege operations is noteworthy. The willingness of such men to give attention to such matters may be seen as a small contribution to their larger military and political successes. Moreover, such involvement affected the development of siege techniques.

[5] OV xii. 36, pp. 340–2.

Whatever the role of individual leaders, the resources of government, so important in organizing and financing siege operations, also increased the level of siegecraft, as well as the frequency of siege warfare. In pursuing their military and political objectives, rulers and commanders sought the best means available in reducing fortified positions and thereby contributed to the growth of twelfth-century poliorcetics.

Communities

Communities as well as rulers provided an impetus to twelfth-century siege warfare and its development. City-states waged wars, particularly in urbanized areas of Latin Europe, in which fortress-taking was central. While the warfare of these communities was much more complex than just capturing rival cities and their dependent fortifications, effective siege warfare was essential in successful campaigns. The concentration of population and technological and material resources of city-states enabled them to conduct complex operations at a sophisticated level. Moreover, the political structures of these communities facilitated the employment of their resources effectively. In pursuing their aims, urban communities also advanced the level as well as the pace of Latin siege warfare.

To a degree forces which participated in some of the great sieges of the crusading movement may be considered in this light. In the course of protracted siege operations communities, albeit temporary ones, were formed which bore a considerable resemblance to some urban centres of western Europe. The communities which participated in the First Crusade, and blockaded Acre during the Third, resembled Latin urban centres as much as they did feudal armies. Of course this observation is not germane to every crusading siege. However, the attack on Lisbon in 1147 involved especially diverse personnel, including inhabitants of urbanized Flanders. In a certain regard, the fleets of Italian maritime states so important in the Latin advance in the Mediterranean brought the military resources of an Italian urban community to a military operation. Whether viewed as composite forces or *ad hoc* communities, these besiegers also played a role in the development of Latin siege technique. Rulers and communities concentrated available resources into siege activities in pursuing their military, political, and dynastic ends. While some leaders and cities were particularly adept in military engineering, most delegated the technical aspects of siege operations to professionals. Yet before turning to these men we might note that employers who sought the best technology available contributed to the spread and development of twelfth-century poliorcetics.

Engineers

Few rulers and military commanders were able to design and build the siege machinery required in their operations, or find allies who could. However, there were professionals who could fulfil these needs, and the military engineer is a key figure in the conduct and development of twelfth-century siege warfare. Although much more evident in the narrative and financial records of the thirteenth century than the twelfth, these men begin to be visible during the period under consideration.[6]

While we have observed these men in Outremer, the cities of northern Italy, and the siege operations of Norman kings, professional experts in siege warfare were active in other areas of western Europe also. Henry II of England employed such men during his great conflict of 1173 and 1174, and doubtless afterwards.[7] While Philip Augustus' crossbowmen are more evident than his engineers, all of these kinds of experts were useful in his siege warfare.[8] While the exact identities of the siege crews of Norman–Sicilian kings remain obscure, their importance in the foundation of the Norman–Sicilian kingdom and its offensive military expeditions is manifest. The military needs of the Lombard city-states required the services of engineers, who enjoyed a high status in the communities in which they served. Alfonso VII of Leon–Castile is reported to have employed his builders and engineers in building machinery for his siege operations throughout much of his reign. It is clear that the military needs of these rulers and states, and doubtless of others, caused them to acquire the services of experts in siege warfare over a considerable period of time. This was another factor in the increased scale and sophistication of twelfth-century siege warfare.

The status experts in siege warfare enjoyed is a worthwhile matter for speculation. Although there is little information as to the size of the payments they could command, there is some evidence that their status was enhanced during the twelfth century. Finó has argued that their appearance in epic literature, and particularly in *La Chevalerie d'Ogier*

---

[6] Contamine, *War*, 105–6; Finó, *Fortresses*, 200–2.   [7] 'Héliot, Chât. Gaill.' 68.
[8] Finó, *Fortresses*, 291.

de Danemarche, where an engineer is rewarded with the equipment befitting a knight after building a machine, suggests this.[9] It is worth recalling in this regard that Marchesius of Crema was rewarded with twelve pounds of silver and a war-horse when he changed sides at the siege of Crema.[10]

Whatever their status, such men enjoyed a wide reputation—at least in certain circles. When the crusader and Venetian siege towers at the siege of Tyre of 1124 were being damaged by defensive artillery, the besiegers knew they lacked the expertise to direct accurate counter-battery fire. Consequently, they sent word to Antioch for a specific Armenian expert, Haverdic, known to be proficient with artillery. It is also worth noting that William of Tyre reports that Haverdic received a good salary to maintain himself in his usual magnificence.[11]

Otto of Morena was well aware of the reputation of Milan's great expert, Guintelmus. That Guintelmus was one of those who handed the keys of Milan over to Barbarossa as part of the city's submission in 1162 is another illustration of the importance of these men in Lombardy.[12]

In some cases, the appearance of engineers in historical accounts has a semi-legendary quality. Orderic Vitalis recounts a story of an expert in fortification who was killed by his employer to prevent his skills from serving a rival.[13] Albert of Aachen uses the story of the Lombard master who came forward with the design for a successful armoured roof after all others had been destroyed to illustrate the unseen hand of God in the Crusade.[14] However, this does not necessarily mean that the story is fictitious, since it may have been cast to fit the author's interpretation of the role of divine influence in the First Crusade. Moreover, as experts in poliorcetics were attracted to major siege operations because of the opportunities they presented, these stories may accurately reflect the manner in which these men made their presence and skills known to potential employers.

While the engineers of epic literature were clearly the result of an author's purposes, their very appearance is itself significant. As with their counterparts in historical writing, engineers are more visible in thirteenth-century works than in those of the twelfth. References to them are generally brief and concentrate on their efforts and rewards. Nevertheless, that they begin to appear at the end of our period illustrates something of the realization of the importance of military engineers in the late twelfth century.

[9] Ibid. 201. [10] See above, pp. 140–1. [11] See above, p. 82.
[12] See above, p. 152. [13] OV, viii. 24, p. 291. [14] See above. pp. 19–21.

Even though a knowledge of the background and training of these professional engineers would be valuable, there is little information about these subjects for men of the twelfth century.[15] Undoubtedly experience was crucial, and the accumulated knowledge of a corps of engineers either in royal service or as part of an urban community must have been vital in refinements and adaptations of techniques. Clearly knowledge in one area of large-scale timber construction was important in building many types of siege engines. Whatever their training and status, professional engineers certainly played an important role in waging and developing the sophistication of twelfth-century siege warfare.

In approximately 1206 a retired *jongleur*, named Guiot of Provins, after a long career entertaining in the courts of western Europe and Outremer, set down his opinions on individuals, institutions, and developments in a vernacular work called his 'Bible'. Guiot was an experienced observer of warfare, and may even have participated in the Third Crusade. Many of the changes he had witnessed had been for the worse, and one was the preponderance of experts and specialists over artists and men of talent in several areas of human endeavour. Warfare had been especially affected by this, and the 'artists' of war, the knights, had ceded pride of place in its waging to the miners, engineers, crossbowmen, and artillerymen of siege warfare.[16] Guiot's words should not be taken as a comprehensive comment on the development of twelfth-century warfare. As we have seen, knights had an important role in siege warfare as assault troops and in guarding devices from defensive sallies. Yet his comments do indicate that, in siege warfare, knights had lost some of their monopoly of pre-eminence in military skills. Moreover, they illustrate the importance of professional experts in siege warfare in the conflicts of Europe and the Mediterranean.

The Development and Dissemination of Siege Techniques

Siege Towers

The increased used of sophisticated machinery, and particularly of mobile siege towers, throughout the Mediterranean and much of

---

[15] J. Harvey, *The Mediaeval Architect* (London, 1972), 87–100.
[16] Guiot of Provins, 'Bible', in *Les Œuvres*, ed. J. Orr (Manchester, 1915), vv. 180–200, pp. 14–16.

western Europe soon after the First Crusade is notable, and the experiences of the expedition have been seen as fundamental in this. It has been suggested that the First Crusade involved an interchange of technological skills from more advanced Near-Eastern cultures through experts who took service with crusader leaders.[17] From this it is possible to explain the increase in the level of poliorcetics associated with the twelfth century by a 'percolation' of knowledge from west to east. While the First Crusade was undoubtedly crucial in this increase, we have encountered little evidence of Near-Eastern influence on the siege warfare of the First or subsequent Crusades. The crusaders drew from their own resources and those of timely reinforcements to solve the particular military and logistical problems which confronted them. The nature and availability of resources, as well as the problems themselves, determined the pattern of siege warfare. It has also been suggested that a western technological superiority and even 'dynamism' at the end of the eleventh century explains the developments in twelfth-century siege warfare.[18] Such a view ignores the particular logistical, topographical, economic, and military factors which were crucial in the course of important sieges. Moreover, such an explanation implicitly assumes that certain machines were in most circumstances superior, and this has been clearly shown not to be the case. Although Italian maritime forces were pre-eminent in much of the Mediterranean siege warfare we have examined, this was not because they enjoyed a monopoly of inherently better siege engines. Mobile wall-dominating siege towers stemmed from Roman military practice and were part of a common tradition of medieval European siegecraft. However, Italian maritime powers concentrated men, materials, and expertise rapidly and economically for those concerned.

Yet the First Crusade played a crucial role in the development of Latin siege technique. That expedition brought together men from all over western Europe and concentrated their abilities in meeting the challenges of the First Crusade. Although the crusaders' responses to their military problems owed something to their European experiences and resources, they also owed much to the particular factors of the expedition. Availability of large-scale manual labour and an overriding need to reduce fortifications rapidly were among the reasons why siege-tower assaults emerged as the principal crusading siege technique. Moreover, the combination of forces involved permitted siege towers built with the assistance of Italian sailors to be manned and

[17] See above, p. 11. [18] See above, p. 88.

guarded by northern European knights. As much the same conditions continued in Outremer after the First Crusade, a similar method of assault was employed, which combined the strengths of the forces of Outremer with their maritime allies. There was a certain flexibility in the logistical and financial preparation for the siege warfare of this expansion which was suited to the forces involved in both the First Crusade and subsequent operations, although in somewhat different ways. In one sense, a siege technique evolved during the First Crusade and first decade of the twelfth century which was well suited to Latin expansion in the twelfth-century Mediterranean.

Moreover, the major sieges associated with the Crusades and Latin expansion in the Mediterranean not only concentrated resources, but also served to instruct some of those who participated in the means of attacking fortified positions employed by their co-religionists. In this context we should note that these endeavours focused the talents of men from all over the Latin west. This reflects ways in which sieges of this scale concentrated the resources and talents of all participants, as well as the probable assistance of Mediterranean expertise in some operations. These major sieges attracted experts in poliorcetics, who came either as part of the contingent or independently, hoping to find an opportunity for their talents. The spread of advanced techniques was closely connected with the Crusades. This was not because they permitted the dissemination of Near-Eastern techniques, but because they focused the talents of western Europe and showed warriors and leaders the capabilities of their military technology.

The skills necessary in building siege towers and the large armoured roofs frequently necessary to support them were closely related to those needed for large-scale wooden constructions, such as fixed defensive fortifications, dwellings, and even bell towers.[19] There was a close relationship between military and other forms of timber engineering which was crucial in the dissemination of siege towers. Men who closely observed the siege towers and armoured roofs on Crusade or in any military operation could draw on a pool of skilled carpenters and joiners to produce siege machinery when they returned home in many areas of western Europe. While these devices may not have been as large or built with the same economy as the great devices of Outremer,

[19] The terms for bell tower and siege tower by Anglo-Norman writers are occasionally the same—*berfredum/beffroi*—and it is not clear whether the siege tower has been named after the bell tower or vice versa. I would like to thank Dr C. Corrie for bringing this to my attention.

they were similar in design. In a sense, the primary means of spreading the employment of siege towers were the sieges in which towers were built.

This may be illustrated from the conduct of an attack on an Obodrite fortification by Henry the Lion, duke of Saxony, in 1163. During the early 1160s the Welf duke augmented his position and power in Saxony by extending his authority into Nordalbingia. He was particularly concerned to maintain and bolster his role as protector of religious establishments east of the Elbe, which embroiled him in conflict with the Obodrite ruler Niklot and his sons Vladislav and Pribislav. Upon their father's death in 1163 his successors began organizing an offensive against German settlers and establishments. The duke, alerted to this impending onslaught, launched a pre-emptive attack early in 1163. Henry concentrated his forces against a fortification at Werle on the Warnow defended by Vladislav. Werle's main defence is described as a large castle. However, it is not clear if this structure was made of wood or stone, or a combination of both. Henry's forces established a close blockade and commenced siege operations, despite harassment by Obodrite warriors under Vladislav's brother Pribislav, who operated in the forests around Werle. Henry supervised the gathering of wood for siege engines, which included rams and a siege tower. The tower was built and moved to the walls, and archers and crossbowmen atop it dominated the defences. Although the position was never stormed, the tower's missile fire and the attackers' attempts to batter and undermine the walls compelled the position's surrender.[20] While such methods, however effectively carried out, may seem unusual for conflicts in the Slavic wars, Henry learnt this siege technique while serving with Barbarossa in northern Italy. According to Helmold, Henry observed the machines built against Crema and Milan and had similar ones made for his siege.[21] The siege tower at Werle was undoubtedly smaller than Barbarossa's great tower at Crema and probably lacked a bridging mechanism. Nevertheless, the story of Henry the Lion's siege tower and how he came to build such a device illustrates how these machines came to be more widely employed in European siege warfare during the twelfth century.

It is clear from this example that Barbarossa's Lombard sieges served

[20] Helmold of Bosau, *Chronica Slavorum*, ed. J. Lappenberg and B. Schmeidler (MGH SRG; Hanover and Leipzig, 1904), i. 93, pp. 182–3; for Henry's activities in Saxony, see K. Jordan, *Henry the Lion: A Biography*, trans. P. S. Falla (Oxford, 1984), 66–81.

[21] Helmond, *Chron.* i. 93, p. 183.

to spread the use of sophisticated machinery in a fashion similar to crusading sieges, albeit on a smaller scale. However, Henry's siege tower at Werle does not usher in a period when the sieges against Slavic fortifications were dominated by these devices, as the aims and context of warfare in this region were obviously very different from those of the twelfth-century Mediterranean. Nevertheless, it shows that large-scale siege operations which were conducted by men from different areas of the Latin west were themselves a basic factor in the spread of techniques.

Lever Artillery

By the beginning of the thirteenth century lever artillery was more effective in destroying fortifications than it had been in earlier periods. Although the full destructive potential of pre-gunpowder artillery was only realized during the thirteenth and fourteenth centuries, a significant phase of its development was completed in the twelfth century. It has not been possible to identify exactly where and when counterweight artillery was first developed. This in part reflects wider problems in the history of medieval pre-gunpowder artillery. As the use of effective artillery depended to a large extent on the painstaking refinement of techniques and weapons, the role of crews and experts employed by commanders who needed their skills is clearly important. Artillery expertise was highly specialized, even within poliorcetics, and the example of the Armenian expert Haverdic at Tyre in 1124 may be noted here.[22] It is likely that more effective artillery was developed by expert siege crews in the regular employment of a commander who could afford them. Some evidence points to the men who employed artillery effectively in the service of Roger II and his successors in this regard. It may be suggested that the artillery men of Norman–Sicilian kings played an important role in the development of counterweight lever artillery. However, there is a strong case that powerful artillery was developed in Islamic lands during the twelfth-century.[23] Yet, it appears, that whatever the origins of this artillery, a powerful incentive for its development was the military rivalry generated by twelfth-century expansion. Wherever counterweight artillery may have undergone its primary development, it spread rapidly at the end of the twelfth

[22] His speciality seems to have been accuracy rather than the use of heavy artillery (see above, p. 82).
[23] See Appendix III.

century, because of the military needs of rulers and besiegers. This artillery became an important feature of western poliorcetics and considerably affected European siege operations until the advent of a more efficient propellant than leverage changed the nature of artillery. To a degree, the spread and development of 'heavy artillery' during the thirteenth century is an indirect result of Latin siege activity in the twelfth.

Sackings

In concluding a discussion of twelfth-century siege warfare we might reflect on matters unrelated to the development of military technology. Positions which fell to a general assault suffered a sack in which inhabitants were at the mercy of attackers who had frequently suffered appreciable casualties in their arduous operations. However, this fate could be avoided by a last-minute negotiated surrender, in which an indemnity was often paid to besiegers. While the number of full-scale sackings seems to have been greater in the early twelfth century, the threat of such action was never removed from the conflict of the period. We have discussed economic motives in this context and in particular a desire by attackers to be recompensed for the considerable cost of siege operations. While important, economic factors were not the only ones. A reputation for ferocity against those who prolonged resistance as well as for success in the conduct of operations was valuable in campaigns of siege warfare. Moreover, the passions which were aroused during often tedious and frustrating as well as laborious sieges should not be ignored. These emotions, together with the loss of comrades and kinsmen on both sides, engendered a desire for revenge. While this affects siege warfare in any period, it is characteristic of such conflict in the twelfth century and should be noted in understanding twelfth-century sieges and those who waged them.

Appendix I. Evidence for Crusader Siege Towers 1099–1169

GIVEN below is a list of crusader siege towers and the evidence for them. I have noted when towers were destroyed, and, when possible, by what means. William of Tyre has been cited only when his account provides additional information.

1099	Arsuf	
	Two towers, both burnt	AA vii. 3–5, 507–11
	Fulcher mentions only one tower	FC ii. 8, pp. 399–400
		WT ix. 19, pp. 445–6
1100	Haifa	
	One siege tower	AA vii. 22–5, pp. 521–3
		TNV 276–7
		WT x. 13, pp. 468–9.
1101	Caesarea	
	One tower, unfinished when the city was stormed	FC ii. 9, pp. 400–4
		AA vii. 55–6, pp. 543–4
		AI 10–12
		WT ix. 14, 469–70
1103	Acre	
	One tower, damaged by defenders and destroyed by besiegers when siege abandoned	AA ix. 19, pp. 601–2
		WT x. 27, pp. 486–7
1108	Sidon	
	One tower, damaged and destroyed when siege abandoned	AA x. 46, pp. 652
		DC 87
1110	Beirut	
	Three towers	*DC* 99–100
	One destroyed by garrison artillery	FC ii. 42, pp. 534–6
	Two other siege towers	WT xi. 13, pp. 515–16
1110	Sidon	
	One siege tower	AA xi. 33–4, pp. 678–9
		DC 107
		WT x. 14, pp. 517–19

Appendix I

1111–12	Tyre Two towers, both burnt	AA xii. 6, pp. 691–2 FC ii. 46, pp. 558–62 WT xi. 17, pp. 521–2 DC 122–5
1124	Tyre Two towers	WT xiii. 6–10, pp. 593–8
1153	Ascalon One tower	WT xvii. 24–30, pp. 793–805 DC 314–16
1167	Alexandria One tower	WT xix. 28, pp. 903–5
1169	Damietta Two towers, both destroyed by besiegers when siege abandoned	WT xx. 15–16, pp. 929–33

Appendix II. Siege Engines: General Descriptions and Terms

THE purpose of this appendix is to identify terms used by twelfth-century writers to denote siege machinery, and to provide descriptions of types of devices which may facilitate an understanding of events related in the text. Descriptions of machines are necessarily general and the information has been derived so far as it is possible from twelfth-century sources.

Unlike thirteenth-century siege warfare, for which manuscript illumination provides valuable evidence for contemporary machinery, the sources for the devices of the twelfth century are primarily Latin narrative histories. Whatever the relationships between Roman and medieval siege technology, some of those who described medieval siege operations were influenced by Roman writers. The use of phrases from Roman history in describing or summarizing events such as famine, blockade, or acts of bravery does not necessarily invalidate a medieval account. However, the use of Roman material in describing operations and particularly machinery complicates matters. The most influential Roman work on siege machinery and warfare was Book IV of the *De re militari* of Vegetius.[1] This treatise was recopied and consulted throughout the medieval period and widely distributed in western Europe.[2] However, this portion of the work is more of a general guide to offensive and defensive siegecraft than a technical descripton of poliorcetics, and provides relatively little material which could be extracted in constructing a narrative. I have drawn as much as possible from twelfth-century sources which pay close attention to the course of events so that it is hoped that what follows is based closely on twelfth-century descriptions.

While not every sort or variation of a device employed in twelfth-century sieges is mentioned, I have tried to give an account of as many of the major types as possible. Although in some cases the classification may seem arbitrary, I have tried to organize sections according to the kind of functions that machines carried out or made possible. I have also described armour utilized by devices and small groups of combatants according to its main function. I have not

---

[1] Vegetius, *Epitoma rei militaris*, ed. C. Lang (Stuttgart, 1872; repr. 1967), iv. 1–30, pp. 125–50.
[2] Schrader has identified 304 Vegetius manuscripts written between 600 and 1600 (C. Schrader, 'The Ownership and Distribution of Manuscripts of the *De rei militari* of Flavius Vegetius Renatus before 1300', Ph.D. thesis (Columbia, 1976). For a discusssion of the work and its context, see W. Goffart, 'The Date and Purpose of Vegetius' *De re militari*', *Traditio*, 33 (1977), 65–100.

discussed such basics as archery, crossbows, ladders, axes, hammers, and small battering-rams. As the spelling of medieval machines varies, I have selected basic forms.

Armoured roof

Small armoured roofs and portable shields protected small labouring crews, archers, and crossbowmen in areas within missile range of defenders. While basic to siege operations and requiring little description we should not overlook their significance.

Terms *cattus, gattus, testudo, vimineus*.

Large armoured roofs protected labouring crews while they prepared approaches, filled ditches, and lowered mounds. They appear to have been built along a central shaft with timber sloped roofs to aid in deflection, and reinforced with internal frames. They were protected against incendiaries with hides and against projectiles with a range of shock absorbers. They moved along rollers or wheels, depending on their size and the terrain. These devices were particularly useful in twelfth-century operations in preparing a pathway for siege-tower assaults. Battering-rams could be suspended underneath roofs of any size.

Terms: *catus, gatus, testudo*.

Battering-ram

Large battering-rams consisted of a massive beam—sometimes a ship's timber—often capped with an iron head. Such devices were usually operated under the protection of an armoured roof. While some were suspended underneath a roof, it is not certain that all were. Large rams were able to be moved independently of the roof in initial manœuvrings and placement.

Terms: *aries, bercellum*.

Digging machine

Although digging machines could be considered armoured roofs, the tasks they performed caused them to be described by a different terminology. While some were doubtless armoured roofs which protected specialists in mining and sapping, others possessed means for excavating the walls. This usually consisted of a rotating beam occasionally with an iron head which drilled through stone and mortar to permit labourers to remove foundation-stones.

Terms: *ericus, mus, sus, talpa, vulpis*.

Sapping

Although not a machine, this technique should be briefly described. In sapping, foundation-stones were removed and replaced by wooden stakes, which were ignited in the hope that they would be consumed at approximately the same time and so cause the walls to collapse. In this process the rapid loss of support of a whole section of walls at the same time, rather than gradually, foundation-stone by foundation-stone, was crucial in causing masonry to sheer off and collapse. The efficacy of this technique varied with the quality of stone and masonry in each fortification.

Siege tower

Mobile siege towers which over-topped walls provided an elevated platform for men with missile weapons to dominate defences. Many but not all possessed bridging apparatus at the level of the walls which enabled assault troops to make an escalade. Some towers possessed internal stairs so that more assault troops than could be carried in the device could make an escalade. Towers were based on four vertical corner beams slanted inward at the summit, which increased stability. While it was easier to build devices around corner beams which ran the height of the devices, it was possible to scarf together reinforced smaller sections which could be built into a number of storeys. Towers were armoured with layers of osier or wickerwork and other shock-absorbing materials against projectile weapons. While there may have been wooden planking on some devices, this would have added significantly to the tower's weight. Shock-absorbing armour was further protected by armouring designed to ward off incendiaries, which usually consisted of semi-cured animal hides. These could be suspended in such a way that liquid poured from the summit irrigated the whole surface of the device. They could also be suspended in layers so that, should one section be ignited, it could be cut away. Devices moved along rollers or wheels on absolutely level terrain, making prepared approaches vital for the advance of these towers. The largest such devices are said to have had crews of five hundred men, doubtless including men who propelled the tower. Terms: *berfredum, castellum, ligneum castellum, ligneum castrum, turris ambulatoria*.

Cratis

This was not a machine but rather a term for armour which was used in several ways by twelfth-century writers. It was used to denote armouring against projectile weapons employed on siege towers and armoured roofs, as well as portable shields which protected small numbers of men operating near defences. Crews which operated artillery in range of defensive missile fire were shielded by *crates*. In these cases the term must have embraced any form of protective weapon armouring. In some instances it was used to describe the bridge of a siege tower, indicating that the front of the bridge facing defenders was armoured.

Appendix III. The Problem of Artillery

QUESTIONS about the working and destructive capabilities of rock-throwing devices have dominated one of the more protracted academic debates concerning the development and diffusion of medieval siege technology: the forms of pre-gunpowder artillery. Despite the interest of scholars and artillerymen for more than a century, no consensus about the type and functioning of artillery in the period 600–1200 has emerged. This appendix discusses aspects of artillery and scholarship concerning it pertinent to an understanding of twelfth-century siege warfare. Although aimed at clarifying problems of Latin siege artillery in the twelfth century, this discussion must refer to a wider chronological period.

The most basic issue involved in the debate concerns the continuity of the torsion weapons of the Greeks and Romans against the employment of a form of lever artillery unknown in the ancient world. While it is clear that lever stone-throwers provided the pre-eminent siege artillery of the thirteenth century, the importance of these weapons for earlier periods has not been definitively established.

It should be noted at the outset that this lack of certainty regarding the forms of artillery employed in the early and central Middle Ages reflects fundamental problems with the available evidence. Almost all of our direct knowledge for the artillery of this period derives from narrative accounts which are seldom sufficiently detailed to provide a clear understanding of the artillery mentioned. While medieval writers employed terms for artillery drawn from Vegetius and Roman historians, they are employed in an inconsistent, confusing, and often unspecific fashion. Greek and Roman engineers provided detailed descriptions of their artillery and its development with a precision which makes possible the reconstruction of devices known to have been employed.[1] However, it is not possible to link medieval accounts with material drawn from these treatises dealing specifically with artillery.[2] Thirteenth-century accounts are more numerous and often more detailed, permitting a clearer understanding of the artillery described in them. More importantly, this material is amplified and illustrated by evidence not available for the earlier period—contemporary descriptions of terms and devices, and manuscript illuminations. Conse-

---

[1] Marsden translates and annotates important treatises in the second volume of his study of ancient artillery. For an authoritative discussion of all aspects of this subject, see E. W. Marsden, *Greek and Roman Artillery* (2 vols.; Oxford, 1969–71).

[2] This may reflect not only the brevity of medieval accounts but also major differences between ancient and medieval artillery; see below, pp. 255–6, 266–7.

quently, varying understandings of artillery in the early period have emerged from differing interpretations of a limited number of historical accounts. Abbo's account of the artillery employed during the Viking siege of Paris 885–6 represents one example of this. In his eyewitness account, Abbo uses several terms in describing the projectile weapons of both sides, including *balista*, *catapulta*, and *mangana*.[3] Köhler used this account to demonstrate the continuity of torsion artillery, and as part of his argument equated the term *mangana* with torsion artillery throughout the Middle Ages.[4] However, Schneider, who argued that the term *mangana* denoted lever artillery, consequently used Abbo's account to suggest that the Vikings first developed this form of artillery.[5] Subsequent writers have argued that Abbo's description is not sufficiently detailed to prove the existence of lever artillery in the ninth century.[6] To apply such criteria vigorously, one may only conclude that Abbo's eyewitness account cannot be used to confirm any specific form of artillery at the siege of Paris. If nothing else, this illustrates the source problems involved in ascertaining precisely which kind of artillery was used in the early and central Middle Ages.

Torsion Artillery

Two aspects of torsion artillery are of special interest to those who wish to demonstrate either the continuity or the demise of the ancient tradition of artillery for the medieval period. The first concerns the technology involved in making effective torsion springs essential in this form of artillery. The second involves the degree of mobility and ease of assembly of large siege artillery pieces.

The artillery of Roman military science utilized and refined the sophisticated Hellenistic tradition upon which it was based. The principal siege artillery of the Greeks and Romans cast projectiles from the energy of twisted coils of animal sinew—hence the descriptive term torsion artillery. The main Hellenistic siege piece was a two-armed torsion stone-thrower which employed two vertical columns of twisted sinew and cast projectiles at a high initial velocity and along a relatively flat trajectory.[7] Although Roman armies employed devices of this type, the main siege weapon of the late Empire was the one-armed torsion stone-thrower—the *onager*.[8] Instead of two vertical coils of twisted sinew, the *onager* had one massive torsion spring placed horizontally and at right angles to a strong frame. A sling attached to a single rotating arm,

---
[3] Abbo, *Le Siège de Paris par les Normands*, ed. and trans. H. Waquet (Paris, 1962), i, vv. 87, 137, 364–6, pp. 22, 42.
[4] Köhler, *Entwickelung*, iii. 154, 159.
[5] R. Schneider, *Die Artillerie des Mittelalters* (Berlin, 1910), 60–1.
[6] Finó, *Fortresses*, 97.
[7] For the development and nomenclature of this artillery, see Marsden, *Artillery*, i. 5–31, 86–98.
[8] For changes in terminology and the possible disappearance of the two-armed device, see ibid. ii. 249–54, 263–4.

itself connected to the torsion spring, projected stones on a more parabolic trajectory than the two-armed device. The most important description of the *onager* was that of the historian Ammianus Marcellinus, who wrote in the later fourth century AD.[9] Although capable of casting heavy projectiles, the *onager* was unwieldy, difficult to transport, and probably less accurate than the two-armed stone-thrower. Nevertheless, its simple torsion spring required less maintenance and expertise in co-ordination than the two coils of twisted sinew of the two-armed device.

The effective performance of this form of artillery was contingent on employing appropriate materials for the torsion springs. Animal sinew and tendons—termed *nervi torti*, *nervi*, or *funes*—were the preferred materials, and certain animals were thought to provide better materials than others. Although animal and even human hair could be employed, sinew and tendons provided the best results, and ancient artillery designs enabled artillerymen to get the most out of this material.[10] In any case, questions about the continuation of the technology involved in this aspect of ancient artillery are of considerable importance in the debate over forms of medieval artillery.

Another important aspect of torsion artillery, and in particular the *onager*, was the considerable baggage train that such large devices required. In order to cast heavy projectiles, the *onager* employed a massive torsion spring and strong frame to stabilize its energy. A full-size working replica built in the last century weighed almost two tons.[11] Although devices which used sinew may have been less cumbersome than this reproduction, the one-armed torsion siege piece was unwieldy in transport and operation. Thus, in examinations of medieval narrative accounts, the degree of mobility and ease of erection have been important criteria in attempting to ascertain the type of artillery described.

Torsion vs. Lever Artillery: The Scholarly Debate

Because of the different operating principle of lever artillery, the means for identifying it are quite diverse, although the question of mobility remains important.[12] For the period before 1200 interest has focused on terms for artillery which may be found in Latin from the later ninth century onwards. The key terms are *mangana* in its various forms, *mangonella*, and *petraria*.[13] Although Roman terms for artillery—*ballista*, *catapulta*, and *tormentum*—persist, the first three terms mentioned are those most frequently used for

[9] Ammianus Marcellinus, *Rerum gestarum Libri qui supersunt*, ed. and trans. J. C. Rolfe (3 vols.; London, 1935–63).
[10] J. G. Landels, *Engineering in the Ancient World* (London, 1978), 107–10.
[11] R. Payne-Gallwey, *The Crossbow, with a Treatise on the Ballista and Catapult of the Ancients* (2nd edn., London, 1958), Appendix, p. 18.
[12] For a description of lever artillery, see below, pp. 266–7.
[13] Although there are permutations, these three represent clearly distinguishable terms. *Trabicum* clearly denotes a counterweight artillery device. See below, pp. 267–8. For clarity I have continued to use Köhler's terms.

artillery during the twelfth century. As we have mentioned, narrative descriptions concerning these terms, especially before 1100, are seldom sufficiently detailed to be conclusive as to which form of artillery is being described. Consequently considerable argumentation and reasoning has gone into debates over the meaning of these three terms in relation to each other. In its most simple form, the debate has involved one camp which argues that the change in nomenclature obscures the continuation of the torsion tradition, and, at the other extreme, another which sees the change in terminology as indicative of a change in the form of artillery.

A debate about the forms of medieval artillery as well as practical experimentation of its capabilities developed with the investigations of Napoleon III and Favé. Although their most important contribution was the construction and 'test-firing' of a full-size counterweight lever piece, they suggested a complete break between ancient and medieval artillery traditions. This was based both on a belief that the technology of torsion artillery did not survive the end of the Roman Empire and on their own practical experience.[14]

The subject was treated with a more thorough scrutiny by the Prussian soldier and veteran artilleryman General Köhler, in the third volume of his *Die Entwickelung des Kriegwesens und der Kriegführung in der Ritterzeit von Mitte des 11. Jahrhunderts bis zu den Hussitenkriegen* as part of his account of siege machinery and technology.[15] Köhler argued for the continued employment of torsion artillery throughout the Middle Ages. Although he allowed that medieval torsion weapons may not have been equal in materials or design to their Greek and Roman antecedents, he strove to demonstrate that medieval besiegers drew upon and continued the torsion artillery tradition.

In his chronology of the development of medieval artillery, Köhler saw the thirteenth century as the great age of lever artillery, largely due to the development of counterweight devices which dated to around 1200. Although many of Köhler's judgements have been modified or abandoned, this dating of the advent of counterweight lever artillery has remained one point upon which there is much agreement.[16]

Köhler believed that western Europeans came into contact with traction lever artillery during the twelfth century, and consequently saw that period as a transitional one for artillery traditions. Although lever artillery became dominant in the thirteenth century, torsion artillery continued to be utilized.

In developing his arguments about medieval artillery, Köhler gave clear definitions for the principal Latin terms for artillery in the period 1050–1200: *mangana*, *petraria*, and *mangonella*. Köhler understood *petraria* to refer to the traction lever artillery, itself most significant as a forerunner of the counterweight pieces of the thirteenth century. As the earliest unambiguous reference

---

[14] L. N. Bonaparte and I. Favé, *Études sur le passé et l'avenir de l'artillerie* (6 vols.; Paris, 1848–71), ii. 26–40.
[15] Köhler, *Entwickelung*, iii. 140–221.   [16] See below, pp. 263–4.

to such artillery was at the siege of Lisbon in 1147, he suggested that Latins first came into contact with this artillery in the twelfth century via Islam.[17] The term *mangana* and its meaning was as central to Köhler's typology of medieval artillery as to his argument about the continued employment of torsion artillery. He believed that it referred to a one-armed torsion stone-thrower similar to if not the same as the Roman *onager*. His understanding was based on a complicated analysis of terms from a very wide range of primary material, not all of it treated critically. However, his interpretation of fifteenth-century manuscript illustrations and accounts of Barbarossa's artillery in Lombardy were important props of his argument.[18] In any case, Köhler believed that both the *petraria* and the *mangana* cast their projectiles along a high, parabolic trajectory.

The trajectories of these devices were unlike that of the third kind of device, the *mangonella*. This term gave Köhler difficulty, as he modified his understanding of its meaning between the publication of the first and third volumes.[19] Although he understood *mangonella* to refer to a device which cast rocks smaller than those projected by the *mangana*, he did not believe that the former term referred to a smaller one-armed torsion device. Rather he took it to mean a small two-armed torsion weapon. The principal significance of this device according to Köhler's understanding was that it cast its stone projectile along a flatter trajectory than the other two kinds of artillery.[20] Köhler also suggested that the *mangonella* was rediscovered by the Latin West through the military experience of the First Crusade.[21]

Before turning to Köhler's chief critic, we may reflect generally on aspects of his work. His efforts, particularly in collecting examples from narrative sources, were considerable and provided the basis for further research and discussion. Köhler was aware of some of the limitations of his primary source material, most notably a lack of military and artillery experience on the part of the largely clerical historians upon whom he drew. However, his work suffered from his uncritical use of these sources and an urge to find a clarity and coherence in a set of terms characterized by the absence of just those features. Moreover, on occasion conjecture and a firm belief in the continuity of ancient tradition provided the principal basis for his opinions.[22] In any case, Köhler's work also considerably influenced Sir Charles Oman, as much of what Oman wrote on siege warfare, and in particular artillery, was drawn from Köhler's discussion.[23]

[17] Köhler, *Entwickelung*, iii. 164–6.   [18] Ibid. iii. 154, 164–6.
[19] In an account of the siege of Carcassonne in 1240 he had described this device as a giant crossbow (ibid. i. 444).
[20] A distinction between flat and high trajectories, which he compared with cannons and mortars of his own period, is fundamental to his classification of ancient and medieval artillery (ibid. iii. 154–9).
[21] Ibid. iii, p. 157 n. 4.
[22] This is particularly true of his treatment of *mangonella*.
[23] The structure and choice of examples in Oman's discussion of artillery and its terminology is related to Köhler's material (Oman, *Art of War*, i. 136–40; ii. 43–6).

Appendix III

The main criticism of Köhler's discussion of artillery came in 1910 with the publication of Schneider's *Die Artillerie des Mittelalters*. In this relatively short work Schneider attacked Köhler's thesis concerning the continuation of torsion artillery; he collected texts and manuscript illustrations important to understanding lever artillery, and argued for a major revision of the history of the development of medieval artillery. Schneider, who saw much to admire in the work of Napoleon III, argued for a complete break between ancient and medieval artillery forms in the course of the upheaval of the *Völkerwanderung*. He challenged Köhler's evidence for the continued employment of torsion artillery, including one of the General's translations of a passage from Otto of Freising's *Gesta Friderici I Imperatoris* concerning artillery employed at the siege of Tortona in 1155. More importantly, he challenged Köhler's interpretation of a fifteenth-century illustration as a torsion piece similar to the *onager*. At the very least, Schneider was able to show that the artillery described by Köhler was unlike the *onager* of Ammianus Marcellinus in several important ways.[24] The principal thrust of Schneider's attack on the continuation of torsion artillery was his belief that the employment of twisted sinew was an essential of torsion artillery and his claim that evidence for such material could not be found in Latin accounts of artillery during the Middle Ages.[25]

Schneider believed that lever devices were the characteristic artillery of the Middle Ages and assembled key technical and narrative texts which illustrated the working of this form of rock-thrower. Schneider also reproduced some important manuscript illuminations then available to illustrate his interpretation. He argued that the great virtue of lever artillery was its simplicity, which allowed its construction from timber and iron fastenings available in most military encampments. Thus forces employing it required little or no encumbering baggage train, as artillery was built on site. Thus it was well suited to military forces lacking the logistical and engineering support of the Roman army.

In arguing for a complete break between ancient and medieval forms of artillery, Schneider described the period from the fifth to the ninth centuries as an age without artillery, basing his understanding on the lack of specific references to twisted animal sinew, and on narrative accounts of early medieval siege operations.[26]

Schneider argued that artillery once again became a factor in European siege warfare in the later ninth century, and that this form of lever artillery remained Europe's siege artillery until the development of gunpowder weapons. Central to his thesis is an argument that the term *mangana* and its various forms denotes lever artillery. Although Schneider formulated this interpretation from a range of material, he examined the artillery of twelfth-century crusaders and Byzantines, as reflected through the works of Albert of Aachen, William of Tyre,

---

[24] Schneider, *Artillerie*, 14–15, figs. 3, 6.   [25] Ibid. 10–16.
[26] Ibid. 1–26.

and Anna Comnena.[27] Without reference to the possible interrelationship of the texts of Albert and William, Schneider argued that the artillery described by a range of terms—including *petraria, mangana, mangonella*, and *tormenta*— was exclusively lever artillery.[28] The lack of reference to baggage trains and logistical support, as well as the apparent ease with which crusaders erected and employed their artillery, were important in Schneider's interpretation. He also noted that artillery was able to be employed in areas where wood was scarce in that it was made from parts of ships broken up for use in machinery construction. Schneider also argued that lever artillery was the main artillery form for the Byzantines and the crusaders' Muslim opponents. However, his principal concern was to demonstrate that lever artillery was already established as an important component of Latin siege warfare by the time of the First Crusade. Thus Schneider's age of transition was the period between the late ninth century and the First Crusade, rather than the twelfth century, as Köhler had suggested.

Although Schneider suggested meanings for the key terms *mangana, petraria*, and *mangonella*, he was less adamant in attaching specific and exclusive meanings to them. Schneider maintained a critical attitude towards source material, and made useful observations about the difficulties presented by ancient, medieval, and some modern narrative accounts of artillery employment and especially terminology.[29] Schneider suggested that *petraria* might refer to a counterweight piece, *mangana* to traction lever artillery, and *mangonella* to very small weapons. He also suggested that the different terms might simply refer to different-size traction devices. In any case, his principal concern was not to delineate precisely the types of lever weapons of the twelfth century and earlier, but to demonstrate that the artillery of the Middle Ages was lever artillery.

Schneider's work was an important development in the study of medieval artillery, as it permitted a clearer understanding of the working and importance of lever artillery. Although Schneider's thesis for the total eclipse of torsion artillery and ascendancy of lever artillery was not conclusively proven because of the nature of the evidence available, his criticisms of Köhler's evidence for the continuation of torsion artillery were telling. Aspects of Schneider's interpretation, however, were based on little evidence and much conjecture, most particularly his suggestion that lever artillery was first developed by the Vikings.[30] However, this should not detract from the importance of his work or his main conclusions about the significance of lever artillery for the Middle Ages.

Whatever its merits, Schneider's work prompted considerable research and debate, with much of it appearing in the pages of *Zeitschrift für Historische*

[27] Ibid. 50–60.   [28] For a discussion of these texts, see above, pp. 14–15.
[29] Schneider, *Artillerie*, 12.
[30] Schneider seems to have believed in the correlation of nautical and lever artillery technologies, as his comments on crusading artillery illustrate (ibid. 53–4).

Waffenkunde, as more evidence came to the attention of those interested in medieval artillery.[31]

Rathgen, another veteran soldier, attempted to defend Köhler's interpretation of the importance of torsion artillery and added some evidence for torsion artillery in the fourteenth century. However, this material is not unambiguous, and did not affect Schneider's criticisms of Köhler's evidence for the continuation of torsion artillery. Rathgen also tried to calculate the hurling power of lever pieces from scale models. Unfortunately his efforts were of little value, because of the energy dynamics of lever artillery.[32]

The next notable development in the scholarship of medieval artillery came with a study by a Finnish orientalist Huuri in 1941.[33] Huuri's study analysed medieval artillery and its terminology from Latin, Arabic, and Byzantine sources and was thus able to discuss the problem of artillery in a wider context. This also permitted concentration upon new material instead of the examples available in the Latin West, which were already becoming overworked. Although Huuri also discussed artillery in medieval Persia, China, and India, these sections were ancillary to his main area of study. His most important conclusions were that traction lever artillery, which had been initially developed in China, was in the regular employment of Byzantine and Muslim armies from the seventh century and evident in the Mediterranean from the eighth century. Secondly, Huuri confirmed Köhler's dating of the development and ascendancy of counterweight lever devices to the thirteenth century and suggested that there was some evidence for such pieces at the end of the twelfth century. Thus, for Huuri, the whole period from the seventh to the twelfth centuries was an age of transition.[34]

Regarding torsion artillery, Huuri made several points. In the first place he found no evidence for two-armed torsion devices and thereby rejected Köhler's analysis of the *mangonella*.[35] He also suggested that one-armed torsion artillery continued to be employed and coexisted with lever artillery during the transitional period. However, as Huuri noted, there was no clear evidence for such artillery between the seventh and fourteenth centuries, and consequently suggested that changes in terminology obscured the employment of such devices.[36]

Huuri's discussion of artillery terminology in the Latin West is complicated

---

[31] Attention was drawn to the important manuscript illustrations of Peter of Eboli in the *Liber ad honorem Augusti* by Erben; Rathgen published his evidence for the employment of torsion artillery in the fourteenth century in the same volume (W. Erben, 'Beiträge zur Geschichte des Geschützwesens im Mittelalter'; B. Rathgen and K. H. Schäfer, 'Feuer und Fernwaffen beim päpstlichen Heere im 14. Jahrhundert', *Zeitschrift für Historische Waffenkunde*, 7 (1915–17).

[32] 34. B. Rathgen, *Des Geschütz in Mittelalter* (Berlin, 1927), 631–9; D. R. Hill, 'Trebuchets', *Viator*, 4 (1973), 106
[33] Huuri, 'Geschichte'.   [34] Ibid. 56–65, 212–27.   [35] Ibid. 52.
[36] While a discussion of this follows, it should be noted that Huuri never addressed the problem of the relationship of this putative torsion artillery to the *onager* (ibid. 53–62).

by attempts to defend the continued use of torsion artillery. He believed that the terminology from the eighth to the eleventh centuries was particulary inconsistent, partially reflecting the primitive level of Latin artillery science. However, *mangana* and *petraria* were associated with the new form of lever artillery and their use reflects the advent of traction-lever artillery.

The centre-piece of Huuri's analysis of Latin artillery and its terms was a discussion of light and heavy artillery. Huuri believed that, by the time of the First Crusade, two types of artillery were at the disposal of Latin besiegers. Light artillery cast projectiles up to five kilograms in weight and was identified by the term *mangonella* and its variants. *Petraria* and occasionally *mangana* referred to heavy artillery which was capable of hurling weights of 50–75 kilograms. While all heavy artillery pieces were traction trebuchets, light artillery devices were both small-lever and one-armed torsion weapons. One-armed torsion weapons were subsumed under this generic term for light artillery, because their projectile trajectory and weight were similar to those of small levers. For Huuri, the twelfth century was an age of notable artillery development which culminated in the counterweight 'super-heavy' devices at the end of the century.[37]

While scholars after Huuri have found his classification of terminology too rigid and his calculations unsatisfactory, they have accepted the broad outlines of his work. Although accepting that both torsion and lever systems coexisted, Huuri's account of the initial diffusion and development of lever artillery has received most attention.

Despite its shortcomings and confusing structure, Huuri's work remains important. His survey of the sources and especially the Arabic material remains fundamental.[38] Moreover, his attempt to synthesize not only diverse primary material but also modern scholarship with manuscript illumination evidence commands respect.

In addition, it is appropriate to draw attention to particular aspects of Huuri's work in connection with Latin siege artillery in the twelfth century. Huuri's distinction between light and heavy artillery of all types is important as it illustrates two different functions of artillery in siege operations—the first being the neutralization of defenders and their artillery, and the second the destruction of fortifications. Although Huuri's attempts to calculate shot weights of different devices is suspect, he underscored the role of traction artillery in debilitating—or at least attempting to debilitate—fortifications.

---

[37] Huuri's focus on the appearance together in the same text of two or more terms for artillery, as well as his defence of torsion artillery, is reminiscent of Köhler's approach (ibid. 57–65).

[38] For comments on Huuri's work, see C. Cahen, 'Les Changements techniques militaires dans le Proche Orient médiéval et leur importance historique', in V. J. Parry and M. E. Yapp (eds), *War, Technology and Society in the Middle East* (London, 1975), 118–19; Cahen's review of Huuri in the *Journal asiatique* (1946), 168–70; Hill, 'Trebuchets', 107.

Earlier analysts tended to overlook the role of this device and placed greatest emphasis on the more effective counterweight devices.[39] Thus, Huuri's attempt to depict the role of traction artillery in siege operations is another important contribution to an understanding of artillery in the twelfth century.

Little specifically concerned with artillery in the Latin West has been written since Huuri's time. However, studies of artillery in other areas provide interesting information on the subject of medieval artillery. Cahen edited and partially translated a treatise on weaponry written for Saladin dated to approximately 1180 which discusses artillery.[40] Cahen drew two main conclusions from this important text.

In the first place, lever artillery was the principal form of artillery with which the author and his informants were familiar; this is another indication of the dominance of this form of artillery. Secondly, the treatise gives a somewhat confusing description of a counterweight lever device. In analysing this description, Cahen reviewed the problem of the development of counterweight artillery and confirmed Köhler and Huuri's dating for the advent and initial development of counterweight lever artillery to the period 1180–1220.[41]

Hill has provided an engineer's discussion of the energy dynamics of lever artillery which is particularly useful to the historian. While principally interested in Arabic material and large counterweight pieces, Hill's discussion of the importance of a sling in effective casting is especially useful for all forms of lever artillery.[42] Finó has reviewed artillery as part of his discussion of developments in the art of fortification. Although his survey says little about the problems of torsion artillery, he collected useful information about the capabilities and developments of lever artillery.[43]

Studies of the artillery of medieval China provide useful material to those interested in the rock-throwers of the medieval West and Near East. Sir Joseph Needham has now given a clear Chinese pedigree for lever artillery from the ancient period with a suggested date of 480 BC.[44] Although the pattern and chronology of diffusion from China to the Mediterranean is not yet clear, Needham has put beyond question China's ancient claim to be the originator of this form of artillery.[45] Moreover, Needham and Franke have published in-

---

[39] Payne-Gallwey did not believe that traction artillery existed or could be operated (Payne-Gallwey, *Crossbow*, 309 n. 2).
[40] Murdā ibn 'Ali Al-Tarsūsī, *Tabsirat arbab al-albab fi kayfiyyat al-najat fi'e-hur-ub*, ed. and trans. C. Cahen, as 'Un traité d'armurerie composé pour Saladin', *Bulletin d'études orientales*, 12 (1947–8), 157–9.
[41] Ibid.   [42] Hill, 'Trebuchets', 106–14.
[43] Finó, *Fortresses*, 53, 97, 150–8; 'Machines de jet médiévales', *Gladius*, 10 (1972), 25–43.
[44] J. Needham, 'China's Trebuchets, Manned and Counterweighted', in B. S. Hall and D. C. West (eds.), *On Pre-modern Technology and Science: Studies in Honor of Lynn White Jr.* (Malibu, 1976).
[45] For the vexed question of the possible diffusion of counterweight lever pieces into China, see ibid. 111–42.

formation from Chinese technical treatises which give details of the employment and performance of lever artillery.[46] Although this information cannot be directly applied to medieval narrative accounts of artillery operations, it does provide some kind of guide-line for the capability of this kind of artillery, useful in its own right as well as a check on the calculations of modern analysts.

Before summarizing this material, some comment on the *ballista* should be made. This term originally described a small two-armed torsion bolt or stone-thrower. By the fourth century it referred to a tension weapon which projected only bolts and arrows. Although this form of giant standing crossbow had a lesser range than torsion weapons, it was easier to maintain and operate.[47] The medieval *ballista* used a composite bow which propelled bolts at a high initial velocity and along a relatively flat trajectory. The history of this projectile weapon in the medieval period is more closely associated with the crossbow than with stone-throwers. *Ballista*, however, was doubtless used generically to refer to any form of projectile weapon. While the term occasionally is used to describe crossbows, large standing devices may have been employed in siege operations. There is evidence from the thirteenth century of large bows drawn by a windlass mounted defensively upon fortifications.[48] If large standing crossbows were used in twelfth-century operations, their role would have been that of an anti-personnel weapon.

It has not been possible to solve the problems of pre-gunpowder artillery in the Middle Ages through an examination of Latin siege warfare in the twelfth century. However, it is appropriate to summarize the above material in reference to my investigation of twelfth-century siege warfare.

Turning first to lever artillery, it does not seem possible to view the twelfth century as the first period of contact between Latins and this form of artillery. Huuri's work in Arabic and Greek sources confirmed Schneider's argument that the term *mangana* was connected with lever artillery and its diffusion. Although the scarcity of evidence makes it difficult to be certain that lever artillery became a regular feature of Latin siege warfare in the ninth century, it seems clear that it was by the eleventh. Those interested in the artillery of the crusading period have focused on lever artillery and its development. It may be noted that White seems not to accept the importance of lever artillery in the Latin West before the twelfth century. While his opinion has changed somewhat over the years, he seems to date the first unambiguous reference to what he calls 'beam-sling artillery' to the middle of the twelfth century.[49]

In this context it is worth noting an anecdote from the early twelfth century

---

[46] Ibid. 112–13; H. Franke, 'Siege and Defense of Towns in Medieval China', in F. Kierman (ed.), *Chinese Ways in Warfare* (Cambridge, Mass., 1974), 167–70.

[47] For changes in terminology, see Marsden, *Artillery*, i. 188–91.

[48] Köhler, *Entwickelung*, iii. 173–5; Huuri, 'Geschichte', 47–51; Cahen, 'Changements', 118–19; Finó, 'Machines', 26–7.

[49] L. H. White, *Medieval Technology and Social Change* (Oxford, 1962; repr. 1978), 101–3; 'Thrust', 102.

which suggests a familiarity with traction trebuchets. Peter Tudebode, whose work is dated to approximately 1110, relates the story of a certain defender of Jerusalem captured spying on the crusaders' siege machinery and preparations for attack during the siege of 1099.[50] With his hands and feet bound, the hapless Saracen was placed into the sling of an artillery piece termed a *petrera* and projected by the crusaders unsuccessfully and fatally towards the city whence he had come.[51] Quite apart from its macabre qualities, this anecdote indicates familiarity with traction-lever artillery. That this story is related in such a fashion so soon after the First Crusade strongly suggests that the author and his intended audience were already familiar with the workings of such machinery.[52] In any case, it seems almost certain that Latins were familiar with lever artillery before the First Crusade.

Concerning counterweight lever artillery, I have not found clear and unambiguous references to such artillery in western sources of the twelfth century. In a fashion this may help confirm the chronology discussed above. However, such weapons were in use in the early 1200s, and thus it is likely that Latins developed or acquired the knowledge of this type of artillery during the later twelfth century. The development or perhaps 'reception' of counterweight artillery in the thirteenth century is in part a result of the military experience and requirements of the twelfth.

Turning to the question of the continuation of torsion artillery: while it seems relatively certain that two-armed stone-throwing weapons were not a part of the repertoire of medieval besiegers, it seems difficult to prove conclusively the same for the one-armed torsion devices. However, I have found no evidence which indicates the employment of such devices. In this context something should be said about the methodology of Köhler and to some degree Huuri in focusing on the significance of the appearance of two terms referring to artillery in the same passage. Because such an approach does not evaluate the relative merits of important texts, it can result in utilizing writers whose accounts consist primarily of an almost arbitrary string of terms for siege machinery. It is necessary in confirming conclusions to analyse the terminology of important writers whose relation to the events they describe is close. In this regard, Schneider's concentration upon important texts and thorough investigation of particular writers' uses of specific terms have much to recommend them. Although this makes comparisons more difficult, it allows for the

[50] For a discussion of this source, see above, pp. 13–14.
[51] 'Christiani dixerunt bonum esse, atque eum acceptum, ligatis manibus ac pedibus, posuerunt eum in funda cujusdam ingenii, quod petrera vocatur, atque cum omnibus viribus suis cogitantes eum projicere infra civitatem, nequiverunt. Nam cum tanto impetu venit, quod, ruptis vinculis, antequam ad murum pervenisset civitatis, dilaceratus est' (Peter Tudebode, *His. Hiero. itin.* xv. 136. Albert of Aachen has a similar story which differs only in the Saracen's mission and means of capture. In his case, the projectile weapon is termed 'tormento mangenae' (AA vi. 14, p. 474).
[52] This is a much earlier example of what White calls 'beam-sling artillery' than he has indicated (White, 'Thrust', 102 n. 3).

fact that some twelfth-century writers had different understandings or uses for the same corpus of terminology. The implication of this is that the appearance of two different terms for artillery does not necessarily mean that different forms of artillery are described. Some such usages are almost formulaic. That is, whenever an author refers to artillery, he employs the same two terms, regardless of the context.

Let us now turn to the different possible meaning for the terms *petraria* and *mangana* and *mangonella*. While both *petraria* and *mangana* refer to traction-lever artillery, it is not so much that they were synonymous but that some authors used one, some the other, and some both. Although *manganella* does often seem to refer to a device which projected a smaller rock than other pieces, it is by no means clear that it operated by a principle other than leverage. William of Tyre explained the differences between the *petraria* and *mangana* simply in terms of the size of projectile—not in method of propulsion.[53] While William the Breton noted that a *mangonella* projected a smaller rock than the *petraria*, he used the same verb to describe the action of both devices.[54] Although examples could be multiplied, it should be clear that the difference between the artillery mentioned by the terms was in weight of projectile and by implication the size of the artillery piece. Even if one-armed torsion devices were employed, their significance cannot have been great, as they were used in a manner very similar to small-lever pieces. Thus the clearest hypothesis is that differences in terminology when significant reflect differences in the weight of projectile cast and perhaps design of lever artillery rather than fundamental differences in the type of artillery. In any case it seems clear that the stone-throwing artillery of the twelfth century cast its projectiles in a high trajectory. *Mangonella* is sometimes used in reference to devices which were obviously small—such as those mounted upon siege towers. However, a rigorous differentation between terms used by a majority of those who refer to twelfth-century siege artillery is not usual. Terms: *machina, mangana, mangonella, petraria, tormenta.*

Lever Artillery

This section gives a basic description of lever artillery and its workings, as well as of particular points which have not been emphasized in earlier discussions and which are useful in understanding the operation of traction trebuchets. Lever artillery—so called because the principal means of propulsion is leverage—has been described by a number of modern as well as medieval terms. If trebuchet is taken as a generic term for lever artillery, then one must

[53] 'Alii vero minoribus tormentis, quae mangana vocantur, minores immittendo lapides, eos qui erant in propugnaculis, a nostrorum infestatione compescere satagebant' (WT viii. 13, p. 403).

[54] 'Nunc mangonellus, Turcorum more, minora Saxa rotat; nunc vero minax petraria verso Vi juvenum multa procliviter axe rotatur Retrogrado' (William the Breton, *Phillipid*: Œuvres de Rigord et Guillaume le Breton, ed. H. F. Delaborde (Paris, 1882), ii. 54).

Appendix III

distinguish between traction and counterweight trebuchets, and between fixed- and moving-counterweight devices.

A beam was balanced on an elevating stand which provided a fulcrum and allowed rotation, with most of its length behind the stand and with the end of the beam closest to the target elevated above the frame. As the front of the beam started to move, the beam went through a rotating action which the beam's balance and mass completed. As the front (target end) of the beam reached its maximum downward movement, the back (projectile end) reached the completion of its arc and a projectile shot forward. It should be noted that the projectile end of the lever travelled at a greater velocity than its forward section. With a ratio of 1:5 the projectile end would have travelled five miles an hour for every mile an hour the front end accelerated. The process was further refined by a sling, attached to the projectile end, into which a rock was placed. The sling travelled in its own arc near the completion of the beam's action, thus increasing the projectile's velocity.

All types of lever artillery cast projectiles in a high parabolic 'Howitzer-like' trajectory. Forms of lever artillery were principally differentiated through the means by which the target end of the lever beam started moving at the beginning of its rotating action. Human traction provided one method, and counterweights attached to the target end another. In the first method, crews at the foot of the device's stand pulled on ropes attached to the beam's front until it began moving, and presumably got out of the way as the beam completed its rotation.

Counterweight devices worked by substituting a load of earth, stones, or lead attached to the target end for human traction. Three types of counterweight pieces are described by Aegidio Colonna in what is the most detailed account of this form of artillery in the medieval West.[55]

The fixed-counterweight device—termed *trabucium*—had its counterweight secured directly to the beam and had a means of securing and releasing the projectile end to the ground. Because the load was fixed, and consequently did not shift or jerk during the beam's rotation, this device was the most accurate and consistent. The second kind of device—termed *biffa*—had a mobile counterweight in place of a fixed one, which Colonna describes as moving around the beam's end in the course of its rotation.[56] Hill believes that this must mean that the counterweight was lifted over and above the target end of the beam, and that casting commenced by releasing the ropes which secured the counterweight. Presumably, the counterweight itself revolved around the beam as the initial stage of getting the beam moving. However, a number of manuscript illuminations depict what is

---

[55] Schneider provides a text in his work on artillery: Aegidio Colonna, *De regimine principium*, Bk. III, part iii, in *Artillerie*, iii. 18, pp. 162–4.

[56] 'Aliud genus machinae habet contrapondus mobiliter adhaerens circa flagellum vel virgam ipsius machinae, vertens se circa huiusmodi virgam' (ibid. 163–4).

FIG. 1. Drawing of a thirteenth-century stone carving at the Church of Saint-Nazaire in Carcassonne (Viollet le Duc, *Dictionnaire*, viii. 389; redrawn from Finó, *Fortresses*, 157)

clearly a mobile counterweight with no apparent means or structure for elevating the counterweight in the above fashion.[57] It is possible that the mobile counterweight always remained below the level of the target throughout the beam's movement.[58] In any case, the mobile-counterweight trebuchet was less accurate than the fixed-counterweight device, although it could cast a heavier projectile. The *tripantum* was the third kind of device, and it had both fixed and moving counterweights. Colonna also refers to the traction device, noting that, although it could not cast projectiles as heavy as those of counterweight trebuchets, the traction weapon was more mobile and could cast more rapidly. It should be noted that there is evidence for a 'transitional device' which was both man-pulled and possessed a fixed counterweight. A stone carving in the Church of Saint-Nazaire in Carcassonne dated to the first third of the thirteenth century, and believed to illustrate the death of Simon of Montfort at the siege of Toulouse in 1218, shows a rock-thrower which possesses a fixed counterweight, and is also being pulled by a crew of men and women. This carving also demonstrates that crewmen did not stand only underneath the

[57] e.g. Schneider, *Artillerie*, Figs. 9, 10, 14.
[58] This is something which could perhaps be established by experimentation with full-size reconstruction. As Hill points out, the mechanics of trebuchets are such that experimentation with cale models gives misleading results (Hill, 'Trebuchets', III).

Appendix III 269

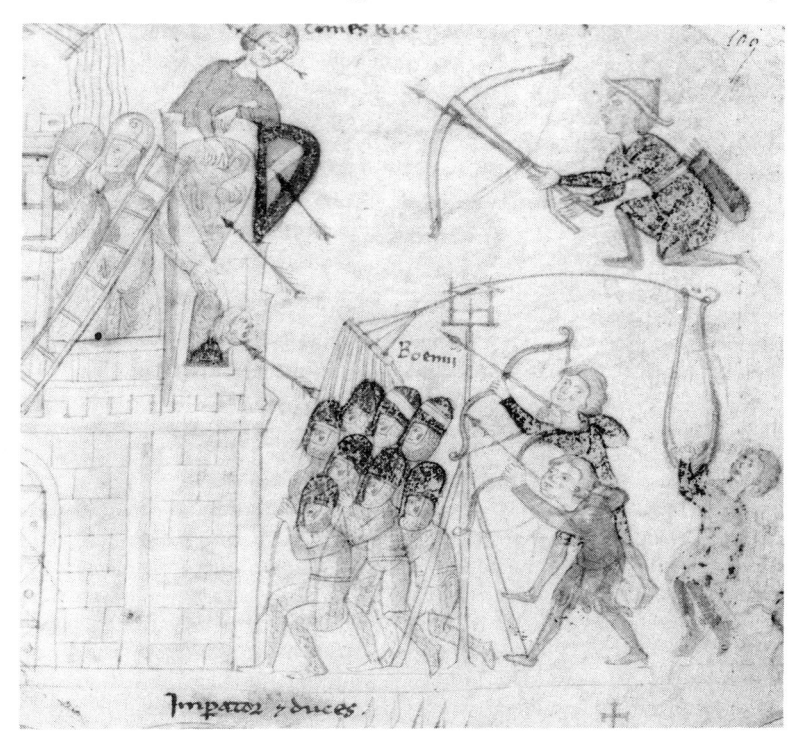

FIG. 2. Traction lever artillery piece with fixed counterweight (from 'Maciejowski Bible', Pierpont Morgan Library, New York, M.638, fo. 23ᵛ)

target end of the beam. By attaching ropes in such a way that men could pull horizontally as well as vertically, the number of pullers could be increased significantly.[59] The lever artillery depicted in the 'Maciejowski Bible' also appears to be both traction and fixed counterweight.[60]

Although artillery of this sort is not clearly described in narrative or technical accounts, it seems likely that it played an important role in the development of counterweight devices as well as in siege operations of the thirteenth and perhaps the late twelfth century.

Turning to aspects of construction and materials, it is clear that the lever-beam was the basic component of all lever pieces. Wood for the frame, ropes, a sling, and iron fastenings were also necessary for all lever pieces. It has been noted, most forcefully by Schneider, that one of lever artillery's advantages

[59] See Fig. 1.   [60] See Fig. 2.

Appendix III

over torsion was its ease of transport and construction. Schneider argued that such artillery could be built on site from materials usually available in most military encampments and so very little indeed needed to be transported from siege to siege. While this may have been so for light artillery, it was clearly not so for the beams of counterweight devices, which needed considerable strength and mass to cast heavy projectiles.

Traction weapons intended to cast heavy projectiles may also have required specially prepared beams. Murda specifies that certain kinds of hardwood for the beam and its axle gave the best results. Chinese treatises mention that the shafts for some weapons required a long period for curing.[61] However, we should not overestimate the difficulties in transporting lever artillery. Although the length and size of beams for devices are not given, it seems that thirty feet was about the maximum length.[62] While transporting counterweights may have added to an artillery train, most of the counterweights could be made up from materials available at the scene of operations. As for ammunition, difficulties in obtaining it could have been related to particular conditions or to the need for especially large projectiles.[63]

Let us now turn to the means by which beams were secured to their fulcrums. While modern commentators have referred to a rotating bolt which attached a beam directly to its frame, as illustrated in later medieval manuscripts,[64] there is some evidence to suggest that a variety of means were used. The artillery illustrated in Peter of Eboli's *Liber ad honorem Augusti* shows beams secured by twisted ropes stretched across a rigid frame.[65] In this context it is appropriate to note the fifteenth-century German illustration of a lever artillery piece with its beam secured to its frame by twisted ropes, over which Köhler and Schneider disagreed.[66] While the energy of twisted ropes may have increased the hurling power of these devices, they cannot be considered torsion weapons—at least in comparison with the torsion artillery of the ancients. Leverage and the flexibility of the beam clearly provided the main force in hurling.

---

[61] Murdā ibn 'Ali Al-Tarsūsī, *Tabsirat arbab al-albab fi kayfiyyat al-najat fi'l-hur-ub*, ed. and trans. C. Cahen, as 'Un traité d'armurerie composé pour Saladin,' *Bulletin d'études orientales*, 12 (1947–8); Eng. trans., *Islam from the Prophet Muhammad to the Capture of Constantinople*, ed. and trans. B. Lewis (2 vols.; London, 1974), i. 219–22; Franke, 'Siege and Defense', 167–8.

[62] Marino Sanudo recommends beams of this size, and this tallies with Chinese data (Marino Sanudo, *Liber secretorum Fidelium Crucis super Terrae Sanctae recuperatione et conservatione*, included in Schneider, *Artillerie*, ii. 4. 22, pp. 94–5; Franke, 'Siege and Defense', 167, 168).

[63] The Norman–Sicilian attack on Alexandria in 1174 is one clear example of the transportation of ammunition in the twelfth century, and this was clearly due to the nature of the expedition, see above, pp. 120–1.

[64] See Fig. 2.

[65] See Fig. 3. This is very similar to Chinese illustrations, as White notes (White, 'Thrust', 101–2).

[66] See above, pp. 259–60.

Appendix III 271

FIG. 3. Traction lever artillery piece (from Peter of Eboli, *Liber ad honorem Augusti*, Burgerbibliothek, Berne)

It seems likely that the large beams necessary in casting heavy projectiles needed to be secured directly, but that this was not necessary with lighter devices. This illustrates the diversity of lever artillery and perhaps something of the differences between light and heavy weapons.

The performance of lever artillery is a subject which has interested scholars for a considerable period and for which there are no definitive answers, especially concerning artillery in the twelfth century. Experimentation and theoretical calculations have involved the large counterweight devices, partially because of the difficulties of calculating the casting power of traction weapons and the dangers involved in experimentation with it.[67]

Huuri estimated that traction trebuchets had a projectile weight of 50–75 kilograms and an effective range of 200 metres. Although he noted one

---

[67] The results of Napoleon III and Favé have been reported in a number of works: Hill relates these data and his own calculations and comments on the problems of such calculations (Hill, 'Trebuchets', III–14; Napoleon and Favé, *Études*, ii. 26 n. 2).

eleventh-century account of a massive device which cast a projectile weighing 175 kilograms, there are difficulties with the precise unit of measurement used in the account.[68] Finó has noted that an account of the death of Simon of Montfort at Toulouse in 1218 confirms a 200-metre range, although it is not clear precisely what kind of artillery piece killed the leader of the Albigensian Crusade.[69] Chinese material collected by Needham and Franke suggests that 100 metres may have been the maximum range for traction trebuchets.[70] It therefore seems appropriate to discuss aspects of traction artillery which have generally been overlooked—particularly as they pertain to the artillery of the twelfth century.

One of the important aspects in the effective use of traction trebuchets was the energy provided by the beam's flexibility. The strength of the cast could be partially determined by the degree of flexibility of the beam. This energy was controlled and carefully timed by the engineer in charge of the pieces. Murda makes this perfectly clear when he notes that the range of traction devices was affected by the degree of dryness, which affected the flexibility of the beam. Closely related to this was the engineer's role in resisting the beam's movement initially, and even where he positioned himself.[71] One of the artillery illustrations from the 'Maciejowski Bible' depicts a man being dragged up into the air while holding on to a sling attached to an artillery piece.[72] Although this has been thought to reflect an artist's fancy, what is depicted is probably much less fanciful. The man hanging on to the end is not only adding to the beam's flexibility, but keeping the sling in a fixed position as well as stabilizing the whole device during the beam's rotation.

That the operation of this kind of artillery was also dangerous is well attested in primary material. Jordan Fantosme relates a story of King William the Lion of Scotland's siege of Wark in 1174 in which a Scottish knight operating a *periere* was saved by his armour from the first rock projected by the piece, which tumbled out of its sling and on to him.[73] Henry of Livonia described a bombardment by Russian besiegers who had only recently become acquainted with artillery in 1205. Their first attempts entertained the defenders, as rocks were projected backwards on to the Russians' own troops.[74] A thirteenth-century

[68] The account itself seems exaggerated in several respects (Huuri, 'Geschichte', 13–14, 91, 149).
[69] Finó, *Fortresses*, 156–7.
[70] Needham, 'China's Trebuchets', 112–13; Franke, 'Siege and Defense', 167–8.
[71] 'If the operator stands directly under the cup in a straight line, the stone is very high and the range is short, and it often falls on the crew; if the operator moves from the cup toward the end of the shaft by about a span, the range is far; the furthest one should move from the shaft is two spans, for if one goes beyond this the launch will not work' (Murdā, *Islam from the Prophet Muhammad*, ed. and trans. Lewis, i. 218–19).
[72] See Fig. 2.
[73] Jordan Fantosme, *Chronicle*, ed. R. C. Johnston (Oxford, 1981), 127–8, vv. 1235–49, pp. 92–3.
[74] Henry of Livonia, *Chronicon Lyvoniae*, ed. W. Arndt (MGH SRG; Hanover, 1874), iii. 12, pp. 36, 37.

Appendix III

Franciscan preaching *exemplar* has a story of the Virgin appearing to an artilleryman at the siege of Seville, admonishing confession because of imminent death. According to the story he was killed by a rock which spun backwards from his weapon.[75] Such examples not only illustrate the difficulties and risks involved in operating this artillery, but also help explain why modern researchers have so far not conducted practical experiments into the capabilities of lever artillery.

One final conjecture may be made concerning the employment of lever artillery at close range in reference to projectile weight, trajectory, and accuracy. It is usually assumed that lever artillery was employed so that projectiles struck their targets near the end of their parabolic trajectory. But not every device need have been used in precisely this fashion. Weapons projecting a heavy projectile, with a consequent short range, could have been situated in certain terrain conditions—such as in ditches—so that rocks struck walls on the initial half of their trajectory. This would have compensated for short range and perhaps been more accurate. Moreover, this would have allowed one section of the walls to be subjected to bombardment from several different positions. Unfortunately, unless more evidence is uncovered, the only way we may resolve how this artillery was actually used is by experimentation.

[75] *Liber ad usum praedicantium*, ed. A. G. Little (Aberdeen, 1907), 26–8.

BIBLIOGRAPHY

PRIMARY SOURCES

ABBO, *Le Siège de Paris par les Normands*, ed. and trans. H. Waquet (Paris, 1962).
ALBERT OF AACHEN, *Historia Hierosolymitana* (RHC Occ. 4).
ALEXANDER OF TELESE, *De rebus gestis Rogerii Sicilae regis libri quatuor*, in *Cronisti e scrittori sincroni Napoletani*, i. *Storia della Monarchia*, ed. G. del Re (Naples, 1845).
AMATUS OF MONTECASSINO, *Ystorie de li Normant: Storia de' Normanni*, ed. V. de Bartholomaeis (FSI 76; 1935).
AMBROISE, *L'Estoire de la Guerre Sainte*, ed. G. Paris (Collection des Documents Inédits sur l'Histoire de la France; Paris, 1897).
—— *The Crusade of Richard Lion-Heart*, ed. and trans. M. J. Hubert and J. L. La Monte (New York, 1941; repr. 1976).
AMMIANUS MARCELLINUS, *Rerum gestarum Libri qui supersunt*, ed. and trans. J. C. Rolfe (3 vols.; London, 1935–63).
Annales Barenses (MGH SS 5).
——'The *Annales Barenses* and the *Annales Lupi Protospatharii*: Critical Edition and Commentary', ed. W. J. Churchill, Ph.D. thesis (Toronto, 1979).
Annales Ceccanenses (MGH SS 19).
Annales Herpibolenses (MGH SS 16).
Annales Sancti Disibodi (MGH SS 17).
Annali Genovesi di Caffaro e de' suoi continuatori, ed. L. T. Belgrano and C. Imperiale di Sant'Angelo (5 vols.; FSI 10–15; 1890–1929).
Annalista Saxo (MGH SS 6).
ARNULF, 'Epistula ad Milonem', Recueil des historiens de Gaules et de la France (24 vols.; Paris, 1869–94).
BALDRIC OF DOL, *Historia de Peregrinatione Jerosolimitana* (RHC Occ. 4).
BONCOMPAGNO, *Liber de obsidione Ancone*, ed. G. Zimolo (RIS 6; Bologna, 1937).
BOSO, *Vita Alexandri III, Liber Pontificalis*, ed. M. L. Duchesne (2 vols.; Paris, 1892).
CAFFARO, *De liberatione civitatum orientis*, in *AG*.
—— *Ystoria captionis Almarie et Turtuose*, in *AG*.
—— OBERTO, and OTTOBUONO, *Annales Ianuenses*, in *AG*.
Carmen in victoriam Pisanorum, ed. H. E. J. Cowdrey, in 'The Mahdia Campaign of 1087', *English Historical Review*, 92 (1977).
Chronica Adefonsi Imperatoris, ed. L. Sanchez Belda (Madrid, 1956).

Bibliography

Chronica monasterii Casinensis (MGH SS 7).
Chronica Najerense, ed. A. Ubieto Arteta (Valencia, 1966).
CODAGNELLUS, JOHN, *Annales Placentini*, ed. O. Holder-Egger (MGH SRG; Hanover and Leipzig, 1901).
Codice diplomatico di Republica di Genova, ed. C. Imperiale di Sant'Angelo (3 vols.; FSI 77–79; 1936–8).
COLONNA, AEGIDIO, *De regimine principium, De re militari*, ed. D. S. F. Hahn (Brunswick, 1724); Book III, part iii, ed. R. Schneider, *Die Artillerie des Mittelalters* (Berlin, 1910).
COMNENA, ANNA, *Alexiade*, ed. and trans. B. Leib (3 vols.; Paris, 1941–5).
DANDOLO, ANDREA, *Chronica*, ed. E. Pastorello (RIS 12; Bologna, 1938).
De bello et excidio urbis Comensis, ed. J. M. Stampa (Rerum Italicarum Scriptores, 5; Milan, 1724).
De expugnatione Lyxbonensi, ed. C. W. David (New York, 1936).
De expugnatione Scalabis, Portugaliae monumenta historica (Scriptores, 1; Lisbon, 1856).
De ruina civitatis Terdonae, ed. A. Hofmeister (Neues Archiv der Gesellschaft für altere deutsche Geschichtskunde, 44; 1920).
Die Lateinische Fortsetzung Wilhelms von Tyrus, ed. M. Salloch (Leipzig, 1934).
DUODECHINUS, 'Epistula Cunonis', *Annales Sancti Disibodi* (MGH SS 17), 27–8.
EADMER, *Vita Anselmi*, ed. R. W. Southern (Oxford, 1962).
L'Estoire de Eracles empereur et la conqueste de la Terre d'Outremer (RHC Occ. 2).
ERNOUL, *Chronique d'Ernoul et de Bernard le Trésorier*, ed. M. L. de Mas-Latrie (Paris, 1871).
EUSTATHIOS OF THESSALONICA, *Narratio de capta Thessalonica*, ed. B. G. Niebuhr (Corpus Scriptorium Historiae Byzantinae; Bonn, 1842).
FALCANDUS, HUGO, *Liber de regno Sicilie*, ed. G. B. Siragusa (FSI 22; 1897).
FALCO OF BENEVENTO, *Chronicon de rebus aetate sua gestis*, in *Cronisti e scrittori sincroni Napoletani*, i. *Storia della monarchia*, ed. G. del Re (Naples, 1845).
FANTOSME, JORDAN, *Chronicle*, ed. R. C. Johnston (Oxford, 1981).
Fiscal Accounts of Catalonia under the Early Count-Kings (1151–1213), ed. T. N. Bisson (2 vols.; Berkeley and Los Angeles, 1984).
FULCHER OF CHARTRES, *Historia Hierosolymitana*, ed. H. Hagenmeyer (Heidelberg, 1913).
——*A History of the Expedition to Jerusalem, 1095–1127*, ed. H. S. Fink, trans. F. R. Ryan (New York, 1969).
GALBERT OF BRUGES, *The Murder of Charles the Good*, trans. J. B. Ross (New York, 1967).
GEOFFREY OF VILLEHARDOUIN, *La Conquête de Constantinople*, ed. E. Faral (2 vols.; Paris, 1938–9).
Gesta comitum Barcinonensium, ed. L. Barrau-Dihigo and J. Masso-Torrents (Barcelona, 1925).
Gesta di Federico I in Italia, ed. E. Monaci (FSI 1; 1887).

Gesta Federici I Imperatoris in Lombardia, auctore cive Mediolanensi, ed. O. Holder-Egger (MGH SRG; Hanover, 1892).
Gesta Francorum et aliorum Hierosolimitanorum, ed. R. Hill (London, 1962).
Gesta triumphalia per Pisanos facta de captione Hierusalem et civitatis Maioricarum et aliarum civitatem et de triumpho habito contra Ianuenses, ed. M. L. Gentile (RIS 6; Bologna, 1936).
GODFREY OF VITERBO, *Gesta Friderici I et Heinrici VI*, ed. G. H. Pertz (MGH SRG; Hanover, 1870).
GUIBERT OF NOGENT, *Gesta Dei per Francos* (RHC Occ. 4).
GUIOT OF PROVINS, 'Bible', in *Les Œuvres*, ed. J. Orr (Manchester, 1915).
HAGENMEYER, H. (ed.), *Die Kreuzzugsbriefe aus den Jahren 1088–1108* (Innsbruck, 1901).
HAYMAR, *De expugnatione civitatis Acconensis*, in Roger of Howden, *Chronica*, ed. W. Stubbs, iii (RS 51; 1870).
HELMOLD OF BOSAU, *Chronica Slavorum*, ed. J. Lappenberg and B. Schmeidler (MGH SRG; Hanover and Leipzig, 1904).
HENRY OF LIVONIA, *Chronicon Lyvoniae*, ed. W. Arndt (MGH SRG; Hanover, 1874).
Historia Compostelana, ed. H. Florez (Espana Sagrada, 20; Madrid, 1765).
Historia Ducum Veneticorum (MGH SS 14).
Historia de expeditione Friderici Imperatoris, in *Quellen zur Geschichte des Kreuzzuges Kaiser Friedrichs I*, ed. A. Chroust (MGH SRG NS 5; Berlin, 1928).
Historia Peregrinorum, in *Quellen zur Geschichte des Kreuzzuges Kaiser Friedrichs I*, ed. A. Chroust (MGH SRG NS 5; Berlin, 1928).
Historia Roderici Campiodocti Rodrigo el Campeador, ed. M. Malo de Molina (Madrid, 1857).
Historia Silense, ed. J. Perez de Urbel and A. Gonzalez Ruiz-Zorilla (Madrid, 1959).
Itinerarium peregrinorum et gesta regis Ricardi: Chronicles and Memorials of the Reign of Richard I, ed. W. Stubbs (RS 38, i; London, 1864).
Itinerarium peregrinorum, ed. H. E. Mayer (MGH Schriften, 18; 1962).
JOHN OF MARMOUTIER, *Historia gaufredi ducis Normanorum et comitis Andegavorum: Chroniques des comtes d'Anjou et des seigneurs d'Amboise*, ed. L. Halphen and R. Poupardin (Paris, 1913).
KINNAMOS, JOHN, *Deeds of John and Manuel Comnenus*, trans. C. M. Brand (New York, 1976).
LANDULF OF ST PAUL, *Historia Mediolanensis* (MGH SS 20).
Liber Maiolichinus de gestis Pisanorum illustribus, ed. C. Calisse (FSI 29; 1904).
Liber ad usum praedicantium, ed. A. G. Little (Aberdeen, 1907).
MALATERRA, GEOFFREY, *De rebus gestis Rogerii comitis et Roberti Guiscardi ducis et fratris eius*, ed. E. Ponteri (RIS 5; Bologna, 1927).
MARAGONE, BERNARD, *Annales Pisani*, ed. M. L. Gentile (RIS 6; Bologna, 1930).

Bibliography

MATTHEW OF EDESSA, *Chronicle*, ed. and trans. E. Doustourian, Ph.D. (Princeton, NJ, 1978).
Narratio itineris navalis ad Terram Sanctam, in *Quellen zur Geschichte des Kreuzzuges Kaiser Friedrichs I*, ed. A. Chroust (MGH SRG NS 5; Berlin, 1928).
ODO OF DEUIL, *De Profectione Ludovici VII in Orientem*, ed. and trans. V. G. Berry (New York, 1948).
OLIVER OF PADERBORN, *Historia Damiatina: Die Schriften des Kölner Domscholasters, späteren Bischofs von Paderborn und Kardinalbischofs von S. Sabina Oliverius*, ed. H. Hoogeweg (Tübingen, 1894).
ORDERIC VITALIS, *Historia ecclesiastica*, ed. M. Chibnall (6 vols.; Oxford, 1967–80).
OSBERN, *De Expugnatione Lyxbonensi*, *Itinerarium peregrinorum et gesta regis Ricardi*, *Chronicles and Memorials of the Reign of Richard I*, ed. W. Stubbs (RS 38, London, 1864).
OTTO OF FREISING, *Chronica*, ed. A. Hofmeister (MGH SRG; Hanover and Leipzig, 1912).
—— and RAHEWIN, *The Deeds of Frederick Barbarossa*, trans. C. Mierow (New York, 1966).
—— —— *Gesta Friderici I Imperatoris*, ed. B. von Simson (MGH SRG; Hanover and Leipzig, 1912).
OTTO OF MORENA AND ACERBUS, *Historia Frederici I: Das Geschichtswerk des Otto Morena und seiner Fortsetzer über die Taten Friedrichs I in der Lombardei*, ed. F. Guterbock (MGH SRG NS 7; Berlin, 1930; repr. 1964).
PETER OF EBOLI, *Liber ad honorem Augusti*, ed. G. B. Siragusa (FSI 33; 1906).
PETER OF VAUX-CERNAY, *Historia Albigensis*, ed. P. Guébin and E. Lyon (Société de l'Histoire de France; Paris, 1882).
RALPH OF CAEN, *Gesta Tancredi in expeditione Hierosolymitana* (RHC Occ. 3).
RALPH OF DICETO, *Ymagines Historiarum: The Historical Works*, ed. W. Stubbs (2 vols.; RS 68; 1876).
RAYMOND OF AGUILERS, *Historia Francorum qui ceperunt Iherusalem (The History of the Frankish Conquerors of Jerusalem)*, ed. and trans. J. H. Hill and L. L. Hill (Philadelphia, 1968).
—— *Le Liber*, ed. J. H. Hill and L. L. Hill (Paris, 1969).
Regesta Regni Hierosolymitani, 1097–1291, ed. R. Röhricht (Innsbruck, 1904).
RICHARD OF DEVIZES, *Chronicle*, ed. J. T. Appleby (Oxford, 1963).
RICHARD THE PILGRIM, *La Chanson d'Antioche*, ed. S. Duparc-Quioc (Paris, 1976).
RIGORD, *Gesta Philippi Augusti*, *Œuvres de Rigord et Guillaume le Breton*, ed. H. F. Delaborde (Société de l'Histoire de France; 2 vols.; Paris, 1882).
ROBERT OF CLARI, *La Conquête de Constantinople*, ed. P. Lauer (Paris, 1924).
ROBERT THE MONK, *Historia Hierosolymitana* (RHC Occ. 3).
RODERIGO XIMENEZ DE RADA, *De rebus Hispaniae*, ed. A. Wechel (Rerum Hispanicarum Scriptores; Frankfurt, 1579).
ROGER OF HOWDEN, *Chronica*, ed. W. Stubbs (4 vols.; RS 51; London, 1868–71).

ROMUALD OF SALERNO, *Chronicon sive annales*, ed. C. A. Garufi (RIS 7; Città di Castello, 1935).
SUGER, *Vie de Louis VI le Gros*, ed. and trans. H. Waquet (Paris, 1964).
Translatio Sancti Nicolai in Venetiam (RHC Occ. 5).
TUDEBODE, PETER, *Historia de Hierosolymitano itinere*, ed. and trans. J. H. Hill and L. L. Hill (Philadelphia, 1974).
—— *Historia de Hierosolymitano itinere*, ed. J. H. Hill and L. L. Hill (Paris, 1977).
VEGETIUS, FLAVIUS RENATUS, *Epitoma rei militaris*, ed. C. Lang (Stuttgart, 1872; repr. 1967).
La Vengance Nostra Seigneur, ed. L. A. T. Gryting (Michigan, 1952).
VINCENT OF PRAGUE and GERLACH, *Annales* (MGH SS 17).
VITRUVIUS, *De architectura*, trans. F. Granger (2 vols.; London and Cambridge, Mass., 1931, repr. 1970).
WILLIAM OF APULIA, *La Geste de Robert Guiscard (Gesta Roberti Wiscardi)*, ed. M. Mathieu (Palermo, 1961).
WILLIAM THE BRETON, *Philippid: Œuvres de Rigord et Guillaume le Breton*, ed. H. F. Delaborde (Société de l'Histoire de France; Paris, 1882), ii.
WILLIAM OF TYRE, *Chronicon*, ed. R. B. C. Huygens, identification des sources historiques et détermination des dates par H. E. Mayer et G. Rösch (2 vols.; Turnhout, 1986).
—— *A History of Deeds Done Beyond the Sea*, trans. E. A. Babcock and A. C. Krey (2 vols.; New York, 1943).

ARABIC SOURCES

ABU SHAMAH, *Le Livre des deux jardins* (RHC Orient. 4–5).
BEHA-ED-DIN, *The Life of Saladin, or What Befel Sultan Yusuf*, ed. and trans. C. W. Wilson and C. R. Conder (Palestine Pilgrims' Text Society, 13; London, 1897).
IBN AL-ATHIR, *Extrait de Kamel-Altevarykh* (RHC Orient. 1, 2).
IBN AL-QALANISI, *The Damascus Chronicle of the Crusades*, ed. and trans. H. A. R. Gibb (London, 1932; repr. 1967).
KEMAL ED-DIN, *Extraits de la Chronique d'Alep* (RHC Orient. 3).
MURDĀ IBN ʿALI AL-TARSŪSĪ, *Tabsirat arbab al-albab fi kayfiyyat al-najat fi'l-hur-ub*, ed. and trans. C. Cahen, as 'Un traité d'armurerie composé pour Saladin', *Bulletin d'études orientales*, 12 (1947–8); English translation of sections of material regarding artillery: *Islam from the Prophet Muhammad to the Capture of Constantinople*, ed. and trans. B. Lewis (2 vols.; London, 1974).
USAMAH IBN MUNQIDH, *Memoirs of an Arab-Syrian Gentleman and Warrior in the Period of the Crusades*, trans. P. K. Hitti (New York, 1929).

Bibliography

SECONDARY SOURCES

ABULAFIA, D., *The Two Italies* (Cambridge, 1977).
AMARI, M., *Storia dei Musulmani di Sicilia* (2nd edn., 3 vols., Catania, 1937).
ANDRESSOHN, J. C., *The Ancestry and Life of Godfrey of Bouillon* (Bloomington, 1947).
APPELT, H. 'Friedrich Barbarossa und die Italienischen Kommunen', in G. Wolf (ed.), *Friedrich Barbarossa* (Darmstadt, 1975).
ARNOLD, B., *German Knighthood* (Oxford, 1985).
AUDOIN, E., 'Essai sur l'armée Royale au temps de Phillipe-Augustus', *Le Moyen Âge*, 25–6 (1912, 1913).
BACH, E., *La Cité de Gênes au XIIe siècle* (Copenhagen, 1955).
BEAUMONT, A., 'Albert of Aachen and the County of Edessa', in L. Paetow (ed.), *The Crusades and Other Historical Essays Presented to D. C. Munro* (New York, 1928).
BEELER, J., *Warfare in Feudal Europe, 730–1200* (Ithaca, NY, and London, 1971).
BERNHARDI, W., *Geschichte Lothar III von Supplinburg: Jahrbücher der Deutschen Geschichte* (Leipzig, 1879).
BESTA, E., *La Sardegna Medioevale* (Palermo, 1905).
BIGGS, A. G., *Diego Gelmirez, First Archbishop of Compostela* (Washington, 1949).
BISHKO, C. J., 'The Spanish and Portuguese Reconquest', in H. Hazard (ed.), *A History of the Crusades*, iii. *The Fourteenth and Fifteenth Centuries* (Madison, Wis., 1975).
BLAKE, E. O., and MORRIS, C., 'A Hermit Goes to War: Peter and the Origins of the First Crusade', in W. J. Shiels (ed.), *Monks, Hermits and the Ascetic Tradition* (Oxford, 1985).
BLEIBURG, G. (ed.), *Diccionario de historia de Espana* (Madrid, 1968).
BLOCH, R. H., *Medieval French Literature and Law* (Berkeley, Calif., 1976).
BOASE, T. S. R., *Kingdoms and Strongholds of the Crusaders* (London, 1971).
BOEHM, L., 'Die "Gesta Tancredi" des Radulf von Caen. Ein Beitrag zur Geschichtsschreibung der Normannen um 1100', *Historisches Jahrbuch*, 75 (1956).
BONAPARTE, L. N., and FAVÉ, I., *Études sur le passé et l'avenir de l'artillerie* (6 vols.; Paris, 1848–71).
BONNASSIE, P., *La Catalogne du milieu de x^e à la fin du xie siècle* (2 vols.; Toulouse, 1975–6).
BOSISIO, A., 'Crema al tempi di Federico Barbarossa', *Archivio storico lombardo*, 10 (1960).
BOSL, K., *Die Reichsministerialität der Salier und Staufen* (MGH *Schriften*, 10; Stuttgart, 1950–1).
BOUSSARD, J., 'Les Mercenaires au XIIe siècle. Henri II Plantegenet et les origines de l'armée de métier', *Bibliothèque de l'École des Chartes*, 106 (1945–6).

BROWN, R. A., 'Royal Castle Building in England, 1154–1216', *English Historical Review*, 70 (1955).
—— *English Medieval Castles* (3rd edn., London, 1976).
BRUHN-HOFFMEYER, A., 'Military Equipment in the Byzantine Manuscript of Scylitzes', *Gladius*, 5 (1966).
BRUNDAGE, J., 'The Army of the First Crusade and the Crusade Vow', *Medieval Studies*, 33 (1971).
BUTLER, W. P., *Lombard Communes* (London, 1906).
BYRNE, E. H., 'The Genoese Colonies in Syria', in L. Paetow (ed.), *The Crusades and Other Historical Essays Presented to D. C. Munro* (New York, 1928).
—— *Genoese Shipping in the Twelfth and Thirteenth Centuries* (Cambridge, Mass., 1930; repr. 1970).
CAHEN, C., 'Les Changements techniques militaires dans le Proche Orient médiéval et leur importance historique', in V. J. Parry and M. E. Yapp (eds.), *War, Technology, and Society in the Middle East* (London, 1975).
—— 'Le Premier Cycle de la croisade', *Le Moyen Âge*, 63 (1957).
—— *La Syrie du Nord à l'époque des croisades* (Paris, 1940).
CARDINI, F., 'Profilo di un crociato, Guglielmo Embriaco', *Archivio storico italiano*, 136 (1978).
CARTELLIERI, A., *Philip II. August, König von Frankreich*, ii. *Der Kreuzzug* (Leipzig, 1906).
CASPAR, E., *Roger II und di Gründung der Normanisch-Sicilischen Monarchie* (Innsbruck, 1904).
CHALANDON, F., *Histoire de la domination normande en Italie et en Sicilie* (2 vols.; Paris, 1907).
—— *Jean II Comnène (1118–1143) et Manuel I Comnène (1143–1180)* (Paris, 1912).
CITAPELLA, A., 'The Relations of Amalfi with the Arab World before the Crusades', *Speculum*, 43 (1967).
CLAUSEWITZ, C. von, *On War*, ed. and trans. M. Howard and P. Paret (New Jersey, 1976).
CLEMENTI, D. R., 'Some Unnoticed Aspects of the Emperor Henry VI's Conquest of the Norman Kingdom of Sicily', *Bulletin of the John Rylands Library*, 36 (1954).
COHN, W., *Die Geschichte der Normannisch-Sizilischen Flotte unter der Regierung Rogers I und Rogers II, 1060–1154* (Breslau, 1910).
CONSTABLE, G., 'A Note on the Route of the Anglo-Flemish Crusaders of 1147', *Speculum*, 28 (1953).
—— 'The Second Crusade as Seen by Contemporaries', *Traditio*, 9 (1953).
CONTAMINE, P., *War in the Middle Ages*, trans. M. Jones (Oxford, 1984).
COWDREY, H. E. J., 'The Genesis of the Crusades: The Springs of Western Ideas of Holy War', in T. P. Murphy (ed.), *Holy War* (Columbus, 1976).
—— 'The Mahdia Campaign of 1987', *English Historical Review*, 92 (1977).
—— *Popes, Monks, and Crusaders* (London, 1984).

CRESWELL, K. A. C., 'Fortification in Islam before A.D. 1250', *Proceedings of the British Academy*, 38 (1952).
CURTIS, E., *Roger of Sicily and the Normans in Southern Italy* (London, 1912).
DAVID, C. W., 'The Authorship of the "De Expugnatione Lyxbonensi"', *Speculum*, 7 (1932).
—— *Robert Curthose, Duke of Normandy* (Cambridge, Mass., 1920).
DAVID, P., *Études historiques sur la Galice et le Portugal du VIe au XIIe siècle* (Paris and Lisbon, 1947).
DEFOURNEAUX, M., *Les Français en Espagne aux XIe et XIIe siècles* (Paris, 1949).
DELBRÜCK, H., *Geschichte der Kriegskunst im Rahmen der Politischen Geschichte* (7 vols.), iii. *Das Mittelalter* (Berlin, 1907).
DEVIC, C., and VAISSETE, J., *Histoire générale de Languedoc* (10 vols., Toulouse, 1874–1914), iii.
DOUGLAS, D. C., *The Norman Achievement* (London, 1969).
—— *The Norman Fate* (London, 1976).
DOWNEY, G., *A History of Antioch in Syria* (Princeton, NJ, 1961).
DOZY, R., *Recherches sur l'histoire et la littérature de l'Espagne pendant le Moyen Âge* (2 vols.; 3rd edn., Leyden, 1881).
DUBY, G., *Le Dimanche de Bouvines* (Paris, 1973).
DUPARC-QUIOC, S., *Le Cycle de la Croisade* (Paris, 1955).
DUSSAUD, R., *Topographie historique de la Syrie antique et médiévale* (Paris, 1927).
EHRENKREUTZ, A. S., 'The Place of Saladin in the Naval History of the Mediterranean Sea in the Middle Ages', *Journal of the American Oriental Society*, 75 (1955).
EDBURY, P. W. (ed.), *Crusade and Settlement: Papers Read at the First Conference of the Society for the Study of the Crusades and the Latin East and Presented to R. C. Small* (Cardiff, 1985).
—— and ROWE, J. G., *William of Tyre: Historian of the Latin East* (Cambridge, 1988).
ELDERKIN, G. (ed.), *Antioch on the Orontes* (Princeton, NJ, 1934).
ENGELS, D. W., *Alexander the Great and the Logistics of the Macedonian Army* (Berkeley, Calif., 1978).
ENLART, C., *Manuel d'archéologie français depuis les temps merovingiens jusqu'à la Renaissance* (3 vols.; Paris, 1927–32).
ERBEN, W., 'Beiträge zur Geschichte des Geschützwesens im Mittelalter', *Zeitschrift für Historische Waffenkunde*, 7 (1915–17).
—— 'Kriegsgeschichte des Mittelalters', *Historische Zeitschrift*, 16 (1929).
FACE, R., 'Secular History in Twelfth Century Italy: Caffaro of Genoa', *Journal of Medieval History*, 6 (1980), 166–80.
FASOLI, G., 'Friedrich Barbarossa und die Lombardischen Städte', in G. Wolf (ed.), *Friedrich Barbarossa* (Darmstadt, 1975).
FERREIRO, A., 'The Siege of Barbastro, 1064–1065: A Reassessment', *Journal of Medieval History*, 9 (1983).

FINÓ, J.-F., *Fortresses de la France médiévale* (3rd edn., Paris, 1977).
—— 'Le Feu et ses usages militaires', *Gladius*, 9 (1970).
—— 'Machines de jet médiévales', *Gladius*, 10 (1972).
FISHER, C. B., 'The Pisan Clergy and the Awakening of Historical Interest in a Medieval Commune', *Studies in Medieval and Renaissance History*, 3 (1966).
FLETCHER, R. A., 'Reconquest and Crusade in Spain, c.1050–1150', *Transactions of the Royal Historical Society*, 5th ser. 37 (1987).
—— *Saint James's Catapult: The Life and Times of Diego Gelmirez of Santiago de Compostela* (Oxford, 1984).
FOLDA, J., *Crusader Manuscript Illumination at Saint Jean d'Acre, 1275–1291* (Princeton, NJ, 1976).
FOREY, A. J., 'The Military Orders and the Spanish Reconquest in the Twelfth and Thirteenth Centuries', *Traditio*, 40 (1984).
—— 'The Failure of the Siege of Damascus in 1148', *Journal of Medieval History*, 10 (1984), 13–23.
FOSTER, S. M., 'Some Aspects of Maritime Activity and the Use of Sea Power in Relation to Crusading States, 1096–1169', D.Phil. thesis (Oxford, 1978).
FOURNIER, G., *Le Château dans la France médiévale* (Paris, 1978).
FRANCE, J., 'The Crisis of the First Crusade: From the Defeat of Kerbogha to the Departure from Arqa', *Byzantion*, 40 (1970).
—— 'The Departure of Tatikios from the Crusader Army', *Bulletin of the Institute of Historical Research*, 44 (1971).
—— 'An Unknown Account of the Capture of Jerusalem', *English Historical Review*, 87 (1972).
FRANKE, H., 'Siege and Defense of Towns in Medieval China', in F. Kierman (ed.), *Chinese Ways in Warfare* (Cambridge, Mass., 1974).
FRY, P. S., *British Medieval Castles* (London, 1974).
FULLER, J. F. C., *A Military History of the Western World* (3 vols.; New York, 1954–6).
GABRIELI, F., 'The Arabic Historiography of the Crusades', in B. Lewis and P. M. Holt (eds.), *Historians of the Middle East* (London, 1962).
GIBB, H. A. R., 'The Achievement of Saladin', in *Studies in the Civilization of Islam* (London, 1962).
—— 'The Armies of Saladin', in *Studies in the Civilization of Islam* (London, 1962).
GIESEBRECHT, W. von, *Geschichte der Deutschen Kaiserzeit* (6 vols.; Brunswick and Leipzig, 1880–95), v.
GILLINGHAM, J. B., 'Why did Rahewin Stop Writing the *Gesta Frederici*?', *English Historical Review*, 83 (1968), 294–303.
—— 'Frederick Barbarossa, a Secret Revolutionary?', *English Historical Review*, 86 (1971).
—— *Richard the Lion Heart* (London, 1978).

—— 'Richard I and the Science of War in the Middle Ages', in J. B. Gillingham and J. C. Holt (eds.), *War and Government in the Middle Ages: Essays in Honour of J. O. Prestwich* (Cambridge, 1984).
GOFFART, W., 'The Date and Purpose of Vegetius' *De re militari*', *Traditio*, 33 (1977), 65–100.
GOHLKE, W. VON, 'Das Geschützwesen des Altertums und des Mittelalters', *Zeitschrift für Historische Waffenkunde*, 5–6 (1909–14).
GOITEIN, S. D., 'Contemporary Letters on the Capture of Jerusalem by the Crusaders', *Journal of Jewish Studies*, 3 (1952).
GROUSSET, R., *Histoire des croisades et du royaume franc de Jerusalem* (3 vols.; Paris, 1934).
GRUNDMANN, H., 'Rotten und Brabanzonen: Söldner-Heere im 12. Jahrhundert', *Deutsches Archiv für Erforschung des Mittelalters*, 5 (1941–2).
GUILMARTIN, J. F., Jun., *Gunpowder and Galleys: Changing Technology and Mediterranean Warfare at Sea in the Sixteenth Century* (Cambridge, 1974).
HAGENMEYER, H., 'Chronologie de la Première Croisade', *Revue de l'orient latin*, 6, 7 (1898, 1900).
—— *Peter der Eremite* (Leipzig, 1871).
HALL, J. B., 'The "Carmen de Gestis Frederici Imperatoris in Lombardia"', *Studi Medievali*, 26 (1985), 969–76.
HAMBLIN, W. J., 'The Fatimid Army during the Early Crusades', Ph.D. thesis (Michigan, 1984).
HAMPE, K., *Germany under the Salian and Hohenstaufen Emperors*, trans. R. Bennet (Oxford, 1973).
HARVEY, J., *The Mediaeval Architect* (London, 1972).
HÉLIOT, P., 'Le Château-Gaillard et les fortresses des XIIe–XIIIe siècles en Europe occidentale', *Château Gaillard*, i (1962).
HEYD, W., *Histoire du commerce du Levant au Moyen Age* (2 vols.; Leipzig, 1885).
HEYWOOD, W., *A History of Pisa* (Cambridge, 1921).
HIESTAND, R., 'Reconquista, Kreuzzug und heiliges Grab: Die Eroberung von Tortosa 1148 im Lichte eines neuen Zeugnisses', *Gesammelte Ausätze zur Kulturgeschichte Spaniens*, 31 (1984).
HILL, D. R., 'Trebuchets', *Viator*, 4 (1973).
—— and AL-HASSAN, A. Y., *Islamic Technology: An Illustrated History* (Cambridge and Paris, 1986).
HILL, J. H., and HILL, L. L., *Raymond IV of St Gilles, Count of Toulouse* (Ithaca, NY, 1962).
HOFFMANN, H., 'Hugo Falcandus und Romuald von Salerno', *Deutsches Archiv für Erforschung des Mittelalters*, 23 (1967).
HOLT, P. M. (ed.), *The Eastern Mediterranean Lands in the Period of the Crusades* (Warminster, 1977).
HOWARD-JOHNSTONE, J. D., 'Studies in the Organisation of the Byzantine Army in the 10th and 11th Centuries', D. Phil. thesis (Oxford, 1971).

HUURI, K. K., 'Zur Geschichte des mittelalterlichen Geschützwesens aus Orientalischen Quellen', *Studia orientalia*, 9 (1941).
JACOBY, D., 'Montmusard, Suburb of Crusader Acre: The First Stage of its Development', in B. Z. Kedar, H. E. Mayer, and R. C. Smail (eds.), *Outremer Studies in the History of the Crusading Kingdom of Jerusalem Presented to Joshua Prawer* (Jerusalem, 1982), 205–13.
JAMISON, E., 'The Norman Administration of Apulia and Capua, more especially under Roger II and William I', *Papers of the British School at Rome*, 6 (1913).
JENAL, A., 'Der Kampf um Durazzo 1107–1108 und Ralph Tortarius', *Historisches Jahrbuch*, 37 (1916).
JORANSON, E., 'The Inception of the Career of the Normans in Italy: Legend and History', *Speculum*, 23 (1948).
JORDAN, K., *Henry the Lion: A Biography*, trans. P. S. Falla (Oxford, 1984).
KEEGAN, J., *The Face of Battle* (Harmondsworth, 1976).
KIERMAN, F. (ed.), *Chinese Ways in Warfare* (Cambridge, Mass., 1974).
KNOCH, P., *Studien zu Albert von Aachen* (Stuttgart, 1966).
KÖHLER, G., *Die Entwickelung des Kriegwesens und der Kriegführung in der Ritterzeit von Mitte des 11. Jahrhunderts bis zu den Hussitenkriegen* (3 vols.; Breslau, 1886–90).
KREBS, F., *Zur Kritik Albert von Aachen* (Munster, 1881).
KREUGER, H. C., 'Postwar Collapse and Rehabilitation in Genoa, 1149–1162', *Studi in Onore di Gino Luzzatto* (2 vols.; Milan, 1950).
LACARRA, M. J., 'La conquista de Zaragoza por Alfonso I', *Al-Andalus*, 12 (1947).
LAMONTE, J. L., *Feudal Monarchy in the Latin Kingdom of Jerusalem, 1100–1291* (New York, repr. 1967).
LANDELS, J. G., *Engineering in the Ancient World* (London, 1978).
LAWRENCE, A. W., *Greek Aims in Fortification* (Oxford, 1979).
LAWRENCE, T. E., *Seven Pillars of Wisdom* (London, 1926).
—— *Crusader Castles* (2 vols.; London, 1936).
LEIGHTON, A. C., *Transportation and Communication in Early Medieval Europe, 500–1100* (Newton Abbot, 1972).
LESTRANGE, G., *Palestine under the Moslems* (London, 1890).
LÉVI-PROVENÇAL, E., *Islam d'occident études d'histoire médiévale* (Paris, 1948).
LEWIS, A., 'The Guillems of Montpellier', *Viator*, 2 (1971).
—— *Naval Power and Trade in the Mediterranean, 500–1100* (Princeton, NJ, 1951).
LEYSER, K. J., 'Early Medieval Canon Law and the Beginnings of Knighthood', in L. Fenske and W. Rösener (eds.), *Institutionen, Kultur und Gesellschaft im Mittelalter: Festschrift für Josef Fleckenstein zu seinem 65. Geburtstag* (Sigmaringen, 1984).
—— 'Some Reflections on Twelfth Century Kings and Kingship', in *Medieval Germany and its Neighbours* (London, 1982).

―― 'The Battle at the Lech, 955', *History*, 1 (1965).
LOMAX, D. W., *The Reconquest of Spain* (London, 1978).
LOPEZ, R., 'The Trade of Medieval Europe: The South', *Cambridge Economic History of Europe*, ii (Cambridge, 1941; 2nd edn., 1952).
LOT, F., *L'Art militaire et les armées au Moyen Âge en Europe et dans le Proche Orient* (2 vols.; Paris, 1946).
LOUD, G. A., *Church and Society in the Norman Principality of Capua, 1058–1197* (Oxford, 1985).
―― 'How "Norman" was the Norman Conquest of Southern Italy?', *Nottingham Mediaeval Studies*, 25 (1981).
LOURIE, E., 'A Society Organised for War: Medieval Spain', *Past and Present*, 35 (1966).
LYONS, M. C., and JACKSON, D. E. P., *Saladin: The Politics of the Holy War* (Cambridge, 1982).
MACKAY, A., *Spain in the Middle Ages: From Frontier to Empire, 1000–1500* (London, 1977).
MAKHOULY, N., and JOHNS, C. N., *Guide to Acre* (2nd edn., Jerusalem, 1946).
MARSDEN, E. W., *Greek and Roman Artillery* (2 vols.; Oxford, 1969–71).
MATTHEW, D. J. A., 'The Chronicle of Romuald of Salerno', in R. H. C. Davis and J. M. Wallace-Hadrill (eds.), *The Writing of History in the Middle Ages: Essays Presented to R. W. Southern* (Oxford, 1981).
MAYER, H. E., *Bibliographie zur Geschichte der Kreuzzüge* (Hanover, 1965).
―― *The Crusades*, trans. J. B. Gillingham (Oxford, 1972).
―― 'Literaturbericht über die Geschichte der Kreuzzüge', *Historische Zeitschrift*, 3 (1969).
―― 'Review of P. Knoch, *Studien zu Albert von Aachen*', *Deutsches Archiv für Erforschung des Mittelalters*, 23 (1967).
MCNEILL, W. H., *The Pursuit of Power Technology, Armed Force, and Society since A.D. 1000* (Chicago, 1982).
MEYER, H., *Die Militärpolitik Friedrich Barbarossa im Zusammenhang mit seiner Italien politik* (Berlin, 1930).
MOLLAT, M., 'Problèmes navales de l'histoire des Croisades', *Cahiers de civilisation medievale*, 10 (1967).
MORGAN, M. R., *The Chronicle of Ernoul and the Continuations of William of Tyre* (Oxford, 1973).
MORRIS, C., 'Policy and Visions: The Case of the Holy Lance at Antioch', in J. B. Gillingham and J. C. Holt (eds.), *War and Government in the Middle Ages: Essays in Honour of J. O. Prestwich* (Cambridge, 1984).
MUNZ, P., *Frederick Barbarossa: A Study in Medieval Politics* (London, 1969).
―― 'Why did Rahewin Stop Writing the *Gesta Frederici*? A Further Consideration', *English Historical Review*, 84 (1969), 771–9.
MURRAY, A., *Reason and Society in the Middle Ages* (Oxford, 1978).
NELSON, L. H., 'Rotrou of Perche and the Aragonese Reconquest', *Traditio*, 26 (1970).

NEEDHAM, J., 'China's Trebuchets, Manned and Counterweighted', in B. S. Hall and D. C. West (eds.), *On Pre-modern Technology and Science: Studies in Honor of Lynn White Jr.* (Malibu, 1976).

NICOLLE, D. C., *Arms and Armour of the Crusading Era, 1050–1350* (2 vols.; White Plains, 1988).

NORGATE, K., *Richard the Lion-Heart* (London, 1922).

O'CALLAGHAN, J. F., *A History of Medieval Spain* (Ithaca, NY, and London, 1975).

OMAN, C., *A History of the Art of War in the Middle Ages* (2 vols.; 2nd edn., London, 1924).

PARRY, V. J., and YAPP, M. E. (eds.), *War, Technology and Society in the Middle East* (London, 1975).

PARTINGTON, J. R., *A History of Greek Fire and Gunpowder* (Cambridge, 1960).

PAYNE-GALLWEY, R., *The Crossbow, with a Treatise on the Balista and Catapult of the Ancients* (2nd edn.; London, 1958).

PIDAL, R. MENENDEZ, *The Cid and his Spain*, trans. H. Sunderland (London, 1934).

PLECHL, H., and PLECHL, S.-C., *Orbis Latinus* (3 vols.; Brunswick, 1972).

PORGES, W., 'The Clergy, the Poor, and the Non-Combatants on the First Crusade', *Speculum*, 21 (1946), 1–20.

POWERS, J. F., 'The Origins and Development of Municipal Military Service in the Genoese and Castillian Reconquest', *Traditio*, 26 (1970).

POWICKE, M., *The Loss of Normandy* (Manchester, 1913; repr. 1963).

PRAWER, J., *Crusader Institutions* (Oxford, 1980).

—— *Histoire du royaume latin de Jérusalem*, trans. G. Nahon (2 vols.; Paris, 1969).

—— 'The Jerusalem the Crusaders Captured: A Contribution to the Medieval Topography of the City', in *CS*.

PRESTWICH, J. O., 'War and Finance in the Anglo-Norman State', *Transactions of the Royal Historical Society*, 5th ser. (1954).

—— 'The Military Household of the Norman Kings', *English Historical Review*, 96 (1981).

—— 'Richard Cœur de Lion: Rex Bellicosus', *Accademia Nazionale dei Lincei: Problemi attuali di scienza e di cultura*, 253 (1981), 3–15.

QUELLER, D. E., *The Fourth Crusade* (New York and Leicester, 1978).

—— and Day, G. W., 'Some Arguments in Defense of the Venetians in the Fourth Crusade', *American Historical Review*, 81 (1976), 717–38.

RATHGEN, B., *Das Geschütz in Mittelalter* (Berlin, 1927).

—— and Schäfer, K. H., 'Feuer und Fernwaffen beim päpstlichen Heere im 14. Jahrhundert', *Zeitschrift für Historische Waffenkunde*, 7 (1915–17).

REILLY, B. F., 'The "Historia Compostelana": The Genesis and Composition of a 12th Century Spanish "Gesta"', *Speculum*, 44 (1969).

—— *The Kingdom of Leon–Castilla under King Alfonso VI, 1065–1109* (Princeton, NJ, 1988).

—— *The Kingdom of Leon–Castilla under Queen Urraca, 1109–1126* (Princeton, NJ, 1982).
REY, E. G., *Étude sur les monuments de l'architecture militaire des croisés en Syrie et dans l'île de Chypre* (Paris, 1871).
RILEY-SMITH, J. S. C., 'The Motives of the Earliest Crusaders and Settlement of Latin Palestine', *English Historical Review*, 98 (1983).
—— *The First Crusade and the Idea of Crusading* (London, 1986).
RÖHRICHT, R., 'Die Belagerung von ʿAkkā (1189–1191)', *Forschungen zur deutschen Geschichte*, 16 (1876).
—— *Geschichte des Königsreichs Jerusalem (1100–1291)* (Innsbruck, 1898).
ROWE, J. G., 'Pascal II, Bohemond of Antioch and the Byzantine Empire', *Bulletin of the John Rylands Library*, 49 (1966).
RUNCIMAN, S., 'The Holy Lance Found at Antioch', *Analecta Bollandiana*, 68 (1950).
—— *A History of the Crusades* (3 vols.; Cambridge, 1951–4).
SCHLIGHT, J., *Monarchs and Mercenaries* (Bridgeport, 1968).
SCHLUMBERGER, G., *Nicéphore Phocas: Un empereur byzantin au dixième siècle* (Paris, 1890).
SCHNEIDER, A. M., and KARNAPP, W., *Die Stadtmauer von Iznik-Nicea* (Berlin, 1938).
SCHNEIDER, R., *Die Artillerie des Mittelalters* (Berlin, 1910).
SCHRADER, C. R., 'The Ownership and Distribution of Manuscripts of the *De re militari* of Flavius Vegetius Renatus before 1300', Ph.D. thesis (Columbia, 1976).
SHELBY, L. R., 'Geometrical Knowledge of Medieval Master Masons', *Speculum*, 47 (1972).
SIMONSFELD, H., *Jahrbücher des Deutschen Reichs unter Friedrich I, Jahrbücher der Deutschen Geschichte* (Leipzig, 1908).
SMAIL, R. C., *Crusading Warfare* (Cambridge, 1956).
—— *The Crusaders in Syria and the Holy Land* (London, 1973).
—— 'The International Status of the Latin Kingdom of Jerusalem, 1150–1192', in P. M. Holt (ed.), *The Eastern Mediterranean Lands in the Period of the Crusades* (Warminster, 1977).
SOUTHERN, SIR RICHARD, 'Aspects of the European Tradition of Historical Writing', *Transactions of the Royal Historical Society*, 5th ser. 20–3 (1970–3).
—— *The Making of the Middle Ages* (London, 1953; repr. 1967).
Storia di Milano, iii (Fondazione Treccani degli Alfieri per la Storia di Milano; Milan, 1953).
SUMBERG, L., 'The Tafurs and the First Crusade', *Medieval Studies*, 21 (1959).
SUMPTION, J., *The Albigensian Crusade* (London, 1978).
SYBEL, H. VON, *The History and Literature of the Crusades*, trans. Lady Duff Gordon (London, 1861).
TOUT, T. F., 'The Fair of Lincoln and the Histoire de Guillaume le Marechal', *Collected Papers of Thomas Frederick Tout* (3 vols.; Manchester, 1932–4).

TOY, S., *Castles* (2nd edn., London, 1966).
UNGER, R., *The Ship in the Medieval Economy* (London, 1980).
VAN CREVELD, M., *Supplying War: Logistics from Wallenstein to Patton* (Cambridge, 1977).
VERBRUGGEN, J. F., *The Art of Warfare in Western Europe during the Middle Ages*, trans. S. Willard and S. C. M. Southern (Amsterdam, 1977).
VITALE, V., *Breviario della storia di Genova* (Genoa, 1955).
VIOLLET LE-DUC, E., *Dictionnaire raisonné: L'Architecture française du XIe au XVIe siècle* (10 vols.; Paris, 1858–68).
WALEY, D., 'Combined Operations in Sicily, A.D. 1060–78', *Proceedings of the British School at Rome*, 22 (1954).
WATTENBACH, W., HOLTZMAN, R., and SCHMALE, F. J., *Deutschlands Geschichtsquellen im Mittelalter* (3 vols.; Cologne, 1971).
WHITE, L. H., 'The Crusades and the Technological Thrust of the West', V. J. Parry and M. E. Yapp (eds.), *War, Technology, and Society in the Middle East* (London, 1975).
—— *Medieval Technology and Social Change* (Oxford, 1962; repr. 1978).
YEWDALE, R., *Bohemond I, Prince of Antioch* (Princeton, NJ, 1917).
WASSERSTEIN, D., *The Rise and Fall of the Party-Kings: Politics and Society in Islamic Spain, 1002–1086* (Princeton, NJ, 1985).

INDEX

Acre 72
  siege of 1103 67, 71, 76, 249
  siege of 1189–91 212–36
Adhemar of Le Puy 16, 19, 32–3, 35
Afonso I of Portugal 159, 182, 183
Agrigento:
  siege of 1091 98
Alfonso I of Aragon 159, 169–72, 237
Alfonso VI of Leon-Castile 163, 170
Alfonso VII of Leon-Castile 158, 162, 172–7, 177–9, 189, 241
Alessandria:
  siege of 1174 146, 147–50
Alexandria:
  siege of 1167 83–4, 250
  siege of 1174 83–4, 86
Alexius I Comnenus, Byzantine Emperor 18, 19, 21–2, 23–4, 24–5, 93, 102–3, 103–5
Almeria:
  siege of 1147 146, 177, 178–9, 196
  retaken by Almohads 1157 189
Almohads 161, 189, 190
Almoravids 160–1, 168–9, 172
Alvora:
  siege of 1189 190
Amalfi:
  siege of 1131 109–10
  siege of 1135 198–9
Amalric I of Jerusalem 83–6
Ancona:
  siege of 1173 147
Antioch:
  battle of December 1097 31
  battle of February 1098 35
  battle of June 1098 39
  siege of 1097–8 25–39, 66, 87
armoured roof(s) 19–20, 22–3, 49, 57, 58, 130, 131, 139, 168, 173, 174, 176–7, 179, 210, 222–3, 252
  see also battering rams
Arquah:
  siege of 1099 45–75, 87
Arsuf:
  siege of 1099 73, 249

siege of 1101 74–5, 249
artillery 38, 43, 49, 58–9, 75, 76, 83, 89, 109, 110, 112–15, 123, 129, 130, 131, 134–5, 137–8, 139, 140, 143, 148–50, 163, 166–7, 168, 169, 174–5, 179, 182, 184, 185, 218, 225, 227–8, 234–5, 247–8, 254–72
Ascalon:
  battle of 1099 66
  siege of 1153 66, 67, 71, 72, 77, 78, 250
Aubord:
  siege of 1167 198

Baldwin I of Jerusalem 66–7, 69, 73, 74–5, 75, 76, 79–82, 99
Balearic Islands:
  campaign of 1113–15 165–9, 194
  campaign of 1146 177, 196, 205
Banyias:
  siege of 1139 89
Barbastro:
  siege of 1064 164, 165
Bari:
  siege of 1068–71 96–7
  siege of 1133 112
  siege of 1137 116, 146
  siege of 1139 118
Barletta:
  siege of 1133 111
Bartholomew, Peter 38–9, 40, 44, 45, 46–7
battering ram(s) 56, 57, 58, 81, 130, 131, 140, 166, 168, 184, 196, 270, 222–3, 252
  see also armoured roof
Beirut:
  siege of 1110 72, 75, 249
Bohemond of Taranto 18, 26, 28, 29, 31, 33, 36, 38, 41, 43–4, 103, 103–5
Brindisi:
  siege of 1129 109
Byzantines, logistical assistance in:
  the First Crusade 18, 19
  possible poliorcetic assistance 24–5
  operations at Damietta 1169 84–6

Caesarea:
  siege of 1101 74, 197, 206, 249

Capua:
    siege of 1057 95
    siege of 1098 101–2
Castrogiovanni:
    siege of 1091 98
Cid El, Rodrigo Diaz de Vivar 163–4
Comibra:
    siege of 1064 163
Como:
    siege of 1127 125
    war with Milan 1118–27 131–2
Coria:
    siege of 1138 174–5
    siege of 1142 176–7
Crema:
    siege of 1159–60 145–6, 151
Cosenza:
    siege of 1091 101
counterforts 30–1, 33–4, 69–70, 95, 98, 164, 170

Damascus:
    siege of 1148 89–98
Damietta:
    siege of 1169 84–6, 250
    siege of 1218–21 233
Durazzo:
    siege of 1081–2 96, 102–3
    siege of 1107–8 103–5

Embriaco, William 50, 51, 73, 74, 196–7, 202, 206
engineers/experts, in siege warfare 19–20, 21, 22, 23, 61, 77, 81, 83, 89, 131–2, 138–9, 140–1, 155, 173–4, 174, 175–6, 186–7, 202, 241, 241–3, 247
    see also Embriaco, William; Gaston IV of Béarn

Fatimids 82–3
Ferdinand I of Leon 163
Frederick I:
    Barbarossa, Emperor 94, 124, 125–6, 132–52
Fraga:
    battle of 1130 161
    siege of 1130 161

Gaston IV of Béarn 151, 171–2, 237–8
Gelmirez, Diego 158, 173–4
Genoa *see also* Genoese 33, 50, 62, 67–8, 69, 73, 75, 78, 172, 177–82, 200–1, 205–6, 209–11, 214, 215, 220–2, 225

Gerald 'the Fearless' of Portugal 189
Godfrey of Bouillon 16, 29, 48, 50, 52, 53–6, 58, 59–60, 61, 73, 76
Guiscard, Robert 92, 94, 95, 96–8, 100, 102–3, 104
Guy of Lusignan 214, 215–16, 220–1

Haifa:
    siege of 1100 72, 249
Harenc:
    siege of 1153 89
Henry I of England 239
Henry the Lion, Duke of Saxony 134, 136, 246–7
Hisn al-Akrad:
    stormed by First Crusaders 49
Holy Lance of Antioch 34, 38–9, 40
Huesca:
    siege of 1096 170

Jerusalem:
    siege of 1099 47–62, 77, 87, 265

Lerida:
    siege of 1149 161, 182
Lisbon:
    siege of 1147 161, 182–9, 192
Lodi:
    siege of 1111 129–30
Lothar III, Emperor 105, 116, 117, 198, 199
Louis of Thuringia 220

Mahdia:
    siege of 1087 194, 208–11
Majorca:
    campaign of 1114–15 167–9
Manerius of Acerenza 100
Marrat-an-Noman
    siege of 1098 39–44, 87, 108
Matera:
    siege of 1139 111
Mequinenza:
    siege of 1133 161
Milan:
    siege of 1158 143–4
    siege of 1161–2 144–6
Mileto:
    siege of 1062 95
Montepeloso:
    siege of 1137 107, 111–12
Motrone:
    battle of 1170 200–1
Naples:

Index

siege of 1135–6 108, 116, 195
surrender, 1131 107
Nicea:
  siege of 1097 17–25, 87
Nocera:
  battle of 1132 115
  siege of 1132 110
  siege of 1134 113
Oreja:
  siege of 1139 175

Palmero:
  siege of 1071–2 96–8, 100
Pantelleria:
  siege of 1089 209
Palma:
  siege of 1114–15 167–9
Peter the Hermit 18, 34, 52
Philip II Augustus of France 224, 225–6, 226–9
Pisa:
  see also Pisans
Pisans 67–8, 84, 88, 116–17, 131–2, 143, 165–9, 198–211, 203, 209–11, 214, 215, 223, 225, 230, 233
Plantagenet, Geoffrey 238–9
Plunder:
  importance of in financial operations 24, 177, 179, 182, 190, 192, 206–7
Pont Audemer:
  siege of 1123 239
Pontirilo:
  siege of 1160 129

Ravello:
  siege of 1131 111
  siege of 1135 199
Raymond-Berenguer III of Barcelona 165–9
Raymond-Berenguer IV of Barcelona 157, 159, 165–6, 177–82, 191
Raymond of St Gilles 16, 19, 26, 32–33, 37, 39, 40, 41–5, 45–7, 48, 50, 51–2, 53–4, 56–9, 59–61, 69
Richard I of England 224, 225–6, 228, 229–30, 231–2, 235, 237
Rignagno:
  battle of 1137 115, 117–18
Roger I of Sicily 92, 95, 96–8, 99, 100, 101, 101–2, 211
Roger II of Sicily 86, 91, 93, 94, 105–20, 151, 198, 199–200, 247

Saladin 215, 217, 219, 221, 229
St Agatha:
  siege of 1135 107
Salerno:
  siege of 1077–8 95, 96
  siege of 1137 116–17, 199–200
Santarem:
  siege of 1147
seapower 2, 35–6, 67–8, 72, 79, 106, 132–3, 155, 165–9, 173–4, 178–9, 196–201, 201–4
Sidon:
  siege of 1107 66–7, 72, 73
  siege of 1108 67, 71, 249
  siege of 1110 75, 76, 249
siege techniques and technology:
  development and possible diffusion 3–7, 10–11, 88–9, 243–8
  historiography of 3–4
siege tower(s) 22, 39, 41–2, 49, 56, 58–66, 71–9, 79–82, 84, 85–6, 88–9, 96–7, 100, 102–4, 131–2, 138–9, 140, 141–2, 148, 149, 157, 167–8, 171, 174, 175, 177, 178, 180–1, 184, 185, 186–8, 204, 220–2, 223, 233–4, 238, 239, 243–8, 249–50, 253–4
Silves:
  siege of 1189 189–90
Squillace:
  siege of 1060 95
Strator at Tortona 133
Syracuse:
  siege of 1085 98–9

Tabeiros:
  siege of 1126 173–4
Tafurs 40–1
Taormina:
  siege of 1079 106
Tatikios 24, 35
Thessalonica:
  siege of 1185 121–2
Tancred of Conversano 37–8, 45, 50, 59, 61
Tocco:
  siege of 1138 113–14
Toledo:
  operations against, 1139 175
  siege of 1085 163
Tortona:
  siege of 1155 133, 134–5, 153, 259
Tortosa:
  siege of 1148 161, 179–82, 196

Toulouse:
  siege of 1174–5 198
  siege of 1218 262, 268
Tramonti:
  siege of 1131 110, 112
Trani:
  siege of 1135 112
Trapani:
  siege of 1077 98, 99
Trencatrala:
  siege of 1154 157
Tripoli:
  operations against, 1102
  siege of 1109 67, 69–70
Troia:
  siege of 1139 118
  siege of 1042 96–7
Tyre:
  siege of 1111–12 73, 76, 77, 79–82, 250
  siege of 1124 68, 76, 78, 82–3, 250

Vegetius 141, 213, 215, 238, 239
Venetians 67–8, 73, 82–3, 147, 214
Venice, *see* Venetians
Venossa:
  siege of 1133 111

Wark:
  siege of 1174 272
William II of Sicily 93
William VI of Montpellier 42, 166
William VIII of Montpellier 198

Zarragoza:
  siege of 1118 169–72, 237